T0302192

Economic Analysis of Oil and Gas Engineering Operations

Economic Analysis of Oil and Gas Engineering Operations

By

Hussein K. Abdel-Aal

CRC Press is an imprint of the
Taylor & Francis Group, an **informa** business

First edition published 2021
by CRC Press
6000 Broken Sound Parkway NW, Suite 300, Boca Raton, FL 33487-2742
and by CRC Press
2 Park Square, Milton Park, Abingdon, Oxon, OX14 4RN

Library of Congress Cataloging-in-Publication Data

Names: Abdel-Aal, Hussein K., author.
Title: Economic analysis of oil and gas engineering operations / by Hussein K. Abdel-Aal.
Description: First edition. | Boca Raton, FL: CRC Press, 2021. | Includes bibliographical references and index. | Summary: "This book focuses on economic treatment of petroleum engineering operations and serves as a helpful resource for making practical and profitable decisions in oil and gas field development. This work will be of value to practicing engineers and industry professionals, managers, and executives working in the petroleum industry who have the responsibility of planning and decision making, as well as advanced students in petroleum and chemical engineering studying engineering economics, petroleum economics and policy, project evaluation and plant design"— Provided by publisher.
Identifiers: LCCN 2020043355 (print) | LCCN 2020043356 (ebook) | ISBN 9780367684716 (hbk) | ISBN 9781003137696 (ebk)
Subjects: LCSH: Petroleum industry and trade. | Oil fields—Economic aspects. | Gas fields—Economic aspects.
Classification: LCC HD9560.5.A257 2021 (print) | LCC HD9560.5 (ebook) | DDC 338.2/728—dc23
LC record available at https://lccn.loc.gov/2020043355
LC ebook record available at https://lccn.loc.gov/2020043356

ISBN: 978-0-367-68471-6 (hbk)
ISBN: 978-1-003-13769-6 (ebk)

Typeset in Times
by KnowledgeWorks Global Ltd.

Contents

SECTION III.IV Downstream Operations: Petroleum Refining of Crude Oil Into Useful Products (Three Parts) and Oil and Gas Transportation (One Chapter)

Preface

Economic Analysis of Oil and Gas Engineering Operations

Considered being the biggest sector in the world in terms of dollar value, the oil and gas sector is a global powerhouse. Many economists agree that crude oil was and remains the single most important commodity in the world as it is the primary source of energy production. Oil and gas organizations compete for resources and reserves around the globe.

The fact that the application of engineering principles in the field of oil operations, especially in exploration and production (E&P) such as the design, manufacture, and operation of efficient and economical plants, and processes, exemplifies the leading role of technology in this regard. Drilling, completing wells and producing oil and gas should be considered an extremely complex business.

In its general scope, economic analysis of petroleum operations would imply the application of economic techniques and analysis to the evaluation of design and engineering alternatives encountered in the petroleum industry. Basically, it involves the systematic evaluation of the economic merits of proposed solutions to engineering problems. What good is the best technological improvement, if the costs to apply these technologies cannot be recovered?

The text focuses on the fact that engineers seek solutions to problems, and the economic viability of each potential solution is normally considered along with the technical merits. This is typically true for the petroleum sector which includes the global processes of exploration, production, refining, and transportation (often by oil tankers and pipelines). Decisions on investment in any oil or gas field development will be made on the basis of its value. This "value" is judged by a combination of a number of economic indicators as presented in Section 2 that consists of 9 chapters, and shown next:

This new book, "Economic Analysis of Oil and Gas Engineering Operations", aims to reflect major changes over the past decade or so in the oil and gas industry. This book is by no means a complete description on the detailed design of any part of oil operations, and many details have been omitted in order to summarize a vast subject.

The subject matter and manner of presentation are such that the book should be of value to senior and graduate students majoring in petroleum engineering, chemical engineering, and economics. It is a helpful resource for practicing engineers and

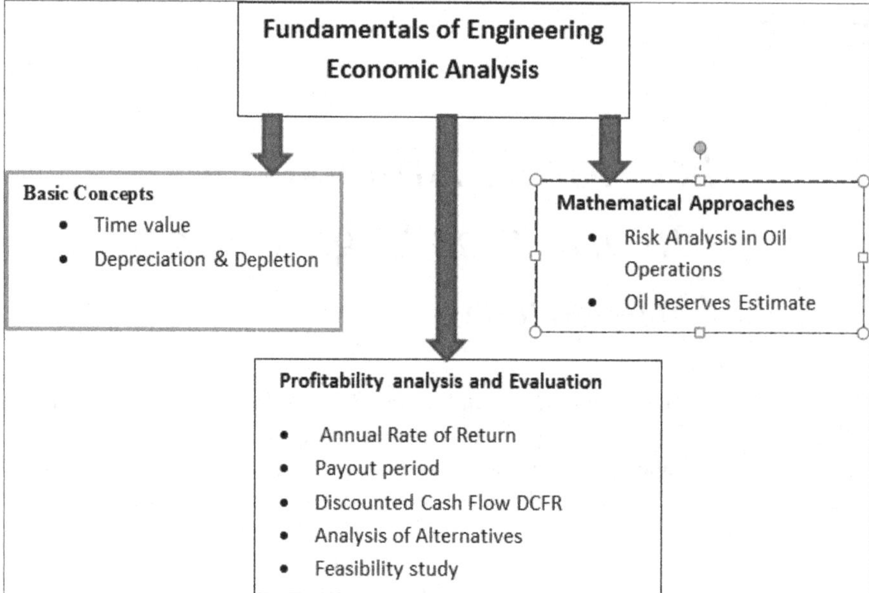

FIGURE 1 Classification of economic tools.

production people working in the petroleum industry, who have the responsibility of planning and decision-making in oil or gas field development. It should serve also as a reference volume for managers, executives, and other personnel engaged in this field. As the title implies, the book is focused on economic treatment of *petroleum engineering* operations, but most of its contents should be equally applicable to other engineering disciplines. The text can be adopted, accordingly, as a principal or supplemental resource book in allied courses such as engineering economics, petroleum economics and policy, project evaluation, and plant design.

The text is basically divided into three sections:

Section 1: Facts about Crude Oil and Gas Industry. It consists of three chapters.

Section 2: Fundamentals of Engineering Economic Analysis. This section represents a thorough coverage on the use of economic analysis techniques in decision-making, in particular, in petroleum-related projects. This section, as shown above, is covered in *9* chapters: Chapters 4–12.

Section 3: Engineering Decisions through Economic Impact Analysis: Applications and Real World Examples.

This section could be considered the backbone of our text. It represents the application of economic analysis to many engineering problems encountered in various sectors of petroleum operations. In addition to introduction, this section is divided into three parts, or subsections as shown next:

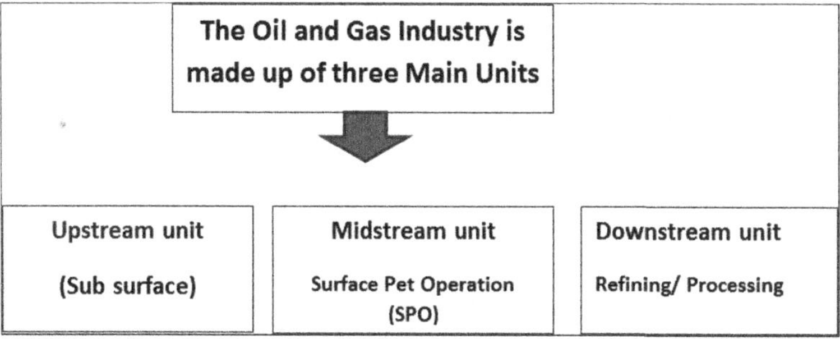

- Introduction
- Upstream operations (subsurface operations), covered in *two* comprehensive chapters: Chapters 13 and 14: Exploration/Drilling & Production (E&P). The arrangement of these two chapters is unique, since both exploration and production are key subsurface operations.
- Middle stream operations: "Surface operations", covered in *4* chapters: Chapters 15–18. Once oil and gas are produced underground, phase separation of oil from gas is accomplished using gas/oil separators. Operations Handling Crude Oil: Treatment, Dehydration, and Desalting come next. Followed by handling gas for treatment and conditioning.
- Downstream operations: *Petroleum refining*, conversion of crude oil into useful products, covered in *3* chapters: Chapters 19–21. Background on Modeling and Computer Applications is devoted in Part 3. Finally, Oil and Gas Transportation is presented in Chapter 22.

It is worth mentioning that Section 1 is a new addition to be found in a text of this merit (Facts about Crude Oil and Gas Industry). It represents a bird's eye view on this strategic industry. Exploration & Production (E&P) is a new addition as a combined subject, following the traditional presentation in oil industry.

Section 3.4, on the other hand, downstream operations involve, *Petroleum refining*, conversion of crude oil into useful products. This section is completely expanded to include oil fractionation (distillation), design of distillation columns, types of distillation schemes, and computations covering flash calculations along with solved examples (spread sheet).

The following chapters of this book contain updated information from chapters that originally appeared in *Petroleum Economics, Third Edition* 9781466506664, eds. Hussein K. Abdel-Aal and Mohammed A. Alsahlawi, Taylor & Francis, 2013.

Chapters 6, 7, and 8 are based on updated materials from "Financial Measures and Profitability Analysis" by *Maha Abd El-Kreem.*

Chapter 9 is based on updated materials from "Analysis of Alternative Selections and Replacements" by *Khaled Zohdy.*

Chapter 11 is based on updated materials from "Risk, Uncertainty, and Decision Analysis" by *Jamal A. Al-Zayer, Taqi N. Al-Faraj, and Mohamed H. Abdel-Aal.*

Chapter 12 is based on updated materials from Exploration and Drilling" by *Hussein K. Abdel-Aal.*

Chapter 14 is based on updated materials from "Production Operations" by *Mohamed A. Aggour and Hussein K. Abdel-Aal.*

Chapter 22 is based on updated materials from "Oil and Gas Transportation" by *M.A. Al-Sahlawi.*

A Guide to Case Studies and Real World Solved-Problems

This guide is a kind of road-map that will lead the reader to the right location.

- It starts first with Case studies:
  1. Recovery of Butane Using Lean Oil Extraction, Egypt, Chapters **19**, 21 Senior Project Committee, KFUPM.
  2. Utilization of Natural Gas Recovered from Gas Plant, Dhahran, SA, Chapters **19**, 23, plant visit to Abqauiq oil field.
  3. Causes of Tight Emulsions in Gas Oil Separation Plants ***Ghawar* field in Saudi Arabia**, Chapter **16**, Plant visit.
  4. To choose between two Alternatives for Drilling Well in a Reservoir, Chapters **13**, 25, **Exhibition, Dhahran, SA.**
  5. The Economic Evaluation of a Gas Lift, thesis for BS students, **HTI, Egypt**, **14**, Chapter **18**, Senior Project, Dept. Board.
  6. Oil and gas industry Case Study (source Orit Mynde, Tullow Oil plc), Chapters **14**, 22, the Republic of Ireland.

  www.cimaglobal.com › Documents › GBC › Case-Study.

- Next, the guide refers the reader to Chapter 10. It is a kind of "treasure", full with many solved real examples, particularly related to Section 2 of the book.
- Going all the way from **UPSTREAM operations through MIDDLE STREAM OPERATIONS to DOWNSTRAM OPERATIONS,** the reader will encounter many selected cases, all the way from Chapter 13 up to Chapter 21, covering material on Section 3. As mentioned in the text, Section 3 is the heart of the book.

Author Biography

Dr. Hussein K. Abdel-Aal is Emeritus Professor of Chemical Engineering & Petroleum Refining, NRC, Cairo, Egypt (retired). Dr. Abdel-Aal worked in the oil industry (1956-1960) as a process engineer in Suez and Homs (Syria) oil refineries before obtaining his PhD from Texas A. & M. University. On returning to Cairo, he joined NRC for the period 1965-1970, followed by one year at UMIST, England, as a postdoctoral scholar.

He joined the department of Chemical Engineering at KFUPM, Dhahran, Saudi Arabia, as a Professor of Chemical Engineering (1971-1985) and (1988-1998). Dr. Abdel-Aal conducted and coordinated projects involving a wide range of process development, feasibility studies, industrial research problems, and continue-education programs for many organizations.

He has contributed to over 90 technical papers and is the editor of Petroleum Economics & Engineering 3rd edition, 2014, the main author of Petroleum and Gas Field Processing, 2nd edition, 2016, the author of Chemical Engineering Primer with Computer Applications, 2016, and the author of Magnesium from Resources to Production, 2019, all books are published by Taylor & Francis, CRC Press.

He is listed in "Who is who in the World", 1982, Sigma Si, Phi Lambda Upsilon. He is fellow and founding member of the board of directors of the International Association of Hydrogen Energy, FL, USA.

Acknowledgments

All praise goes to **Almighty Allah,** who is the creator of this universe. I am grateful to God for the good health and wellbeing that are necessary to complete this book, and to overcome my eye problem.

When authors like me get involved in writing a book, they rather being optimistic.

Optimism is the essential gradient for success. Getting help and support from many individuals provides inspiration that fuel the author's trip throughout the writing process.

I would like to thank my wife for standing beside me throughout my career and for writing my fifth book. She has been my inspiration and motivation for continuing to move my career forward.

I am especially indebted to a group of my students, at KFUPM and H T I, for including some of their homework-solved-problems in the text.

Last but not least, a word of appreciation is to be said to the group of Taylor & Francis for their cooperative support and dedication in getting this product in the hands of the readers. This includes primarily Allison Shatkin and Gabrielle Vernachio.

Section I

Facts about Crude Oil and Gas Industry

1 Oil and Gas Industry Overview

1.1 INTRODUCTION

The oil and gas industry is one of the largest sectors in the world in terms of dollar value, generating an estimated $3.3 trillion in revenue annually. Oil is crucial to the global economic framework, especially for its largest producers: the United States, Saudi Arabia, Russia, Canada, and China.

Oil price increases are generally thought to increase inflation and reduce economic growth. In terms of inflation, oil prices directly affect the prices of goods made with petroleum products. In economics terminology, high oil prices can shift up the supply curve for the goods and services for which oil is an input. As always, there are headwinds and tailwinds, risks and opportunities, uncertainties, and foreseeable trends. We will now explore the oil and gas trends through the year 2020.

To avoid uncertainties, risks, and other opportunities that may include weakening economic growth and intensifying trade tensions all the way to political risks, one must explore global oil and gas trends. To set an example over the past decade, we have seen the heights of bullish optimism and seemingly limitless investment during the years of the $100 per barrel world, from 2011 to mid-2014, and the lows of the price crash and extended oil downturn, from mid-2014 to 2017. How do we assess the oil and gas performance in coming years and its prospects in the future?

As 2018 ended, it's the time where it can be analyzed to figure out the status of both the oil and gas sector as well as the chemicals sector. The oil and gas sector recovered, especially the oil markets from the depths of the post 2014 downturn. Since 2016, oil prices have recovered from $40, reaching $67 in September, 2018. The recovery happened due to several factors. One of them is the success of the production restraint agreement between Organization of the Petroleum Exporting Countries (OPEC) and non-OPEC countries, which is in force since the beginning of first half of 2017. Other factors that influenced the recovery of the oil and gas markets are the less oil coming to the market from challenged producers and the ongoing global demand growth estimated by the Energy Information Administration (EIA); the global demand growth for 2018 was estimated at 1.6 million barrels per day (mb/d).

1.2 CORONAVIRUS (COVID-19)

The impact of the new coronavirus (COVID-19) on demand, slowing supply growth in the United States and other non-OPEC countries, has to be considered. At the same time, global energy transitions are affecting the oil industry. Following decarburization, refiners face a big challenge from weaker transport fuel demand.

In other words, the outbreak of the COVID-19 has added a major consideration of uncertainty to the oil market forecast. Global oil demand is expected to contract for the first time since the global recession of 2009.

In addition, it is anticipated to say that a contraction in 2020 and an expected sharp rebound in 2021, global oil demand growth is set to weaken as consumption of transport fuels increases more slowly. Between 2019 and 2025, global oil demand is forecast to grow at an average annual rate of just below 1 mb/d.

It is worth to state the following observations:

- Exploration and production (E&P) in some parts of the world looking for oil may require complex and cutting-edge technology making exploration difficult, expensive, and highly risky. Large companies, who are well-equipped technologically and are strong financially, will venture in such cases.
- The prospects of the petroleum industry will be increasingly competitive in the future; which may be in line with the super majors as they possess more technical know-how and are well-prepared financially. Nevertheless, the nationally owned companies (NOCs) may be a strong match for them as they are supported by their state governments and also have the wherewithal to seek for concession.
- The strong desire of some countries to create their own national oil companies and the concern about energy security is likely to increase resource nationalism in the near future.
- From the popular saying, "change is the only constant thing in nature", the petroleum industry has had its fair share of structural and organizational changes over the past years.

1.3 GLOBAL OIL AND GAS ANALYSIS

This we may call a bird's eye view to dig out some information on global oil and gas production and demand as well (Figure 1.1).

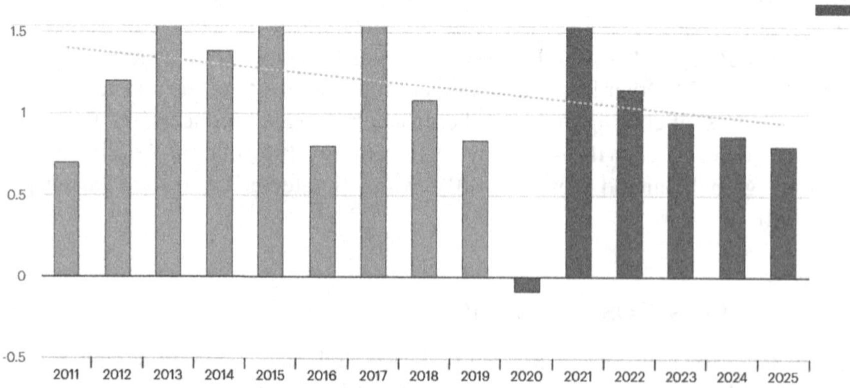

FIGURE 1.1 Global oil demand growth, 2011–2020.

Source: Oil 2020 IEA (International Energy Agency).

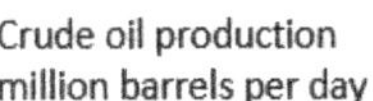

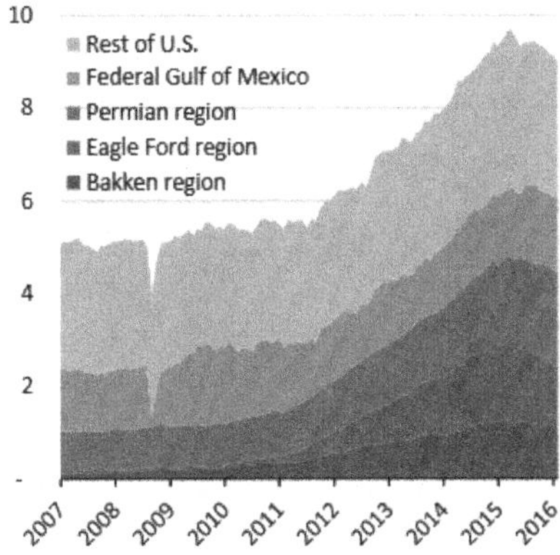

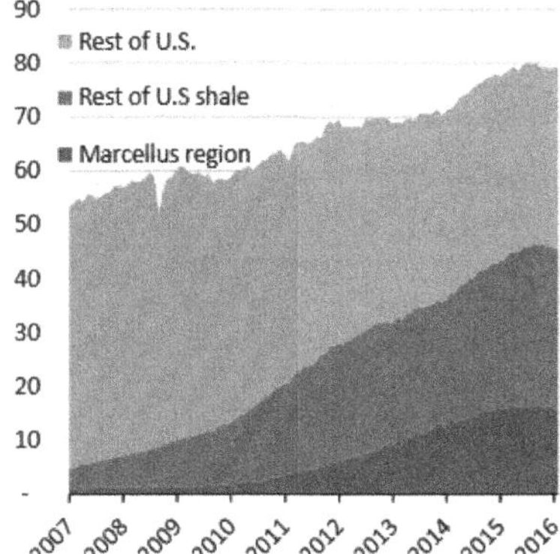

Source: U.S. Energy Information Administration *Drilling Productivity Report* regions, *Petroleum Supply Monthly*, *Natural Gas Monthly*

Note: Shale gas estimates are derived from state administrative data collected by DrillingInfo Inc. and represent the U.S. Energy Information Administration's shale gas estimates, but are not survey data.

FIGURE 1.2 Regional shale development has driven increases in U.S. crude oil and natural gas production.

Source: U.S. Energy Information Administration, March 2016.

1.3.1 Liquid Fuel Production

EIA estimates that global liquid fuels production averaged 91.8 mb/d in the second quarter of 2020, down 8.6 mb/d year over year (Figure 1.2).

1.4 GLOBAL ENERGY PRODUCTION

World energy production continued growing in 2019 (+1.5%), but stayed below its historical trend (2%/year). This is given by the Global Energy Statistical Yearbook, as shown next (Figures 1.3 and 1.4).

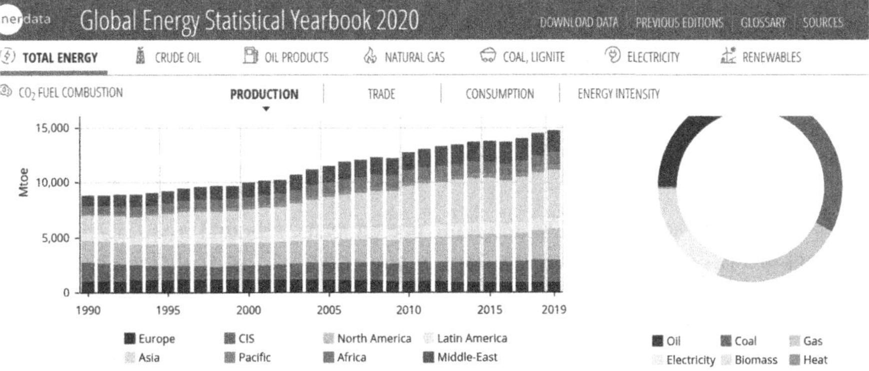

FIGURE 1.3 Global energy production.

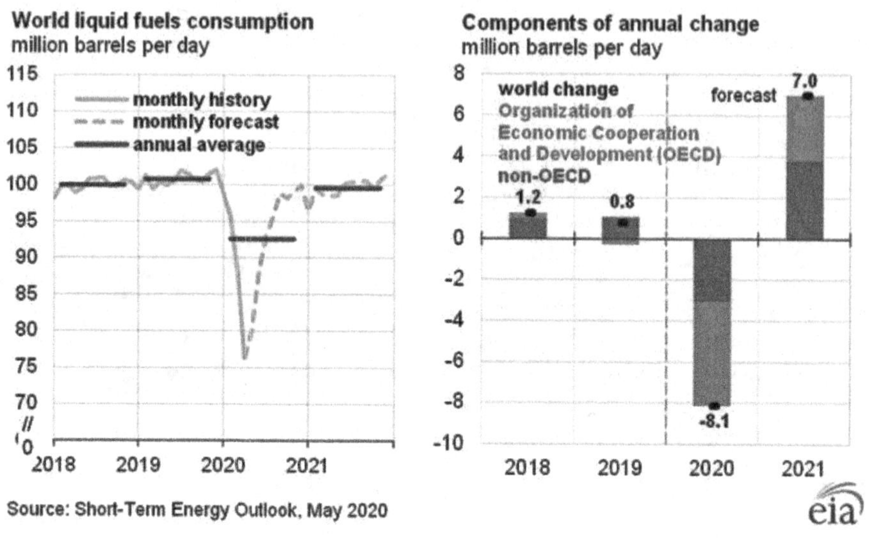

FIGURE 1.4 Global liquid fuel consumption and components of annual change.

1.5 OIL PRODUCTION, 2019

List of countries by oil production

	Country	Oil production 2019 (bbl/day)
01	United States	15,043,000
02	Saudi Arabia (OPEC)	12,000,000
03	Russia	10,800,000
04	Iraq (OPEC)	4,451,516

93 more rows

Following a contraction in 2020 and an expected sharp rebound in 2021, global oil demand growth is set to weaken as consumption of transport fuels increases more slowly. Between 2019 and 2025, global oil demand is forecast to grow at an average annual rate of just below 1 mb/d.

1.5.1 The Top Ten in Oil Reserves

Here are the top 10 largest oil reserves in the world by country.

- **Kuwait**: 101.5 Billion Barrels. ...
- Russia: 103.2 Billion Barrels. ...
- Iraq: 150 Billion Barrels. ...
- **Iran**: 157.8 Billion Barrels. ...
- Canada: 172.9 Billion Barrels. ...
- **Saudi Arabia**: 267 Billion Barrels. ...
- Venezuela: 298.3 Billion Barrels.

More items... • Jun 27, 2019

Venezuela has the largest amount of oil reserves in the world with 300.9 billion barrels.

1.5.2 Top Ten in Natural Gas Production

Country Comparison > Natural gas - production > TOP 10

Rank	Country	Natural gas - productioı (cubic meters)
1	United States	772,799,987,712
2	**Russia**	665,600,000,000
3	**Iran**	214,499,999,744
4	Qatar	166,400,000,000

6 more rows

1.6 STRUCTURE AND ORGANIZATION OF THE OIL AND GAS INDUSTRY

There are basically three phases of oil production:

1. Pre-drilling activities (upstream)
2. Drilling (upstream)
3. Production (midstream and downstream)

It is noticed that the longest and typically most expensive phase is pre-drilling activities. Drilling itself occurs in two phases: drilling down to below the water table and then encasing the well hole in cement to prevent groundwater and soil contamination; and then drilling to the required depth and taking the necessary steps to stimulate upward oil flow.

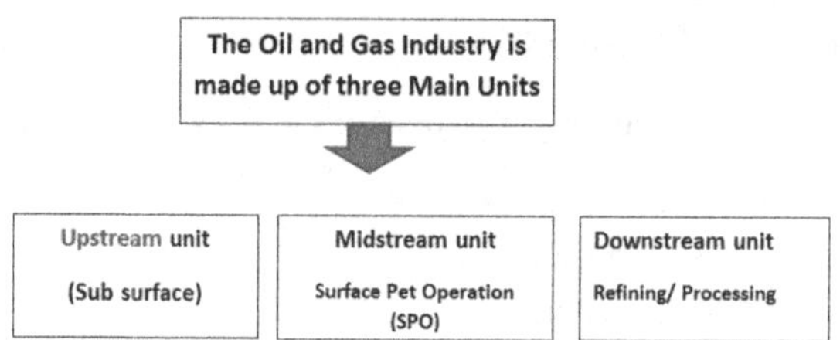

FIGURE 1.5 The oil and gas industry involves three stages.

The oil and gas industry is best illustrated by Figure 1.5.

The stages are described as follows:

a. **Upstream Unit** – It consists of companies involved in the exploration and production of oil and gas. These are the firms that search the world for reservoirs of the raw materials and then drill to extract that material. These companies are often known as "E&P" for "exploration and production".

The upstream segment is characterized by high risks, high investment capital, extended duration as it takes time to locate and drill, as well as being technologically intensive. Virtually all cash flow and income statement line items of E&P companies are directly related to oil and gas production.

b. **Midstream Unit** – Activities are focused on transportation, and involve moving the (separated fluids) extracted raw materials to refineries to process the oil and gas. The midstream segment is also marked by high regulation, particularly on pipeline transmission, and low capital risk. The segment is also naturally dependent on the success of upstream firms.
c. **Downstream Unit** – This is the refining stage which involves upgrading the products by treatment to bring them to specifications such as gasoline, jet fuel, heating oil, and asphalt.

We conclude this chapter by observing the trends that could shape the rest of 2020, setting the ground for a challenging 2021 and a nascent recovery in the early-to-mid 2020s. How can oil and gas organizations remain competitive and emerge stronger in the wake of the COVID-19 pandemic?

As we said earlier, the spread of COVID-19 has disrupted global financial and commodity markets, as well as the U.S. oil and gas industry. The year 2020 poses great challenges to oil and gas industry.

2 Crude Oil: Origin and Background

2.1 ORIGIN, OCCURRENCE, AND RESOURCES OF CRUDE OIL

A brief description of how oil and gas are formed and accumulated underground is presented in this part. Crude oil is typically obtained through drilling, where it is usually found alongside other resources, such as natural gas (which is lighter and, therefore, sits above the crude oil) and saline water (which is denser and sinks below).

Several theories have been proposed to explain the formation and origin of oil and gas (petroleum); these can be classified as the organic theory of petroleum origin and the inorganic theory of petroleum origin. The organic theory provides the explanation most accepted by scientists and geologists.

It is believed, and there is evidence, that ancient seas covered much of the present land area millions of years ago. The Arabian Gulf and the Gulf of Mexico, for example, are parts of such ancient seas. Over the years, rivers flowing to these seas carried large volumes of mud and sedimentary materials into the sea. The mud and sedimentary materials were distributed and deposited layer upon layer over the sea floor. The buildup of thousands of feet of mud and sediment layers caused the sea floors to slowly sink and be squeezed. As time goes by, heat and pressure began to rise as the organisms get buried deeper and deeper below the surface. Depending on the amount of pressure, heat, and the type of organisms, determines if the organisms will become natural gas or oil. The more heat the lighter the oil. If there is even more heat and the organisms were made up of mostly plants, then natural gas is formed.

Eventually the sedimentary rocks (the sandstone and shale, and the carbonates) are formed, where petroleum is found.

2.1.1 How Oil and Gas Are Accumulated?

Here, we refer to the natural resources of forming crude oil. Accumulation is the primary step in making crude oil available as a relatively "*natural resource*". The oil, gas, and salt water occupied the pore spaces between the grains of the sandstones, or the pore spaces, cracks, and vugs of the limestones and dolomites. Whenever these rocks were sealed by a layer of impermeable rock, the cap rock, the petroleum accumulating within the pore spaces of the source rock was trapped and formed the petroleum reservoir. However, when such conditions of trapping the petroleum within the source rocks did not exist, oil gas moved (migrated), under the effects of pressure and gravity, from the source rock until it was trapped in another capped (sealed) rock.

Because of the differences in density, gas, oil, and water segregated within the trap rock. Gas, when existed, occupied the upper part of the trap and water occupied the bottom part of the trap, with the oil between the gas and water. Complete displacement of water by gas or oil never occurred. Some salt water stayed with the gas and/or oil within the pore spaces and as a film covering the surfaces of the rock grains; this water is known as the connate water, and it may occupy from 10% up to 50% of the pore volume.

The geologic structure, in which petroleum has been trapped and has accumulated, whether it was the source rock or the rock to which petroleum has migrated, is called the petroleum reservoir.

In summary, the formation of a petroleum reservoir involves:

- First, the accumulation of the remains of land and sea life and their burial in the mud and sedimentary materials of ancient seas.
- This is followed next, by the decomposition of these remains under conditions that recombine the hydrogen and carbon to form the petroleum mixtures.

Finally, the formed petroleum is either trapped within the porous source rock when a cap rock exists or it migrates from the source rock to another capped (sealed) structure.

2.2 DISCOVERING PETROLEUM RESERVOIRS

Exploration is the beginning to look for oil and gas. Finding or discovering a petroleum reservoir involves three major activities:

- Geologic surveying,
- Geophysical surveying,
- Exploratory drilling activities.

The data collected from the geologic and geophysical surveys are used to formulate probable definitions and realizations of the geologic structure that may contain oil or gas. However, we still have to determine whether petroleum exists in these geologic traps, and, if it does exist, would it be available in such a quantity that makes the development of the oil/gas field economical? The only way to provide a definite answer is to drill and test exploratory wells.

The exploratory well, known as the wildcat well, is drilled in a location determined by the geologists and geophysicists. As this exploratory well is drilled, samples of the rock cuttings are collected and examined for their composition and fluid content. The data are used to identify the type of formation versus depth and to check on the presence of hydrocarbon materials within the rock. Cores of the formations are also obtained, preserved, and sent to specialized laboratories for analysis.

Whenever a petroleum-bearing formation is drilled, the well is tested while placed on controlled production. After the well has been drilled, and sometimes at various intervals during drilling various *logs* are taken. There are several logging tools or techniques (electric logs, radioactivity logs, and acoustic logs) that are used to gather information about the drilled formations.

These tools are lowered into the well on a wireline (electric cable) and, as they are lowered, the measured signals are transmitted to the surface and recorded on computers. The signals collected are interpreted and produced in the form of rock and fluid properties versus depth.

2.2.1 Types and Classification of Petroleum Reservoirs

Petroleum reservoirs are generally classified into two categories:

1. According to their geologic structure
2. Their production (drive) mechanism

2.2.1.1 Geologic Classification

Petroleum reservoirs exist in many different sizes and shapes of geologic structures. It is usually convenient to classify the reservoirs according to the conditions of their formation as follow:

- Dome-shaped and anticline reservoirs, these reservoirs are formed by the folding of the rock layers as shown in Figure 2.1. The dome is circular in outline, and the anticline is long and narrow. Oil or gas moved or migrated upward through the porous strata where it was trapped by the sealing cap rock and the shape of the structure.
- Faulted reservoirs, these reservoirs are formed by shearing and offsetting of the strata (faulting), as shown in Figure 2.2. The movement of the nonporous rock opposite the porous formation containing the oil/gas creates the sealing. The tilt of the petroleum-bearing rock and the faulting trap the oil/gas in the reservoir.
- Salt-dome reservoirs, these types of reservoir structures, which take the shape of a dome, were formed due to the upward movement of a large, impermeable salt dome that deformed and lifted the overlying layers of

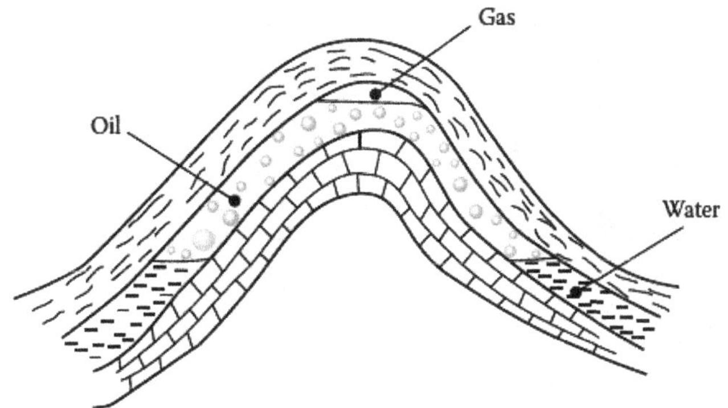

FIGURE 2.1 Dome-shaped reservoirs.

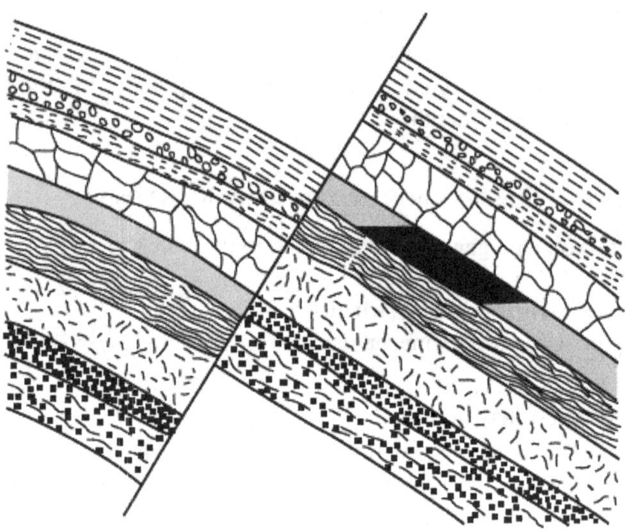

FIGURE 2.2 Faulted reservoirs.

rock. As shown in Figure 2.3, petroleum is trapped between the cap rock and an underlying impermeable rock layer, or between two impermeable layers of rock and the salt dome.

- Combination reservoirs, in these cases, combinations of folding, faulting, abrupt changes in porosity, or other conditions create the trap from these common types of reservoirs.

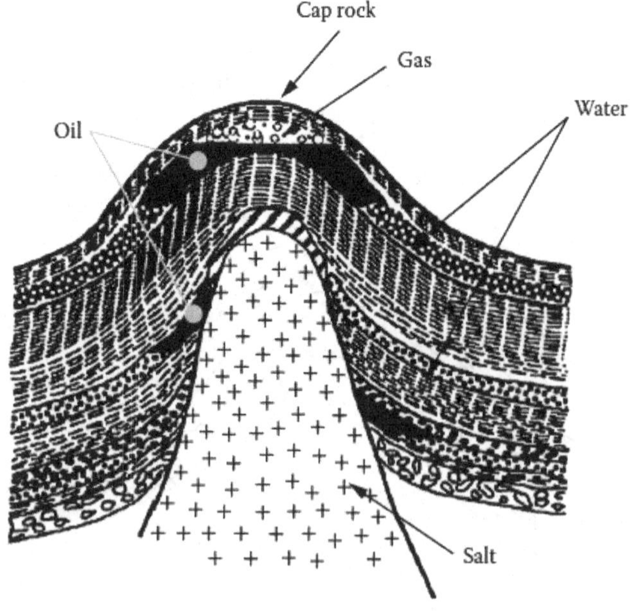

FIGURE 2.3 Salt-dome reservoirs.

2.2.2 Reservoir Drive Mechanisms

At the time oil was forming and accumulating in the reservoir, the pressure energy of the associated gas and water was also stored. When a well is drilled through the reservoir and the pressure in the well is made to be lower than the pressure in the oil formation, it is that energy of the gas, or the water, or both that would displace the oil from the formation into the well and lift it up to the surface. Therefore, another way of classifying petroleum reservoirs, which is of interest to reservoir and production engineers, is to characterize the reservoir according to the production (drive) mechanism responsible for displacing the oil from the formation into the wellbore and up to the surface. There are three main drive mechanisms:

- *Solution gas drive reservoirs*—depending on the reservoir pressure and temperature, the oil in the reservoir would have varying amounts of gas dissolved within the oil (solution gas). Solution gas would evolve out of the oil only if the pressure is lowered below a certain value, known as the bubble point pressure, which is a property of the oil. When a well is drilled through the reservoir and the pressure conditions are controlled to create a pressure that is lower than the bubble point pressure, the liberated gas expands and drives the oil out of the formation and assists in lifting it to the surface. Reservoirs with the energy of the escaping and expanding dissolved gas as the only source of energy are called solution gas drive reservoirs. This drive mechanism is the least effective of all drive mechanisms; it generally yields recoveries between 15% and 25% of the oil in the reservoir.
- *Gas cap drive reservoirs*—many reservoirs have free gas existing as a gas cap above the oil. The formation of this gas cap was due to the presence of a larger amount of gas than could be dissolved in the oil at the pressure and temperature of the reservoir. The excess gas is segregated by gravity to occupy the top portion of the reservoir. In such reservoirs, the oil is produced by the expansion of the gas in the gas cap, which pushes the oil downward and fills the pore spaces formerly occupied by the produced oil. In most cases, however, solution gas is also contributing to the drive of the oil out of the formation. Under favorable conditions, some of the solution gas may move upward into the gas cap and, thus, enlarge the gas cap and conserve its energy. Reservoirs produced by the expansion of the gas cap are known as gas cap drive reservoirs. This drive is more efficient than the solution gas drive and could yield recoveries between 25% and 50% of the original oil in the reservoir.
- *Water drive reservoirs*—many other reservoirs exist as huge, continuous, porous formations with the oil/gas occupying only a small portion of the formation. In such cases, the vast formation below the oil/gas is saturated with salt water at very high pressure. When oil/gas is produced by lowering the pressure in the well opposite the petroleum formation, the salt water expands and moves upward, pushing the oil/gas out of the formation and occupying the pore spaces vacated by the produced oil/gas. The movement of the water to displace the oil/gas retards the decline in oil, or gas pressure, and conserves the expansive energy of the hydrocarbons.

2.3 LIFE CYCLE OF OIL AND GAS FIELDS

Oil and gas field development usually follows the following pattern:

The schematic presentation in Figure 2.4 is detailed as follows:

- Exploration activities are carried out. Geological and geophysical surveys and studies are used to determine the location where a hydrocarbon reservoir may potentially exist. The results of such studies merely provide information about the potential location of the reservoir, its area, depth, and some characteristics such as faults and fractures.
- Based on the information available, what is called a *wild cat well* is drilled. A location (normally at the center of the potential reservoir) is selected to drill this first exploration well.
- As this well is being drilled, rock and fluid properties data for all penetrated formations are collected and analyzed. If hydrocarbons (oil and/or gas) are found, the well is tested to determine the production potential; otherwise, the well is considered a dry well and is abandoned.

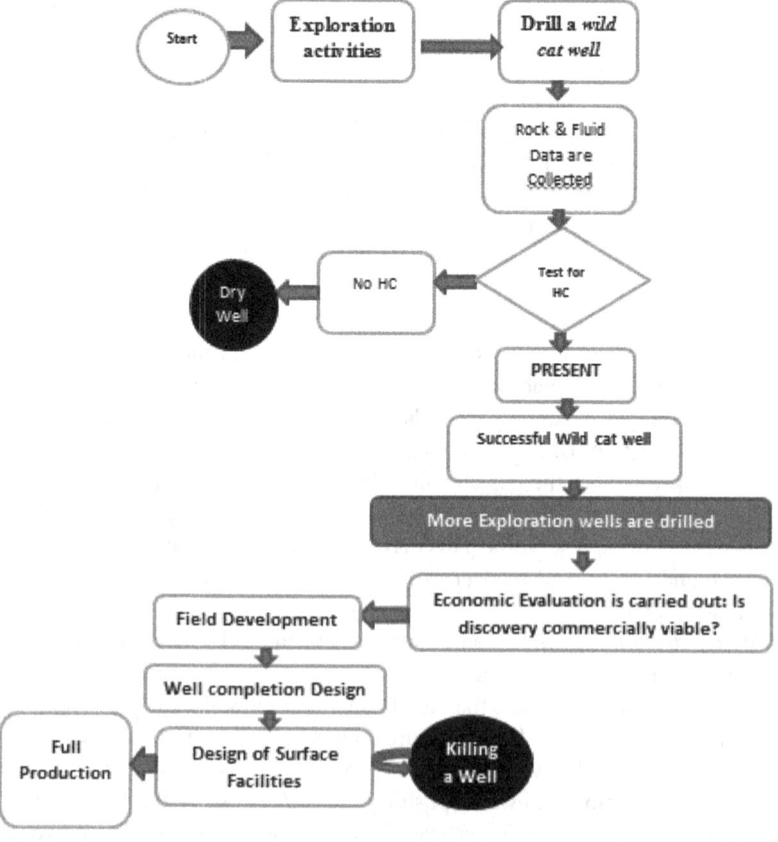

FIGURE 2.4 An outline for oil and gas development.

- If the wild cat is successful, more exploration wells will be drilled and tested. At this stage, the number and locations of these wells are determined to provide as much information as possible about the reservoir volume, the amount of hydrocarbons in place, and the production potential of the discovery. Preliminary reservoir simulation studies coupled with economic evaluations are made in order to determine whether the discovery is commercially viable.
- Once a decision is made to develop the field, extensive simulation studies will be conducted to examine various development and production strategies with the objective of determining the optimum development and production plan, which yield the maximum recovery and best economics.
- Following this, well-completion-designs will be made with the objective of having wells work for the entire life of the field, providing maximum recovery in the most economic and safe manner. Based on the completion designs, the well drilling designs and programs will be developed.
- Based on the production forecast, the surface facilities for separation and treatment of produced fluids are designed. Procurement of materials and equipment is planned and made to secure their availability on time for actual field development and production. To accelerate revenue, all or part of the surface production and processing facility should be on location to produce wells as they are drilled and completed.
- **History matching:** Production data (production rates, pressures, temperatures, gas–oil ratio, and water cut, if any) are collected for a period of time and then compared against the forecasted (predicted) data from reservoir simulations. Normally, no match would be obtained. Then a process called history matching is performed where the reservoir simulations are modified (by changing the data used in developing the original simulations that have the least certainty) until the simulation data match the actual production data. The modified simulations are then used to forecast future production. The history matching process is repeated and this would probably continue until the end of the life of the field. It should be noted that several operations, such as pressure maintenance, improved/enhanced recovery, and artificial lift, might be implemented during the production life of the field. When no more hydrocarbons can be economically produced, the field is abandoned.
- **Killing a well:** When no more hydrocarbons can be economically produced, the well is abandoned and have to be killed, filled with layers of cement and sand, and the surface casing capped. This process is governed by either company or government regulations.

In other words, once the well is shut-in, for drilling, the following well kill procedure is used: Driller's Method, which removes the kick first, then circulates kill fluid throughout the well with two circulations. This method works best for lateral or deviated wells.

3 Composition and Types of Crude Oil and Composition of Natural Gas

3.1 INTRODUCTION: COMPREHENSIVE INFORMATION ABOUT CRUDE OIL

Hydrocarbons make up crude oil and natural gas, known as the organic source, which are naturally occurring substances found in rock in the earth's crust. How these organic materials are formed? They are created by the compression of the remains of plants and animals in sedimentary rocks such as sandstone and limestone. The organic material eventually transforms in sedimentary form into oil and gas after being exposed to specific temperatures and pressure ranges deep within the earth's crust.

Many types of crude oils are produced around the world. The market value of an individual crude stream reflects its quality characteristics. Two of the most important quality characteristics are density and sulfur content. Density ranges from light to heavy, while sulfur content is characterized as sweet or sour.

According to the International Energy Agency (IEA), it's likely to remain so for decades to come, even as the world embarks on a low carbon pathway to help meet climate change commitments. The IEA expects demand for oil to grow to 103.5 million barrels of oil per day (mb/d) in 2040, compared to 92.5 mb/d in 2015—an increase of 11%.

Crude oil is far from being one homogenous substance. Its physical characteristics differ depending on where in the world it is produced, and those variations determine its usage and price. This is due to the fact that oil from different geographical locations will naturally have its own very unique properties. These oils vary dramatically from one another when it comes to their viscosity, volatility, and toxicity.

In its natural, unrefined state, crude oil ranges in density and consistency, from very thin, light weight, and volatile fluidity to an extremely thick, semi-solid heavy weight oil. There is also a tremendous gradation in the color that the oil extracted from the ground exhibits, ranging all the way from a light, golden yellow to the very deepest, darkest black imaginable.

In this chapter, the composition of crude oil is presented using the chemical approach and by applying the physical methods traditionally used. Composition of natural gas is included. Classification and characterization of crude oils based on correlation indexes and crude assays are explained. Different types of well-known crude oils as well as bench-mark are included.

Main crude oil products are briefly described as produced by the backbone distillation operations.

3.2 BACKGROUND

Crude oil, commonly known as petroleum, is a liquid found within the Earth comprised of hydrocarbons, organic compounds, and small amounts of metal. While hydrocarbons are usually the primary component of crude oil, their composition can vary from 50% to 97% depending on the type of crude oil and how it is extracted. Organic compounds like nitrogen, oxygen, and sulfur typically make-up between 6% and 10% of crude oil while metals such as copper, nickel, vanadium, and iron account for less than 1% of the total composition.

Figure 3.1 presents comprehensive information about crude oil. It includes, types of crude oil, identification parameters of crude oil, elemental composition of crude oil, the four main groups of hydrocarbons found in crude oil, and how the quality of crude oil is measured by two parameters:

- The American Petroleum Industry (API) gravity and classified as **Light**: 38 API or higher, **Medium**: 22–38 API, and **Heavy**: 22 API or below
- The sulfur content as **Sweet**: S content<0.5% and **Sour**: S content>0.5

Most of the physical properties of crude oils such as API gravity, viscosity, and coefficient of expansion, depend on reservoir pressures and temperatures, chemical composition of the oil, and sometimes, on the amount of dissolved natural gases.

It is worth mentioning to say that the API gravities of crude oils usually increase with depth. This is because a combination of source and reservoir maturation processes. Associated with slow but continuously increasing geo temperatures, cause the generation of lighter (or high API gravity) oils at greater depths of burial.

This classification is further exemplified by some typical crude oils as shown next in Table 3.1.

3.2 CRUDE OIL COMPOSITION

Crude oil is essentially a mixture of many different hydrocarbons, all of varying lengths and complexities. There are two broad approaches to study and quantify the composition of crude oil:

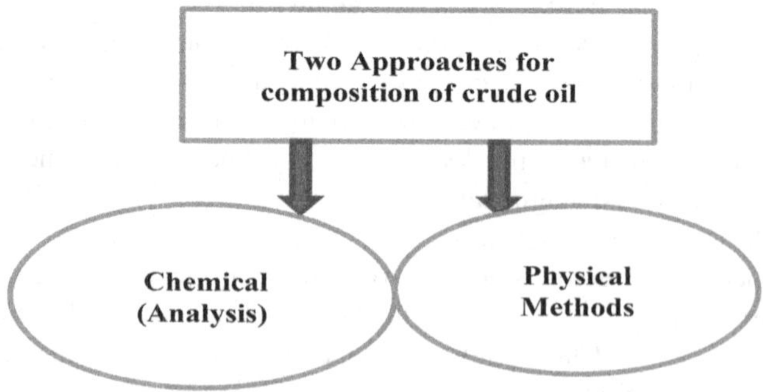

- The chemical approach (analysis)
- The physical methods

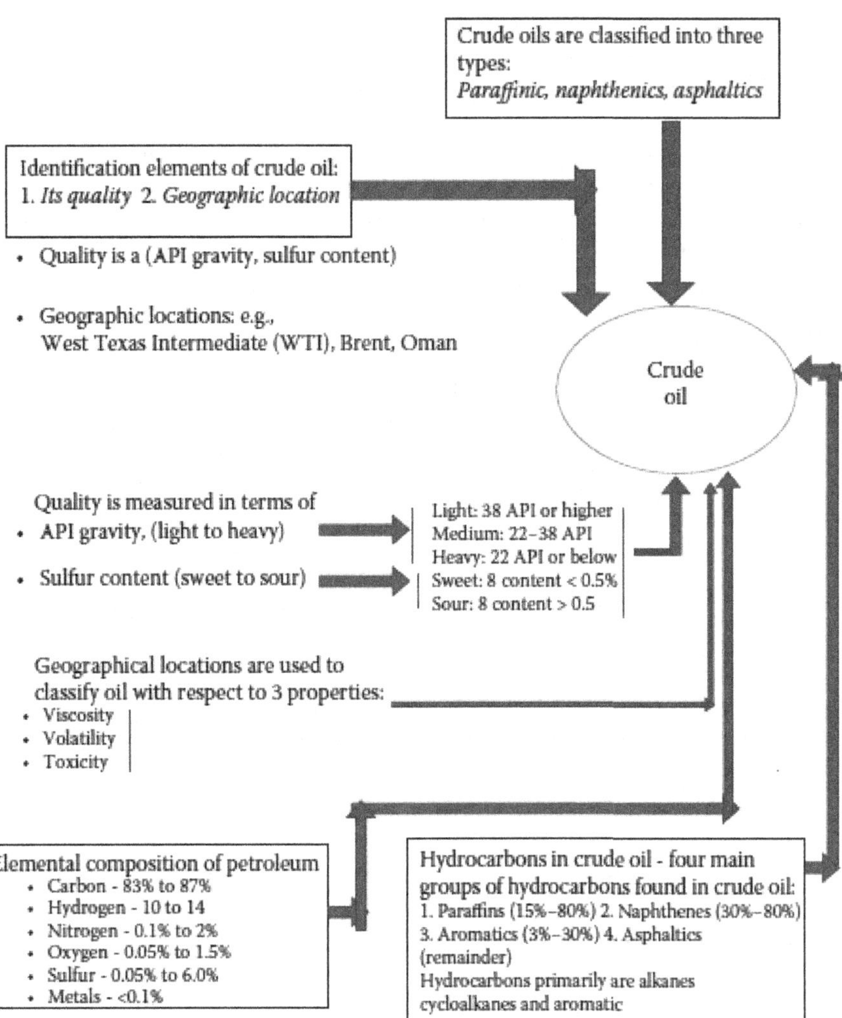

FIGURE 3.1 Facts about crude oil.

TABLE 3.1
Examples of Typical Crude Oils

Classification	API Range	Examples: Crude Name, API
Light	>33	• Saudi Super Light, 39.5 • Nigerian Light, 36 • North Sea Brent, 37
Medium	28–33	• Kuwait, 31 • Venezuela Light, 30
Heavy	<28	• Saudi Heavy, 28 • Venezuela Heavy, 24

Chemical composition describes and identifies the individual chemical compounds isolated from crude oils over the years. *However, no crude oil has ever been completely separated into its individual components*, although many components can be identified. A total of 141 compounds were identified in a sample of Oklahoma crude that account for 44% of the total crude volume. (Source: PetroWiki, published by SPE).

Physical representation, on the other hand, involves considering the crude oil and its products as mixtures of hydrocarbons and describing physical laboratory tests or methods for characterizing their quality.

3.2.1 Chemical Approach

Nearly all petroleum deposits are made up of a mixture of chemical compounds that consist of hydrogen and carbon, known as hydrocarbons, with varying amounts of nonhydrocarbons containing S, N_2, O_2, and other some metals. The composition of crude oil by elements is approximated as shown in Figure 3.1. It could be further stated that these hydrocarbon compounds making up oils are grouped chemically into different series of compounds described by the following characteristics:

- Each series consists of compounds similar in their molecular structure and properties (e.g., the alkanes or paraffin series).
- Within a given series, there exists a wide spectrum of compounds that range from extremely light or simple hydrocarbon to a heavy or complex one. As an example, CH_4 stands for the former group and $C_{40}H_{82}$ for the latter in the paraffinic series.

3.2.1.1 1st Hydrocarbon Series

The major constituents of most crude oils and its products are hydrocarbon compounds, which are made up of hydrogen and carbon only. These compounds belong to one of the following subclasses:

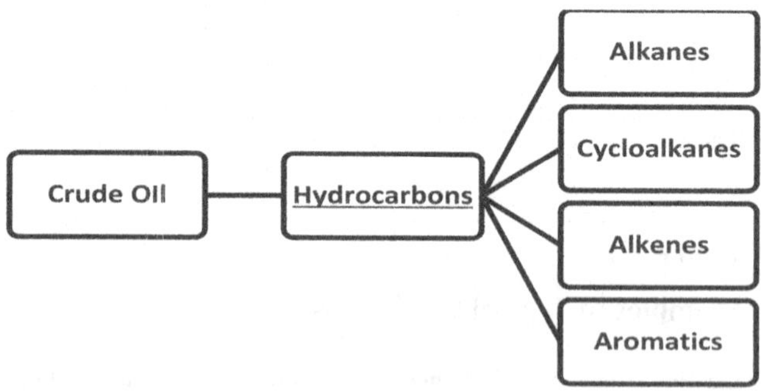

1. *Alkanes or Paraffins:* Alkanes are saturated compounds having the general formula C_nH_{2n+2}. Alkanes are relatively nonreactive compounds in comparison to other series. They may either be straight-chain or branched compounds; the latter are more valuable than the former, because they are useful for the production of high-octane gasoline.

2. *Cycloalkanes or Cycloparaffins (Naphthenes):* Cycloalkanes and bicycloalkanes are normally present in crude oils and its fractions in variable proportions. The presence of large amounts of these cyclic compounds in the naphtha range has its significance in the production of aromatic compounds. Naphtha cuts with a high percentage of naphthenes would make an excellent feedstock for aromatization.
3. *Alkenes or Olefins:* Alkenes are unsaturated hydrocarbon compounds having the general formula C_nH_n. They are practically not present in crude oils, but they are produced during processing of crude oils at high temperatures.

   Alkenes are very reactive compounds. Light olefinic hydrocarbons are considered the base stock for many petrochemicals. Ethylene, the simplest alkene, is an important monomer in this regard. For example, polyethylene is a well-known thermoplastic polymer and polybutadiene is the most widely used synthetic rubber.
4. *Aromatics:* Aromatic compounds are normally present in crude oils. Only monomolecular compounds in the range of C6–C8 (known as B-T-X) have gained commercial importance. Aromatics in this range are not only important petrochemical feedstocks but are also valuable for motor fuels.

Dinuclear and polynuclear aromatic compounds are present in heavier petroleum fractions and residues. *Asphaltenes*, which are concentrated in heavy residues and in some asphaltic crude oils, are, in fact, *polynuclear aromatics* of complex structures. It has been confirmed by mass spectroscopic techniques that condensed-ring aromatic hydrocarbons and heterocyclic compounds are the major compounds of asphaltenes.

3.2.1.2 2nd Nonhydrocarbon Compounds

So far, a brief review of the major classes of the hydrocarbon compounds that exist in crude oils and their products was presented. For completeness, we should mention that other types of nonhydrocarbon compound occur in crude oils and refinery streams. Most important are the following:

- Sulfur compounds
- Nitrogen compounds
- Oxygen compounds
- Metallic compounds

Sulfur Compounds: In addition to the gaseous sulfur compounds in crude oil, many sulfur compounds have been found in the liquid phase in the form of organosulfur. These compounds are generally not acidic. Sour crude oils are those containing a high percentage of hydrogen sulfide. However, many of the organic sulfur compounds are not thermally stable, thus producing hydrogen sulfide during crude processing.

High-sulfur crude oils are in less demand by refineries because of the extra cost incurred for treating refinery products. Naphtha feed to catalytic reformers is hydrotreated to reduce sulfur compounds to very low levels (1 parts per million, ppm) to avoid catalyst poisoning.

The following sulfur compounds are typical:

1. *Mercaptans (H–S–R):* Hydrogen sulfide, H–S–H, may be considered as the simple form of mercaptan; however, the higher forms of the series are even more objectionable in smell. For example, butyl mercaptan (H–S–C_4H_9) is responsible for the unusual odor of the shank.
2. *Sulfides (R–S–R):* When an alkyl group replaces the hydrogen in the sulfur-containing molecule, the odor is generally less obnoxious. Sulfides could be removed by the hydrotreating technique, which involves the hydrogenation of the petroleum streams as follows:

$$\text{R—S—R} + 2\text{H—H} \rightarrow 2\text{R—H} + \text{H—S—H}$$
$$\text{R—S—R} + \text{H—H} \rightarrow \text{R—R} + \text{H—S—H}$$

   The hydrogen sulfide may be removed by heating and may be separated by using amine solutions.
3. *Polysulfides (R–S–S–R):* These are more complicated sulfur compounds and they may decompose, in some cases depositing elemental sulfur. They may be removed from petroleum fractions, similar to the sulfides, by hydrotreating.

Nitrogen Compounds: Nitrogen compounds in crude oils are usually low in content (about 0.1–0.9%) and are usually more stable than sulfur compounds. Nitrogen in petroleum is in the form of heterocyclic compounds and may be classified as basic and non-basic. Basic nitrogen compounds are mainly composed of pyridine homologs and have the tendency to exist in the high-boiling fractions and residues. The nonbasic nitrogen compounds, which are usually of the pyrrole and indole, also occur in high-boiling fractions and residues. Only a trace amount of nitrogen is found in light streams.

During hydrotreatment (hydrodesulfurization) of petroleum streams, hydrodenitrogenation takes place as well, removing nitrogen as ammonia gas, thus reducing the nitrogen content to the acceptable limits for feedstocks to catalytic processes.

It has to be stated that the presence of nitrogen in petroleum is of much greater significance in refinery operations than might be expected from the very small amounts present. It is established that nitrogen compounds are responsible for the following:

- Catalyst poisoning in catalytic processes
- Gum formation in some products such as domestic fuel oils

Oxygen Compounds: Oxygen compounds in crude oils are more complex than sulfur compounds. However, oxygen compounds are not poisonous to processing catalysts. Most oxygen compounds are weakly acidic, such as phenol, cresylic acid, and naphthenic acids. The oxygen content of petroleum is usually less than 2%, although larger amounts have been reported.

Metallic Compounds: Many metals are found in crude oils; some of the more abundant are sodium, calcium, magnesium, iron, copper, vanadium, and nickel. These normally occur in the form of inorganic salts soluble in water—as in the case of sodium chloride—or in the form of organometallic compounds—as in the case of iron, vanadium, and nickel.

The occurrence of metallic constituents in crude oils is of considerably greater interest to the petroleum industry than might be expected from the very small amounts present. The organometallic compounds are usually concentrated in the heavier fractions and in crude oil residues. The presence of high concentration of vanadium compounds in naphtha streams for catalytic reforming feeds will cause permanent poisons. These feeds should be hydrotreated not only to reduce the metallic poisons but also to desulfurize and denitrogenate the sulfur and nitrogen compounds.

Hydrotreatment may also be used to reduce the metal content in heavy feeds to catalytic cracking.

3.2.2 Physical Methods

Having discussed the various chemicals found in crude oils and realizing not only the complexity of the mixture but also the difficulty of specifying a crude oil as a particular mixture of chemicals, we can understand why the early petroleum producers adopted the physical methods generally used for classification.

As may be seen, crude oils from different locations may vary in appearance and viscosity and also vary in their usefulness as producers for final products. It is possible by the use of certain basic tests to identify the quality of crude oil stocks. The tests included in the following list are primarily physical (except sulfur determination):

1. Distillation
2. Density, specific gravity, and API gravity
3. Viscosity
4. Vapor pressure
5. Flash and fire points
6. Cloud and pour points
7. Color
8. Sulfur content
9. Basic sediments and water (B.S.&W.)
10. Aniline point
11. Carbon residue

The details of some of these tests are described next.

3.2.2.1 API Gravity

Earlier, density was the principal specification for petroleum products. However, the derived relationships between the density and its fractional composition were only valid if they were applied to a certain type of petroleum. Density is defined as the mass of a unit volume of material at a specified temperature. It has the dimensions of grams per cubic centimeter.

Another general property, which is more widely, is the specific gravity. It is the ratio of the density of oil to the density of water and is dependent on two temperatures, those at which the densities of the oil sample and the water are measured.

When the water temperature is 4°C (39°F), the specific gravity is equal to the density in the centimetre–gram–second (cgs) system, because the volume of 1 g of water at that temperature is, by definition, 1 mL. Thus, the density of water, for example, varies with temperature, whereas its specific gravity is always unity at equal temperatures. The standard temperatures for specific gravity in the petroleum industry in North America are 60/60°F and 15.6/15.6°C.

Although density and specific gravity are used extensively in the oil industry, the API gravity is considered the preferred property. It is expressed by the following relationship:

$$^{\circ}API = \frac{141.5}{\gamma} - 131.5$$

where, γ is the oil specific gravity at 60°F. Thus, in this system, a liquid with a specific gravity of 1.00 will have an API of 10 degrees. A higher API gravity indicates a lighter crude or oil product, whereas a low API gravity implies a heavy crude or oil product.

3.2.2.2 Carbon Residue

Carbon residue is the percentage of carbon by weight for coke, asphalt, and heavy fuels found by evaporating oil to dryness under standard laboratory conditions. Carbon residue is generally referred to as CCR (Conradson carbon residue). It is a rough indication of the asphaltic compounds and the materials that do not evaporate under conditions of the test, such as metals and silicon oxides.

3.2.2.3 Viscosity

The viscosity is the measure of the resistance of a liquid to flow, hence indicating the "pumpability" of oil.

3.2.2.4 Pour Point

This is defined as the lowest temperature (5°F) at which the oil will flow. The lower the pour point, the lower the paraffin content of the oil.

3.2.2.5 Ash Content

This is an indication of the contents of metal and salts present in a sample. The ash is usually in the form of metal oxides, stable salts, and silicon oxides. The crude sample is usually burned in an atmosphere of air and the ash is the material left unburned.

3.2.2.6 Reid Vapor Pressure

The Reid Vapor Pressure (RVP) is a measure of the vapor pressure exerted by oil or by light products at 100°F.

3.2.2.7 Metals

In particular, arsenic, nickel, lead, and vanadium are potential poisons for process catalysts. Metal contents are reported in ppm.

3.2.2.8 Nitrogen

It is the weight of total nitrogen determined in a liquid hydrocarbon sample (in ppm). Nitrogen compounds contribute negatively to process catalysts.

3.2.2.9 Salt Content

Salt content is typically expressed as pounds of salt (sodium chloride, NaCl) per 1,000 barrels of oil (PTB). Salts in crude oil and in heavier products may create serious corrosion problems, especially in the top-tower zone and the overhead condensers in distillation columns.

3.2.2.10 Sulfur

This is the percentage by weight (or ppm) of total sulfur content determined experimentally in a sample of oil or its product. The sulfur content of crude oils is taken into consideration in addition to the API gravity in determining their commercial values. It has been reported that heavier crude oils may have higher sulfur content.

3.2.2.11 Hydrogen Sulfide

Hydrogen sulfide dissolved in a crude oil or its products is determined and measured in parts per million. It is a toxic gas that can evolve during storage or in the processing of hydrocarbons.

The above tests represent many properties for the crude oils that are routinely measured because they affect the transportation and storage facilities. In addition, these properties define what products can be obtained from a crude oil and contribute effectively to safety and environmental aspects. The price of a crude oil is influenced by most of these properties.

To conclude, it can be stated that light and low-sulfur crude oils are worth more than heavy and high-sulfur ones as illustrated by the next diagram.

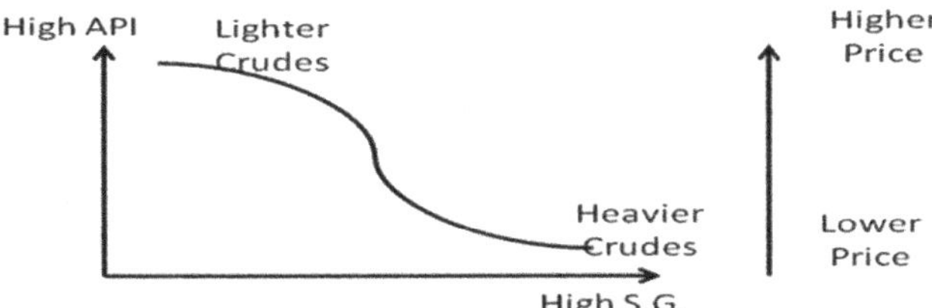

One can summarize the two approaches of examining crude oils as follows:

1. Chemical composition
2. Physical properties:
    a. API, S, salt, metals, nitrogen, and so forth
    b. Distillation: ASTM, TBP, EFV
    c. Correlations: Kw, Ind

where TBP is true boiling point, EFV is equilibrium flash vaporization, Kw is Watson characterization factor, Ind is U.S. Bureau of Mines correlation index.

3.3 CLASSIFICTION OF CRUDE OILS

3.3.1 Broad Classification (Based on Chemical Structures)

Although there is no specific method for classifying crude oils, it would be useful for a refiner to establish some simple criteria by which the crude in hand would be classified. A broad classification of crudes has been developed based on some simple physical and chemical properties. Crude oils are generally classified into three types depending on the relative amount of the hydrocarbon class that predominates in the mixture. These are:

1. *Paraffinic* constituents are predominantly paraffinic hydrocarbons with a relatively lower percentage of aromatics and naphthenes.
2. *Naphthenics* contain relatively a higher ratio of cycloparaffins and a higher amount of asphalt than in paraffinic crudes.
3. *Asphaltics* contain relatively a large amount of fused aromatic rings and a high percentage of asphalt.

3.3.2 Classification by Chemical Composition

Petroleum contains a large number of chemicals with different compositions depending on the location and natural processes involved. Petroleum composition varies (molecular type and weight) from one oil field to another, from one well to another in the same field and even from one level to another in the same well.

A correlation index was introduced to indicate the crude type or class. The following relationship between the midboiling point of the fraction and its specific gravity gives the correlation index, known as "Bureau of Mines Correlation Index":

$$\mathrm{BMCI} = \frac{48640}{\mathrm{K}} + (473.7\mathrm{d} - 456.8)$$

K = midboiling point of a fraction in Kelvin degrees
d = specific gravity of the fraction at 60/60°F

It is possible to classify crudes as paraffinic, naphthenic (mixed) or asphaltic according to the calculated values using the above relationship. A zero value has been assumed for paraffins and 100 for aromatics.

3.3.3 Classification by Density

Density gravity (specific gravity) has been extensively applied to specify crude oils. It is a rough estimation of the quality of a crude oil. Density is expressed in terms of API gravity by the following relationship:

$$\mathrm{API} = [141.5 / \text{specific gravity}] - 131.5$$

Another index used to indicate the crude type is the Watson characterization (UOP) factor. This also relates the midboiling point of the fraction in Kelvin degrees to the density.

TABLE 3.2
General Properties of Crude Oils

Property	Paraffin Base	Asphalt Base
API gravity	High	Low
Naphtha content	High	Low
Naphtha octane number	Low	High
Naphtha odor	Sweet	Sour
Kerosene smoking tendency	Low	High
Diesel-fuel knocking tendency	Low	High
Lube-oil pour point	High	Low
Lube-oil content	High	Low
Lube-oil viscosity index	High	Low

$$\text{Watson correlation factor} = \frac{(K^{1/3})}{d}$$

A value higher than 10, indicates predominance of paraffins while a value lower than 10 indicates predominance of aromatics. Properties of crude oils will thus vary according to their base type, as shown in Table 3.2.

3.4 CRUDE OIL COMPARISONS AND CRUDE OIL ASSAY

In order to establish a basis for the comparison between different types of crude oil, it is necessary to produce experimental data in the form of what is known as an "assay". The assay can be an inspection assay or comprehensive assay. Crude assays are described as the systematic compilation of data for the physical properties of the crude and its fractions, as well as the yield. In other words, a crude assay involves the determination of the following:

The properties of crude oil
The fractions obtained: (a) their percentage yield and (b) properties

Analytical testing only without carrying out distillation may be considered an assay. However, the most common assay is a comprehensive one that involves all of the above-stated parameters.

The basis of the assay is the distillation of a crude oil under specified conditions in a batch laboratory distillation column, operated at high efficiency [column with 14 plates and reflux ratio (RR)]. Pressure in column is reduced in stages to avoid thermal degradation of high boiling components.

A comparison of the characteristics of different types of crude oil over the distillation range could be made via a graph that relates the following:

- The density of distillate fractions
- Their midboiling points

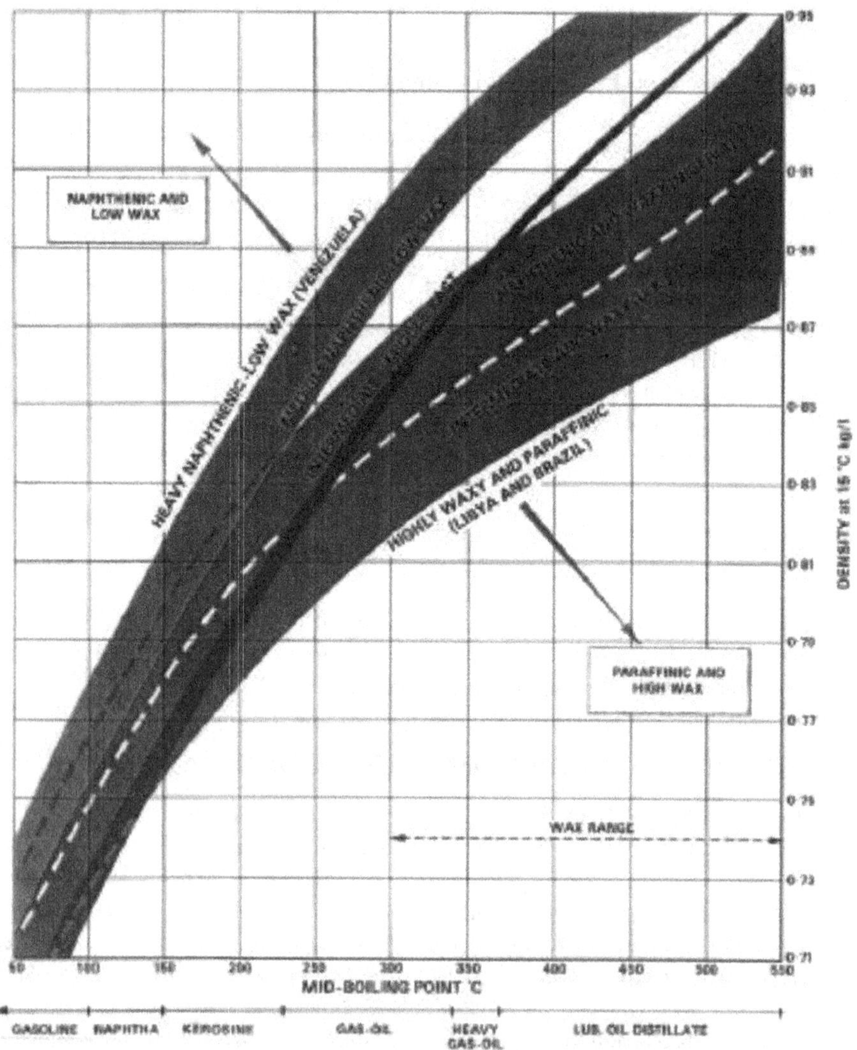

FIGURE 3.2 Comparisons of types of crude oils based on density/midboiling point.

Such a comparison is illustrated in Figure 3.2. The density level of crude at given boiling point on the curve is a function of the relative proportions of the main three hydrocarbon series: aromatics, cycloparaffins, paraffins; their densities decrease in that order.

In addition, a comparison with some standard crude is usually recommended. A review for some of the crudes that are adopted in the oil industry is given next.

3.4.1 Benchmark

The three most quoted oil products are North America's West Texas Intermediate Crude (WTI), North Sea Brent Crude, and the UAE Dubai Crude, and their pricing is used as a barometer for the entire petroleum industry, although, in total, there are 46 key oil exporting countries. Brent Crude is typically priced at about $2 dollars over the WTI Spot price, which is typically priced $5 to $6 dollars above the Energy Information Administration (EIA)'s Imported Refiner Acquisition Cost (IRAC) and OPEC Basket prices.

3.4.2 Benchmark Crude

A benchmark crude or marker crude is a crude oil that serves as a reference price for buyers and sellers of crude oil. There are three primary benchmarks, WTI, Brent Blend, and Dubai. Other well-known blends include:

- Opec Basket used by OPEC
- Tapis Crude, which is traded in Singapore
- Bonny Light used in Nigeria and Mexico's Isthmus

Energy Intelligence Group publishes a handbook which identified 195 major crude streams or blends in its 2011 edition. Benchmark crude oil that is traded so regularly in the spot market that its price quotes are relied upon by sellers of other crude oils as a reference point for setting term or spot prices. Brent, West Texas Intermediate, and Dubai are all benchmark crude oils.

3.4.3 Heavy vs Light and Sweet vs Sour

A concise summary for the properties of heavy crude versus light, and sweet crude versus sour is presented in Figure 3.3.

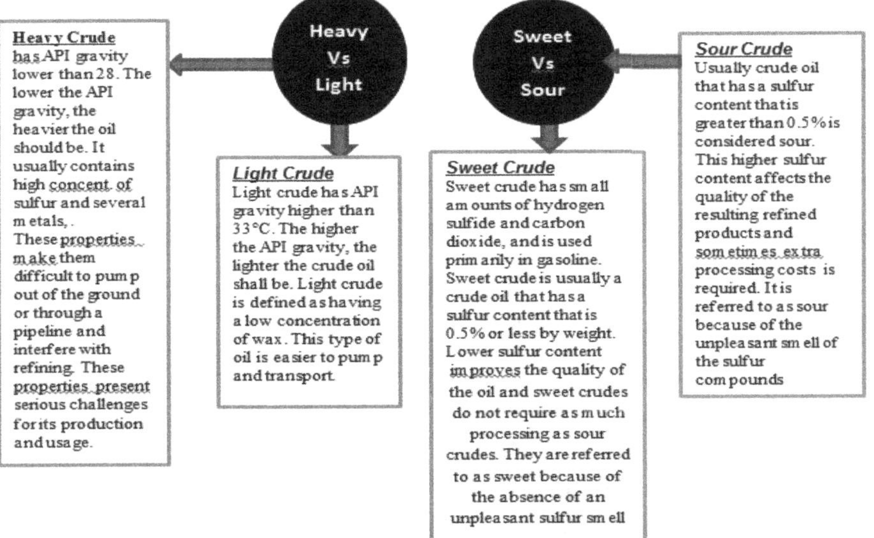

FIGURE 3.3 Summary for the properties of heavy vs light and sweet vs sour.

3.4.3.1 Heavy Crude

Heavy crude has API gravity lower than 28 degrees. The lower the API gravity, the heavier the oil shall be. It usually contains high concentrations of sulfur and several metals, particularly nickel and vanadium (high amount of wax). These are the properties that make them difficult to pump out of the ground or through a pipeline and interfere with refining. These properties also present serious environmental challenges to the growth of heavy oil production and use.

3.4.3.2 Light Crude

Light crude has API gravity higher than 33 degrees Celsius. The higher the API gravity, the lighter the crude oil shall be. Light crude is defined as having a low concentration of wax. This classification of oil is easier to pump and transport.

3.4.3.3 Sweet Crude

Sweet crude has small amounts of hydrogen sulfide and carbon dioxide, and is used primarily in gasoline. Sweet crude is usually a crude oil that has a sulfur content that is 0.5% or less by weight. Lower sulfur content improves the quality of the resulting refined products, and sweet crudes do not require as much processing as sour crudes. They are referred to as sweet because of the absence of an unpleasant sulfur smell.

3.4.3.4 Sour Crude

Usually crude oil that has sulfur content that is greater than 0.5% is considered sour. This higher sulfur content affects the quality of the resulting refined products and sometimes means extra processing is required. It is referred to as sour because of the unpleasant smell of the sulfur.

3.5 COMPOSITION OF NATURAL GAS

3.5.1 INTRODUCTION

Natural gas is a combustible mixture of hydrocarbon gases. While natural gas is formed primarily of methane, it also includes ethane, propane, butane, and pentane. N_2, CO_2, H_2S, Helium are found in natural gas resources. The next picture illustrates the composition. The composition of natural gas can vary widely. A typical composition of natural gas exits a well is shown in Table 3.3.

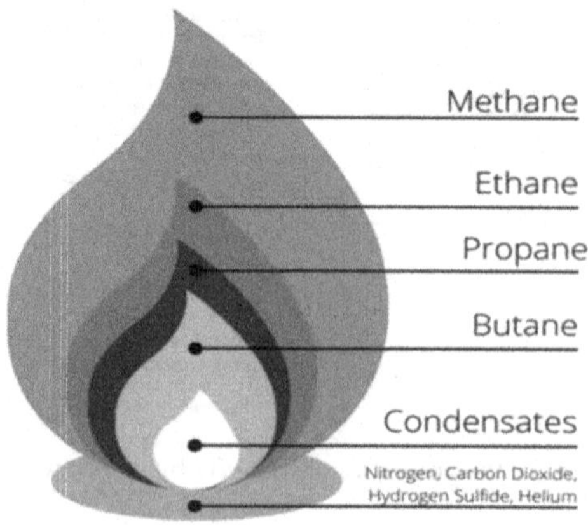

TABLE 3.3
Typical Composition of Natural Gas

Component	Typical Analysis (Mole %)	Range (Mole %)
Methane	94.7	87.0–98.0
Ethane	4.2	1.5–9.0
Propane	0.2	0.1–1.5
Isobutane	0.02	trace–0.3
Normal Butane	0.02	trace–0.3
Isopentane	0.01	trace–0.04
Normal Pentane	0.01	trace–0.04
Hexanes plus	0.01	trace–0.06
Nitrogen	0.5	0.2–5.5
Carbon Dioxide	0.3	0.05–1.0
Oxygen	0.01	trace–0.1
Hydrogen	0.02	trace–0.05
Specific Gravity	0.58	0.57–0.62
Gross Heating Value (MJ/m^3), dry basis*	38.8	36.0–40.2
Wobbe Number (MJ/m^3)	50.9	47.5–51.5

* The gross heating value is the total heat obtained by complete combustion at constant pressure of a unit volume of gas in air, including the heat released by condensing the water vapor in the combustion products (gas, air, and combustion products taken at standard temperature and pressure).

Typical combustion properties of natural gas are as shown here:

- Ignition Point: 564°C*
- Flammability Limits: 4–15% (volume % in air)*
- Theoretical Flame Temperature (stoichiometric air/fuel ratio): 1,953°C*
- Maximum Flame Velocity: 0.36 m/s*

In its purest form, natural gas, such as that used domestically, is almost pure methane. The distinctive "rotten egg" smell that we often associate with natural gas is actually an odorant called "mercaptan", added to the gas before it is delivered to the end user.

3.5.2 Types of Natural Gas

Raw natural gas comes from three types of wells:

1. Oil wells, where natural gas is typically termed *associated gas*. This gas exists separate from oil in the formation (free gas) or dissolved in the crude oil (dissolved gas). Dissolved gas is that portion of the gas dissolved in the

* Information provided is from the Ortech Report No. 26392, Combustion Property Calculations for a typical Union Gas Composition, 2017.
www.uniongas.com › about-us › chemical-composition-o.

TABLE 3.4
South Pars/North Field Gas Reserve *"South Pars Attracts $15b in Domestic Investment"*

	In-place gas reserve		Recoverable gas reserve	
	cu km (km^3)	Trillion cu ft (ft^3)	cu km (km^3)	Trillion cu ft (ft^3)
South Pars	14,000	500	10,000	360
North Dome	37,000	1300	26,000	900
Total	51,000	1800	36,000	1260

Note: 1 km^3 = 1,000,000,000 m^3 = 1 Billion m^3 = 1 Trillion Liters

crude, and associated gas (called gas cap) is free gas in contact with the crude oil. Natural gas extracted from oil wells is called *casinghead gas.*

2. Gas wells, in which there is little or no crude oil, are termed *nonassociated gas.* Gas wells typically produce raw natural gas by itself (dry gas).
3. Condensate wells producing free natural gas along with a semiliquid hydrocarbon condensate (wet gas).

All crude oil reservoirs contain dissolved gas, and may or may not contain associated gas. Whatever the source of the natural gas is, once separated from crude oil (if present) it commonly exists in mixtures of methane with other hydrocarbons, principally ethane, propane, butane, and pentanes. In addition, raw natural gas contains water vapor along with some nonhydrocarbons such as hydrogen sulfide (H_2S), carbon dioxide (CO_2), helium, nitrogen, and other compounds.

The world's largest gas field is the offshore South Pars/North Dome Gas-Condensate field, shared between Iran and Qatar. It is estimated to have 51,000 cubic kilometers (12,000 cu mi) of natural gas and 50 billion barrels (7.9 billion cubic meters) of natural gas condensates. According to the IEA, the field holds an estimated 1,800 trillion cubic feet (51 trillion cubic meters) of in-situ natural gas and some 50 billion barrels (7.9 billion cubic meters) of natural gas condensates, as shown in Table 3.4.

Section II

Fundamentals of Engineering Economic Analysis

The contents of this section are classified as shown next

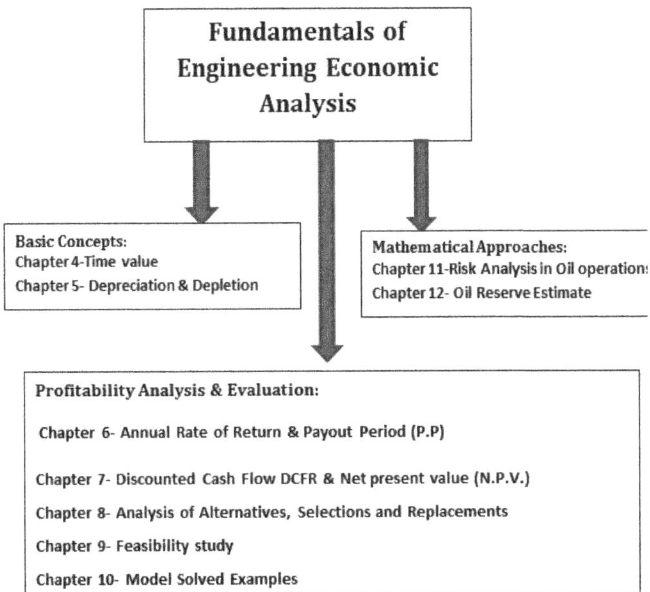

Classification of fundamentals of engineering economic analysis

INTRODUCTION

1st Background

Basic knowledge and techniques for performing investment analysis are presented in this part. This part provides the reader with what we may call it the **Tools** for tackling economic and investment problems in the petroleum sector, keeping in mind that the economic viability of each potential solution is normally considered along with the technical aspects.

Some fundamental principles and concepts are described in the following introductory remarks.

Compound interest tables are helpful in calculation in this section. The tables are provided in Appendix B.

Principle 1: ONE DOLLAR NOW IS WORTH MORE THAN A DOLLAR AT A LATER TIME. THIS EXPLAINED BY THE TIME VALUE OF MONEY (TVM)

Principle 2: Three parameters influence the TVM:

1. inflation
2. risk
3. cost of money (interest)

Of these, the cost of money is the most predictable, and, hence, it is the essential component in our economic analysis.

Principle 3: Additional risk is not taken without the expected additional return.

Next, fundamental engineering economic concepts are discussed in Chapters 4–7 of this part, along with example problems to illustrate the use of various theoretical solutions. This covers the following:

Basic Tools: The mathematical and practical "tools" used in investment analysis for evaluating profitability, and known as economic decision criteria are:

- Annual Rate of Return (ARR)
- Payout or Payback Period (PP)
- Discounted Cash Flow Rate of Return DCFR)
- Net Present Value (NPV)

Basic Concepts

- Cash flow
- Interest rate and time value of money
- Equivalence technique

Cash-Flow Concepts

Cash flow is the stream of monetary (dollar) values, costs (inputs), and benefits (outputs), resulting from a project investment.

Cost of money is represented by

(1) Money paid for the use of a borrowed capital, or (2) A return on investment.

Considering the time value of money is central to most engineering economic analyses. Cash flows are *discounted* using an *interest rate*, *i*, except in the most basic economic studies.

Cost of money is determined by an interest rate.

Time value of money is defined as the time-dependent value of money stemming both from changes in the purchasing power of money (inflation or deflation) and from the real earning potential of alternative investments over time.

Cash-Flow Diagrams[1]

A picture is worth more than one thousand words. It is difficult to solve a problem if you cannot see it. The easiest way to approach a problem in economic analysis is to draw a picture that shows three items:

a. A time interval divided into an appropriate number of equal periods
b. All cash outflows (deposits, expenditures, etc.) in each period
c. All cash inflows (withdrawals, income, etc.) for each period

2nd: Throughout Section 2, we will present the materials in different fashion. Formulas are presented first, without the elaborate derivation, followed next by Applications.

[1] Unless otherwise indicated, all such cash flows are considered to occur at the end of their respective periods.

Section II.I: Basic Concepts

Chapters 4 and 5 come under "**Basic Concepts**". Simply both chapters offer distinctive principles. Chapter 4 emphasizes the role of TVM calculations in our economic. In Chapter 5, on the other hand the concepts of depreciation and depletion are clearly explained.

4 Time Value of Money in Capital Expenditures

TVM in oil operations is signaled in this chapter: One dollar now is worth more than a dollar at a later time.

4.1 INTRODUCTION

Time value of money (TVM) is the value of money figuring at a given amount of interest earned over a given amount of time. TVM is the central concept in finance theory. It is essential to consider the effect time has on capital, as capital must always produce some yield.

Most economic problems in the petroleum industry involve determining what is economical in the long run; that is, over a period of time.

A dollar now is worth more than the prospect of a dollar next year or at some later date.

All of the standard calculations for TVM are based on the most basic algebraic expression for the present value P, of a future sum F, discounted at an interest rate i. This is formulated mathematically by the expression:

$$P = F/(1+i)$$

4.2 BASIC DEFINITIONS

Money invested in oil projects is used, basically, for the following purposes:

1. To purchase and install the necessary machinery, equipment, and other facilities. This is called *fixed capital investment.* Basically this investment is *depreciable.*
2. To provide the capital needed to operate an oil field or a refinery as well as the facilities associated with them. This is what we call *working capital.* Principally it is capital tied up in raw material inventories in storage, in process inventories, finished product inventories, cash for wages, utilities, etc. This is a *non-depreciable* investment.

Working capital must not be ignored in a preliminary estimate of needs for capital investment, because (1) it is usually a sizable amount of any total investment, and (2) no economic picture of oil processing is complete without inclusion of working capital. Investment in working capital is no different than investment in fixed capital except that the former does not depreciate.

Working capital can theoretically be recovered in full when any refinery, or oil field for that matter, shuts down. Yet capital is "tied up" when the refinery is operating and must be considered part of the total investment. Working capital, however, is normally replaced as it is used up by sales dollars the oil company receives for crudes or refined oil products. Therefore, we can safely say that this capital is always available for return to owners. This is not the immediate case with depreciable capital.

Interest may be defined as the compensation paid for the use of borrowed capital. The recognized standard is the *prime interest rate*, which is charged by banks to their customers. In contrast to this definition, which is the one adopted by engineers, the classical definition, on the other hand, describes interest as the money returned to the investors for the use of their capital. This would mean that any profit obtained by using this capital is considered interest, which is not true. Instead a distinction is to be made between interest and the rate of return on capital.

4.3 TYPES OF INTEREST

As shown in Figure 4.1, interest may be described as *simple* or *compound* interest. Simple interest, as the name implies, is not compounded; it requires compensation payment at a constant interest rate based only on the original principal. In compound

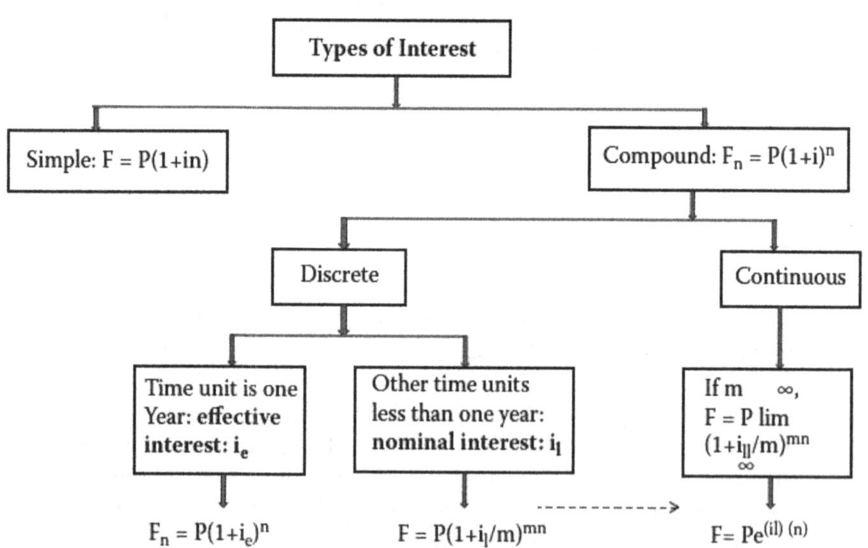

FIGURE 4.1 Mathematical definition and classification of interest.

interest, the interest on the capital due at the end of each period is added to the principal; interest is charged on this converted principal for the next time period. Most oil economics are based on compound interest.

4.4 INTEREST CALCULATION

Now, if P represents the principal (in dollars), n the number of time units (in years) and i the interest rate based on the length of one interest period, then:

Using simple interest: the amount of money to be paid on the borrowed capital P, is given by: (P) (i) (n).

Hence, the sum of capital plus the interest due after n interest periods will be denoted by:

$$F = P + Pin = P(1 + in) \tag{4.1}$$

where F is the future value of the capital P.

Using compound interest: the amount due after any *discrete* number of interest periods can be calculated as follows:

	Principal Capital Available	Interest Earned on P	Principal Plus Interest
For the 1st period	P	Pi	$P(1+i)$
For the 2nd period	$P(1+i)$	$P(1+i)i$	$P(1+i)^2$
	⋮	⋮	⋮
For the nth period	$P(1+i)^{n-1}$	$P(1+i)^{n-1}i$	$P(1+i)^n$

Thus, the general equation is given by:

$$F = P(1 + i)^n \tag{4.2}$$

- Interest may be defined as the compensation paid for the use of borrowed capital. The recognized standard is the *prime interest rate*, which is charged by banks to their customers.
- As shown in Figure 4.1, interest may be described as *simple* or *compound*. In compound interest, the interest on the capital due at the end of each period is added to the principal; interest is charged on this converted principal for the next time period. Most oil economics are based on compound interest.
- *Using compound interest*: the amount due after any *discrete* number of interest periods can be calculated by the following:
- In common engineering practice, 1 year is assumed as the discrete interest period; however, there are many cases where other time units are employed. Thus, the way interest rates are quoted affects the return on investment. For instance, the future value after 1 year of $1,000 compounded annually at 6% is $1,060.

Example 4.1

To illustrate the value of knowledge of the effective interest rate to oil management, assume that a short-term loan for 1 year only could be arranged for an oil company in temporary distress. The company needs \$100,000 for immediate working capital at either a nominal rate of 12% compounded monthly or a nominal rate of 15% compounded semiannually.

The oil company wants to know which arrangement would provide the oil company with the lower debt at the end of the short-term loan period. The use of the effective interest rate formula gives the answer:

SOLUTION

On a nominal 12% rate compounded monthly:

$$\begin{aligned}\text{Effective interest rate} &= \left(1+\frac{0.12}{12}\right)^{12}-1\\ &= (1.01)^{12}-1\\ &= 1.127-1 = 0.127, \text{or } 12.7\%\end{aligned}$$

On a nominal 15% rate, compounded semiannually:

$$\begin{aligned}\text{Effective interest rate} &= \left(1+\frac{0.15}{2}\right)^{2}-1\\ &= (1.075)^{2}-1\\ &= 1.156-1 = 0.156, \text{or } 15.6\%\end{aligned}$$

Conclusion: The loan at 12% compounded monthly has the lower effective interest rate, or 12.7%, while 15.6% for the loan arrangement using a nominal rate of 15% compounded semiannually.

4.5 ANNUITIES AND PERIODIC PAYMENTS

Compound interest and discount factors are defined as follows:

$$C = (1+i)^n$$
$$D = (1+i)^{-n}$$

An annuity is a series of equal payments occurring at equal time intervals, normally at the end of the period. Payments of this type are used to accumulate a desired amount of capital as in depreciation calculations, where engineers face the problem of an unavoidable decrease in value of equipment. The amount of an annuity is the sum of all payments plus interest if allowed to accumulate at a definite rate of interest during the annuity term. Thus, the basic equation (sinking fund factor) is given here.

Finally, the sum of all payments will be F, where,

$$F = A(1+i)^{n-1} + A(1+i)^{n-2} + \cdots + A(1+i) + A$$

Hence, it can be shown that

$$F = \frac{A[(1+i)^n - 1]}{1}$$

The above factor $[(1 + i)^n - 1]/i$ is known as the compound amount factor or sinking fund factor.

4.6 CAPITALIZED COSTS

4.6.1 Calculation of Capitalized Costs of an Asset to Be Replaced Perpetually

Here, we have to establish what is known as "perpetuity". In an annuity, periodic payments were made for a definite number, n years. However, in perpetuity, the periodic payments continue indefinitely:

i.e. Annuity $\rightarrow n$ years
Perpetuity $\rightarrow \infty$ years

To establish a perpetuity based on capitalized costs for an equipment, we should have an accumulated amount of money, K, in order to provide funds for:

1. The capital cost of the new equipment, C_v
2. The capital investment P, the present worth of the same asset, such that at the end of n years, this P should have generated enough money for replacing the equipment, perpetually, i.e., to provide C_R

$$\text{Hence, } K = C_V + P$$

The capitalized cost K is given next:

$$K = V_s + \left[\frac{C_R(1+i)^n}{(1+i)^n - 1}\right]$$

where, $C_v = C_R + V_s$; that is, the cost of new equipment equals the replacement cost plus the salvage value (V_s).

4.6.2 Calculation of the Capitalized Costs of a Perpetual Annual Expense

In determining what the value is at the present time for a perpetual series of annual payments in the future, the equipment "capitalized cost" of annual operating costs, such as repairs and maintenance, that must be paid in an indefinite number of periods in the future—in order to continue the given services—is considered in this section. Thus, if repairs and maintenance in an oil field cost \$300,000 yearly on the average, the capitalized costs of such continuous expenses at an interest rate of, say, 8% will be: 300,000/0.08 = \$3,750,000. This is the equivalent cost of a series of annual operating costs.

To generalize this approach, the capitalized cost in this case is defined as follows:

$$\text{Capitalized cost} = \frac{\text{total annual operating expenses}}{\text{average interest rate}}$$

If the capitalized cost of an asset involves annual operating expenses, the above equation should be written as

$$K = V_s + \left[\frac{C_R(1+i)^n}{(1+i)^n - 1}\right] + \frac{\text{annual operating expenses}}{i}$$

4.7 EQUIVALENCE

The knowledge of equivalent values can be of importance to oil companies. The concept of equivalence is the cornerstone for comparisons of time values of money comparisons. Incomes and expenditures are identified with time as well as with amounts. Alternatives with receipts and disbursements can be compared by use of equivalent results at a given date, thus aiding in decision-making.

The concept that payments that differ in total magnitude but that are made at different dates may be equivalent to one another is an important one in engineering economy.

Specifically, three factors are involved in the equivalence of sums of money:

1. The capital investment involved
2. The time
3. The interest rate

The following examples illustrate the concept of equivalence.

Example 4.2

A sum of $10,000 is borrowed by a refining oil company. Propose four different equivalent plans of money payments for this capital over a period of 10 years assuming the interest rate is 6%.

SOLUTION

As shown in Table 4.1, plan I involves the annual payment of the interest only ($600) until the end. Plans II and III involve systematic reduction of the principal of the debt ($10,000). For plan II this is done by uniform repayment of principal ($l,000/yr) along with diminishing interest, while for plan III a scheme is devised to allow for uniform annual payment for both capital and interest all the way through until the end ($1,359). For plan IV, on the other hand, payment is done only once at the end of the 10th year.

TABLE 4.1
Summary for the Four Plans for Solving Example 4.2

Year	Investment ($)	I ($)	II ($)	III ($)	IV ($)
0					
1		600	1,600	1,359	
2		600	1,540	1,359	
3		600	1,480	1,359	
4		600	1,420	1,359	
5		600	1,360	1,359	
6		600	1,300	1,359	
7		600	1,240	1,359	
8		600	1,180	1,359	
9		600	1,120	1,359	
10		10,600	1,060	1,359	17,908

Example 4.3

Show how $100,000 received by an oil company today can be translated into equivalent alternatives. Assume money is worth 8%.

SOLUTION

Cash flow is translated to a given point in time by determining the present value or the future value of the cash flow. Accordingly, $100,000 today is equivalent to $215,900 10 years from now (using the formula Find F/Given P in the next section or the tables in Appendix A). Also, $100,000 today is equivalent to $25,046 received at the end of each year for the next 5 years (using the formula Find A,/Given P). Many other options can be selected for different periods of time. Figure 4.2 illustrates this concept.

4.8 FORMULAS AND APPLICATIONS: SUMMARY

A list of the fundamental formulas dealing with interest, which express the relationship between the set of variables:

i to represent interest rate per interest period
n to represent number of periods of interest payments (yr, month, … etc.)
P to represent value of principal, $(present)
A to represent annual payments or receipts, $/yr
F to represent future value, $

is given as follows:

1. Find F/Given P $\quad F = P(1+i)^n \quad$ (4.3)

2. Find P/Given F $\quad P = F(1+i)^{-n} \quad$ (4.4)

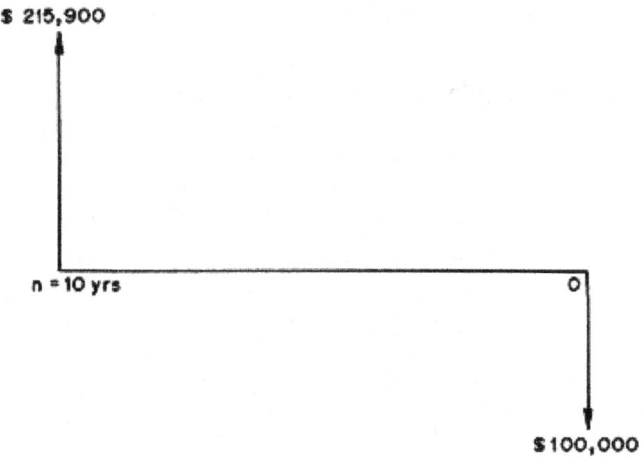

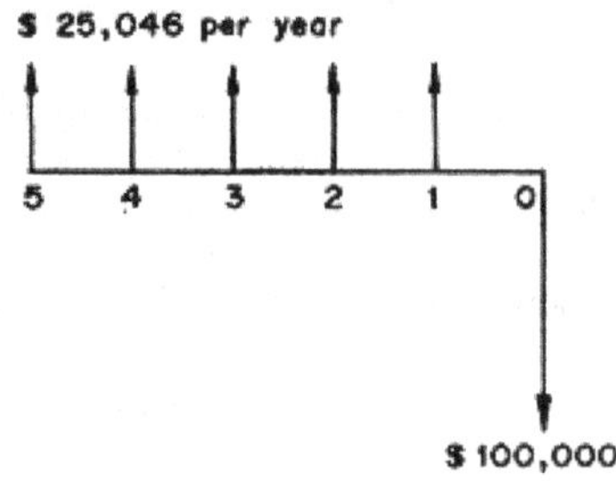

FIGURE 4.2 Solution of example 4.2.

3. Find A/Given F $$A = F\left[\frac{i}{(1+i)^n - 1}\right] \tag{4.5}$$

4. Find A_r/Given P $$A_r = P\left[\frac{i(1+i)^n}{(1+i)^n - 1}\right] \tag{4.6}$$

5. Find P/Given A_r $$P = A_r\left[\frac{(1+i)^n - 1}{i(1+i)^n}\right] \tag{4.7}$$

6. Find A_r/Given A_d $$A_r = A_d(1+i)^n \tag{4.8}$$

4.9 PRACTICAL APPLICATIONS AND SELECTED CASE

Additional examples illustrating the practical applications of each of these interest formulas are presented next:

Example 4.4

In 10 years, it is estimated that $144,860 (future value) will be required to purchase several cooling towers. Interest available at the bank is 8% compounded annually. Calculate the annual annuity payment that will amount to the given fund after 10 years of deposit.

SOLUTION

Using the compound interest tables in Appendix A, and the formula Find A/Given F [Equation (4.25)] for 8% and 10 years, we get:

$$A = (144{,}860)(0.06903)$$
$$= \$10{,}000 \text{ yearly}$$

Thus, each year a payment or deposit of $10,000 should be made into the sinking fund at 8% compounded annually. After 10 years, the fund will contain $144,860 with which the oil company can purchase cooling towers as provided for by the fund. Table 4.2 tabulates the future value at the end of 10 years of $144,860, with total deposits of $100,000. At the end of the second year the fund shows a total of $38,500, and at the end of the fifth year a total of $86,200. Amounts into the fund, including interest, decrease as each year progresses, with no interest being included in the tenth payment.

TABLE 4.2
Tabulation of Results for Example 4.4

Year	Payment into Fund	Compound Interest Factor	Compound Interest	Payment with Interest into Fund (Col. 2 × Col. 3) Fund	Amount in Sinking Fund
1	$10,000	1.999	$(0.08)^9$	$19,990	$19,990
2	10,000	1.851	$(0.08)^8$	18,510	38,500
3	10,000	1.714	$(0.08)^7$	17,140	55,640
4	10,000	1.587	$(0.08)^6$	15,870	71,510
5	10,000	1.469	$(0.08)^5$	14,690	86,200
6	10,000	1.360	$(0.08)^4$	13,600	99,800
7	10,000	1.260	$(0.08)^3$	12,600	112,400
8	10,000	1.166	$(0.08)^2$	11,660	124,060
9	10,000	1.080	$(0.08)^1$	10,800	134,860
10	10,000	0	(0)	$10,000	144,860
Total:	$100,000			$144,860	

Example 4.5

A sinking fund is to be established to cover the capitalized cost of temperature recorders. The recorders cost $2,000 and must be replaced every 5 years. Maintenance and repairs come to $200 a year. At the end of 5 years the accumulated sinking fund deposits are expected to cover the capitalized cost of continuous expense for these recorders. How much money must be deposited each year, at an interest rate of, say, 5%, to cover the capitalized costs at the end of 5 years?

SOLUTION

Two methods are proposed to solve this problem:

1. Using Equation (4.22), where $V_s = 0$ and $C_R = \$2{,}000$:

$$\begin{aligned} \text{K, total capitalized cost} &= 2000\frac{(1.05)^5}{(1.05)^5 - 1} + \frac{200}{0.05} \\ &= 9{,}238.5 + 4{,}000 \\ &= \underline{\$13{,}238.5} \end{aligned}$$

The capitalized cost due to the replacement of the equipment only is 9,238.5 – 2,000 = $7,238.5.

The annual expenditures corresponding to this sum = (7,238.5) (0.05) = $362.

In other words, the *total annual* expenditures to be deposited will be 362 + 200 = $562.

2. Using the formula Find A/Given F [Equation 4.25), we get:

$$A = (2{,}000)(0.18097) = \$362 \text{ per year}$$

Then adding $362 to $200, we get $562.

If these annual deposits of $362 are invested in a sinking fund deposit at 5%, they will be worth exactly $2,000 at the end of 5 years.

Again, the addition of $200 to this $362 will give the required annual deposit of $562 obtained by the 1st method.

A further check is done on the total capitalized cost as follows:

$$562/0.5 = \$11{,}240$$

Hence, 11,240 + 2,000 = $13,240

Example 4.6

In Example 4.5 the annual sum of money of $362 was calculated, which recovers the principal value of the temperature recorders if deposited in a sinking fund ($2,000). What about the cost of the capital (interest on capital)?

Calculate the annual capital recovery costs (A_r) and compare with the annual depreciation cost (A_d).

SOLUTION

In solving Example 4.6, the annual depreciation cost was calculated by: Example 5.5

$$A_d = 2{,}000\left[\frac{0.05}{(1.05)^n - 1}\right]$$
$$= \$362/\text{year}$$

In order to calculate the annual capital recovery costs, use is made of Equation (4.26) as follows:

$$A_r = 2{,}000\left[\frac{0.05(1.05)^5}{(1.05)^5 - 1}\right] = 2{,}000(0.23097)$$
$$= \$462/\text{yr}$$

Now, the difference between A_r and A_d is = 462 – 362 = 100/year.

This $100 accounts for the annual cost (interest) on the capital ($2,000), which makes:

$$(100/2000)(100) = 5\%$$

Also, using Equation (4.28), we can check the value of A_r, given A_d:

$$A_r = (362)(1.05)^5$$
$$= \$462$$

Example 4.7

An oil production company wishes to repay in 10 installments a sum of $100,000 borrowed at 8% annual interest rate. Determine the amount of each future annuity payment A_r required to accumulate the given present value (debt) of $100,000 for a number of payments of 10 years.

SOLUTION

Find A_r/Given P:

$$A_r = (100{,}000)(0.14903)$$
$$= \$14{,}903/\text{year}$$

Thus, for 10 years, $149,030 would have been paid: $100,000 as principal and $49,030 as interest.

The $100,000 is the present value of the 10-year annuity and the $14,903 is the annual payment, or the annual capital recovery by the creditor.

Example 4.8

An oil-exploration company plans to take over offshore operations 7 years from now. It is desired to have $250,000 by that time. If $100,000 is available for investment at the present time, what is the annual interest rate the company should require to have that sum of money?

SOLUTION

Using Equation (4.23), where,

$P = \$100,000$
$F = 25,000$
$n = 7$ years
$i =$ to be found
$250,000 = 100,000(1 + i)^7$

Solving for i: The interest rate = 14%, which is rather high to realize.

Example 4.9

During the treatment of associated natural gas it was decided to install a knockout drum in the feedline of the plant. This vessel can be purchased and installed for $40,000, and will last for 10 years. An old vessel is available and can be used but needs to be repaired. However, the repairing has to be done every 3 years. If it is assumed that the two vessels (the new and the old ones) have equal capitalized costs, how much does the maintenance department have to spend repairing the old knockout drum? Assume interest is 10%.

SOLUTION

Assuming the salvage value, $V_s = 0$. Equation gives: $K = C_R$.
Comparing the *new vessel* with the *old vessel*:

C_R ($)	10,000	Unknown
n (years)	10	3
i	0.1	0.1

Now, on the basis of equal capitalized costs:

$$10,000\left(\frac{(1.1)^{10}}{(1.1)^{10}-1}\right) = C_R\left(\frac{(1.1)^3}{(1.1)^3-1}\right)$$

Solving for C_R, it is found the maximum amount the maintenance department can spend on repairing the old vessel (perpetual service) is $4,047.

In concluding this chapter, steps in the use of compound interest factors or formulas involving F, P, and A for measurement and determination of time values of money for expansion or replacement of older assets are given as follows:

1. Determine what is wanted—F, P, or A.
2. Determine what is given—F, P, or A.
3. Then apply the formula as to what is given and what is desired, or use the appropriate compound factor for the formula (found in Appendix A) with the desired rate of interest (i).

NOTATION

A	Annual payment ($/yr)
$\mathbf{A_d}$	Annual payment, sinking fund depreciation ($/yr)
$\mathbf{A_r}$	Annual capital recovery ($/yr)
C	Compound interest factor $(1 + i)^n$
C_v	Original value of equipment ($), also denoted as ($V_o$)
C_R	Cost of replacement of equipment after "*n*" years of operation ($)
D	Discount factor $(1 + i)^{-n}$
F	Future value of capital ($)
i	Interest rate (%) per time period
i_e	Effective interest rate
i_l	Nominal interest rate for *m* periods
K	Total capitalized cost ($)
n	Number of years
m	Compounding time periods per year

5 Depreciation and Depletion in Oil Projects

Physical assets decrease in value with time, i.e., they depreciated; while Oil resources, like other natural resources, cannot be renewed over the years, they are continuously *depleted.*

5.1 APPROACH

Methods of determining depreciation costs are examined. Depreciation itself as a process is simply defined, on the other hand, as the unavoidable loss in value of a plant, equipment, and materials.

This includes: straight line, declining balance, sum-of-the-digits, and the sinking fund. Comparison between these methods and evaluation of each is presented as well. Depletion allowances are computed next using either the fixed percentage basis or the cost-per-unit basis.

Depreciation (from the accounting point of view): A system which aims to distribute the cost or other basic value of tangible capital assets (less salvage if any), over the estimated useful life of the unit. It is considered a process of *allocation*—not of valuation. Depreciation itself as a process is simply defined, on the other hand, as the unavoidable loss in value of a plant, equipment, and materials.

Depletion (from the accounting point of view): Depletion costs are made to account or compensate for the loss in value of the mineral or oil property, because of the exhaustion of the natural resources.

Depletion is defined as the capacity loss due to materials consumed or produced.

Service life of an asset (equipment): The useful period during which an asset or property is economically feasible to use. The U.S. Bureau of Internal Revenue recognizes the importance of depreciation as a legitimate expense for industrial organizations. It is for this reason that the Bureau publishes an official listing of the estimated service lives of many assets. Table 5.1 includes the service lives of equipment and assets used in different sectors, both manufacturing and nonmanufacturing.

Salvage value/junk (scrap) value: The value of the asset by the end of its useful life service. The term "salvage" would imply that the asset can be of use, and is worth more than merely its scrap or junk value. The latter definition is applicable to cases where assets are dismantled and have to be sold as junk.

The estimation of these values—including the lifetime—is generally based on the conditions of the asset when installed. In many cases, zero values are designated to the salvage and junk values.

Book value, present asset value, or unamortized cost: The value of an asset or equipment as it appears in the official accounting record (book) of an oil organization. It is equal to the original cost minus all depreciation costs made to date.

Market value: The value obtained by selling an asset in the market. In some conditions, if equipment is properly maintained, its market value could be higher than the book value.

Replacement value: As the name implies, it is the cost required to replace an existing asset, when needed, with one that will function in a satisfactory manner.

5.2 VALUATION OF ASSETS USING DEPRECIATION AND DEPLETION: GENERAL OUTLOOK

In oil and gas production operations, one is interested in finding answers to the following questions:

1. Where and how is a capital investment spent?
2. How much does it cost to produce crude oil and natural gas? How do you calculate the total cost of producing one barrel of oil?
3. How do you apply some economic indicators to judge the profitability of an investment in oil and gas projects?

The answer to where and how a capital investment is spent is found through the following discussion. Economic analysis of the expenditures and revenues for oil operations requires recognition of two important facts:

1. Physical assets decrease in value with time, i.e., they *depreciate.*
2. Oil resources, like other natural resources, cannot be renewed over the years and they are continuously *depleted.*

Depreciation or amortization is described as the systematic allocation of the cost of an asset from the balance sheet to a depreciation expense on the income statement over the useful life of an asset. On the other hand, depletion allowance is a depreciation-like charge applied to account for the exhaustion of natural resources.

As shown in Figure 5.1, for oil production operations, we have two phases where capital investment has to be spent. The first phase, called the *pre–oil-production phase*, involves preliminary preparation, exploration, dry-well drilling, and development. The property is now ready for the second phase, where money is spent in providing necessary assets and equipment for the production stage and post–oil production. The question is: How can we recover the capital spent in the pre–oil-production phase and the *production/postproduction phase* as well?

For oil production operation, as seen earlier, capital investment is spent in two consecutive phases: the preproduction oil phase and the production/postproduction oil phase. As far as the second phase, physical assets can be tangibly verified in a property; hence *depreciation accounting* can be applied to recover this capital investment. The *first* phase, on the other hand, exhibits the contrary: intangible costs were invested, because no physical assets can count for them. In this case, *depletion accounting* is introduced to recover the development costs that were spent for exploration and other preliminary operations prior to the actual

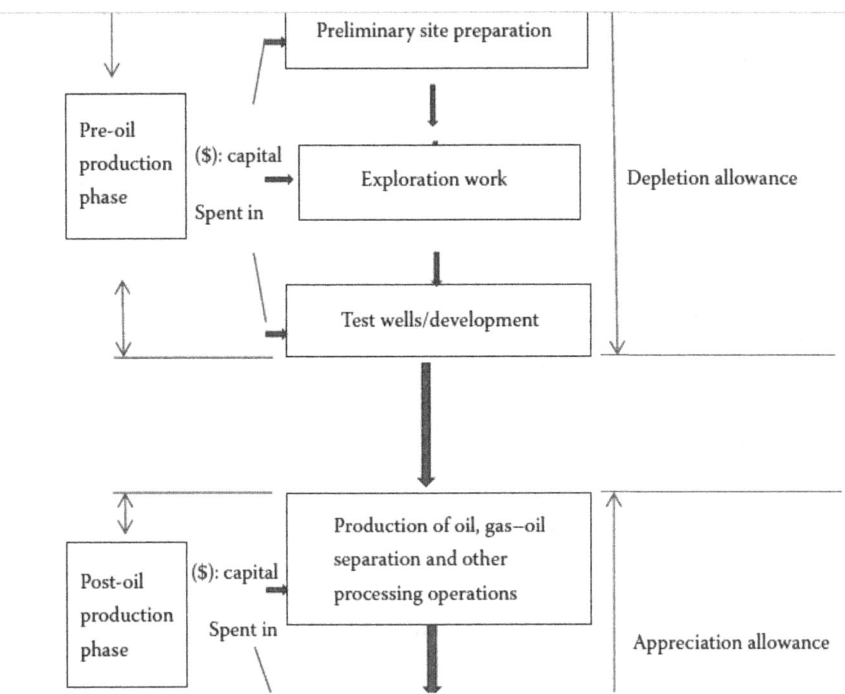

FIGURE 5.1 Contribution of depreciation and depletion allowances in oil production cost.

production of oil and gas. In other words, as mentioned earlier, depletion allowance is a depreciation-like charge applied to an account for the exhaustion of natural resources.

Petroleum company management frequently must determine the value of oil engineering properties. An adequate discussion of the methods used to arrive at the correct value of any property would require at least good-sized volume; so only a few of the principles involved will be considered here—those intimately connected with the subjects of depreciation and depletion.

There are many reasons for determining the value of oil field and refinery assets after some usage. For instance, these values may be needed to serve as a tax base or to establish current value for company statement purposes. Taking depreciation first, the primary purpose of depreciation is to provide for recovery of capital that has been invested in the "physical" oil property. Depreciation is a cost of production; therefore, whenever this production causes the property to decline in value, depreciation must be calculated. Indirectly, depreciation gives a method of providing capital for replacement of depreciated oil equipment. In short, depreciation can be considered as a cost for the protection of the depreciating capital, without interest, over the given period (minimum set by government) during which the capital is used. Finally, the process of valuation is usually an attempt either to make an estimate of present value of future oil profits which will be obtained through ownership of a property, or to determine what would have to be spent to obtain oil property capable of rendering the same service in the future at least as efficiently as the property being valued.

Investment of depreciable capital is used for one of two purposes in the oil fields:

1. As working capital for everyday operating expenses such as wages, materials, and supplies.
2. To buy oil drilling machinery, rigs, etc., used in development and production of oil wells.

Normally, working capital is replaced by sales revenue as it is used up. Thus, this part of investment capital is always available for return to investors.

Investment used for oil drilling machinery, well casings, etc.—that is, fixed capital—cannot be converted directly to original capital invested in oil equipment and machinery, because these physical properties decrease in value as time progresses. They decrease in value because they depreciate, wear out, or become obsolete. Recovery of this investment of fixed capital, with interest for the risks involved in making the investment, must be assured to the investor. The concept of capital recovery thus becomes very important.

The valuation of oil resources in the ground is something else. Oil resources cannot be renewed over a period of years like some other natural resources, such as timber or fish. Also, oil resources cannot be replaced by repurchase as such depreciable physical properties as machinery and equipment can be. Some provision is thus needed to recover the initial investment, or value, of oil reserves and reservoirs, sometimes referred to as an oil lease if purchased by others who are not owners of the land.

One way for investors to recover capital investment in an oil lease—known as depletion capital—is to provide a depletion allowance with annual payments made to the owners of the oil lease. Payments are based on the estimated life of the resource where such an estimate can be made with some degree of accuracy.

Another way to recover capital investment in an oil lease or other depletable capital is to set up a sinking fund with annual deposits based on one interest rate for the depletable capital plus another interest rate or profit on the investment.

In the case of exploration costs and development costs, or money spent for exploration and operations preliminary to actual recovery (production) of oil, such costs are usually recovered by "write-offs" (an accounting term) against other revenues in the year they occur or through a depletion allowance. In the case of foreign oil companies, that is, foreign investors with other outside revenues, these costs can be subtracted from their other revenues along with other expenses in arriving at net income for tax purpose in their own countries. For example, in exploring and developing new leases in the Arabian Gulf area, which could involve millions of dollars before production or perhaps even with little chance of production success, oil companies could write-off these costs against their overall revenues. This would reduce their taxable income, and thereby reduce income taxes they would be liable to pay in their home countries.

To illustrate how both depreciation and depletion costs are calculated, several methods of determining depreciation and depletion are given, with examples of each.

5.3 METHODS FOR DETERMINING DEPRECIATION

There are several ways of determining depreciation for a given period. The following are some of the more popular methods used in most industries. Some are more applicable to the oil industry than others. In general, these methods can be classified into two groups, as shown in Figure 5.2. This classification is based on either neglecting the interest earned on the annual depreciation costs, such that the sum accumulated at the end of the lifetime will equal the depreciable capital, or to take into consideration this interest. Classification of depreciation methods is shown next.

By the end of life time "**The sum of** "$\sum d_y$ = Depreciable cost

5.3.1 Straight-Line Depreciation (S.L.D.)

Mathematically speaking, it is assumed that the value of the asset decreases linearly with time. Now, if the following variables are defined:

d = annual depreciation rate, \$/year
V_o, V_s = original value and salvage values of asset, \$
n = service life, years

$$\text{then the annual depreciation cost} = \frac{\text{depreciable capital}}{\text{n}}, \text{or}$$

$$d = \frac{V_o - V_s}{n}$$

The straight-line method is widely used by engineers and economists working in the oil industry because of its simplicity. Comparison of the different methods is given in Figure 5.3.

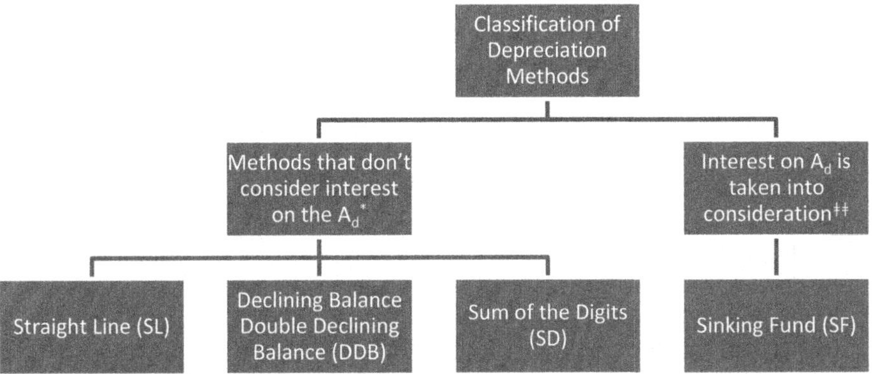

FIGURE 5.2 Methods used to calculate depreciation cost.

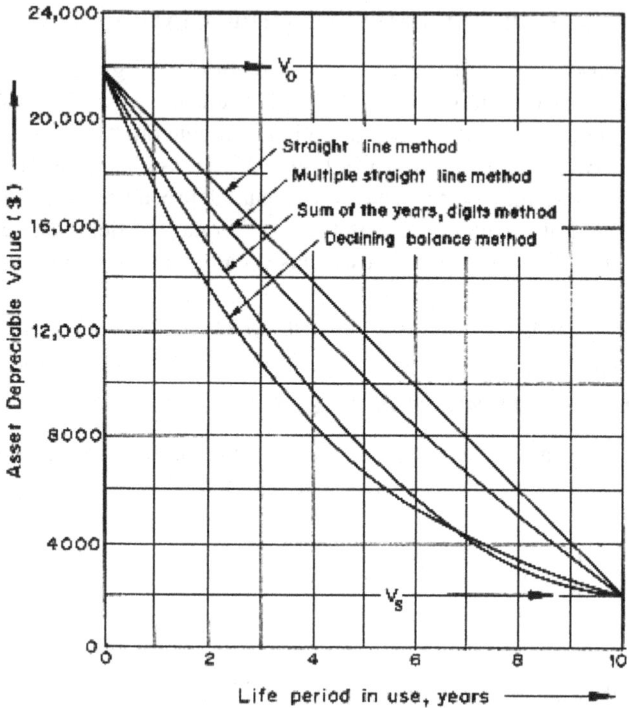

FIGURE 5.3 Comparison of different depreciation methods.

5.3.2 Declining Balance Depreciation (D.B.D.)

The declining balance method assumes that the equipment in question will contribute more to the earning of revenues in the early stage of useful life than it will as the equipment gets older.

A valid use of a declining pattern of depreciation occurs when it is felt that obsolescence will exert a strong influence on the life of the equipment but there is no way of predicting when it will occur. In a simpler way this method is used where utility is higher in the earlier years of life. For example, a computer becomes obsolete within certain period of time due to advancements in technology. In this method, a fixed percentage factor "f" is applied to the new asset value to calculate the annual depreciation costs, which will differ from year to year.

The formula relating "f" to V_a is derived as follows:

By the end of the 1st year: $V_1 = V_o(1 - f)$
By the end of the 2nd year: $V_2 = V_o(1 - f)^2$
By the end of "a" year: $V_a = V_o(1 - f)^a$
By the end of the n year: $V_n = V_o(1 - f)^n$ or $V_s = V_o(1 - f)^n$
since V_n represents value at the end of service life.

Finally solving for the value of f:

$$f = 1 - (V_s / V_o)^{1/n} \tag{5.1}$$

Examining Equation (5.1), one concludes the following:

1. The declining balance method permits the asset investment to be paid off more rapidly during the early years of life. This persuades oil companies starting new ventures to use the D.B.D., because it allows a reduction in income taxes at the early years of their operations.

Example 5.1

An example of how the double declining balance method is calculated is given here. If we assume that an acid injection unit had an original cost of $25,000 and its lifetime is 5 years, it is necessary to calculate the annual depreciation costs and the book value for this unit. The salvage value, V_s, is taken to be $3,000.

SOLUTION

Since n = 5 years, the annual depreciation using S.L.D. will be 20%, and the allowable fixed percentage to be applied using D.B. (doble declining balance) will be (2)(20%) = 40%. The depreciation schedule would be then as shown in Table 5.1.

Example 5.2

A flow/recording control valve installed on the feed line of a caustic-soda treating unit costs $4,000, with a service life of 5 years and scrap value of $400. Calculate the annual depreciation cost using the S.D.D. (Straight line method).

TABLE 5.1
Depreciation Schedule

Year	Depreciation Expense	Book Value	Remaining Depreciable Cost
Start	$25,000	$22,000	(with $3,000 salvage value off)
After first year	$10,000 (40% of $25,000)	$15,000	$12,000 ($22,000–$10,000)
After second year	$6,000 (40% of $15,000)	$9,000	$6000 ($12,000–$6,000)
After third year	$3,600 (40% of $9,000)	$5,400	$2,400 ($6,000–$3,600)
After fourth year	$2,160 (40% of $5,400)	$3,240	$240 ($2,400–$2,160)
After fifth year	$240 (depreciation before salvage value)	$3,000	0

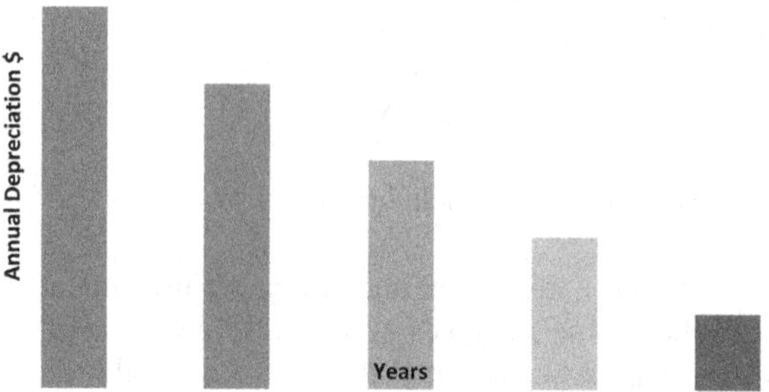

FIGURE 5.4 Annual depreciation per year.

SOLUTION

The sum of arithmetic series of numbers from 1 to 5 = 1 + 2 + 3 + 4 + 5 = 15. Using the bove equqtions, we get:

$$d_1 = \left(\frac{5}{15}\right)(4000 - 400)$$

$$= \$1200$$

$$d_2 = \left(\frac{4}{15}\right)(3600)$$

$$= \$960$$

$$d_3 = \$720$$

$$d_4 = \$480$$

$$d_5 240$$

$$\text{Sum} = \underline{\$3,600}$$

The bar chart clearly depicts that annual depreciation is constantly decreasing as the years pass shown in Figure 5.4

Example 5.3

Assume a petroleum company investment of $10 million for an expansion to a current refinery, allocated $1,000,000 for land and $7,000,000 for fixed and other physical properties subject to depreciation. Additional capital of $2,000,000 is available for operation purposes, but this sum is not subject to depreciation. Investors want a 15% interest rate (or earning rate to investors) on their money for a 10-year period. The sinking-fund method will be used, with depreciation figured at 15% per year. No income taxes are involved in order to simplify the example.

SOLUTION

First-year profit before deducting the sinking-fund depreciation charge made at the earning rate of 15% interest, and assuming no salvage value for the physical properties, is 0.15 × $1,000,000, or $150,000 per year.

But the oil company must earn enough additional money annually to pay for the depreciation occurring on the depreciable capital of $700,000.

Using sinking-fund depreciation and a 15% interest rate for the sinking fund, the annual deposit in the fund is given by:

$$A_d = \frac{\$700{,}000 \times 0.15}{(1.15)^{10} - 1} = \$34{,}440$$

Thus, company profits before depreciation must total $184,440 ($150,000 + $34,440) and not merely $150,000 in the first year. Actually, the $184,440 in the first year represents:

$34,440 = the sum of annual depreciation charge
$105,000 = the 15% interest on the undepreciated part of the depreciable capital which is, in the first year or before any deductions, 0.15 × $700,000
$45,000 = the 15% interest on the non-depreciable capital, or 0.15 × $300,000
$184,440 = the total for the first year

Thus, $139,440 ($105,000 + $34,440) is needed to cover (1) the depreciation deposit in the sinking fund, and (2) the interest on the depreciable capital for that year. This is also calculated by using:

$$A_r = \$700{,}000 \frac{0.15(1.15)^{10}}{(1.15)^{10} - 1} = \$139{,}440$$

In each succeeding year, the book value of the depreciable capital decreases, but the depreciation reserve increases in such a manner that the sum of the two always equals $700,000 and the total annual interest remains constant at $105,000 even though the interest charges on each component vary.

The biggest drawback to the actual use of the sinking-fund method in business is the fact that businesses rarely maintain an actual depreciation sinking fund. The interest rate which could be obtained on such deposits would be small, probably not over 6% in the petroleum business, according to financial experts in the oil industry. An active business, such as an oil company operation, is constantly in need of working capital. This capital will usually earn much more than 6%.

A reasonable rule is that all values should be kept invested in the oil business and not remain idle. As a result, a fictitious depreciation fund is often used: The amounts which have been charged to depreciation are actually left in the business in the form of assets, and a "reserve for depreciation" account is used to record these funds.

Where such a "depreciation reserve" is used, the company is actually borrowing its own depreciation funds. Therefore, there is no place from which interest on these values could be obtained except from the business itself. This would create a situation in which a business pays itself interest for the use of its own money. To accomplish this, the cost of depreciation equal to the sinking-fund deposit has to be charged as an operating expense, and then interest on the accumulated sinking fund has to be charged as a financial expense. Such a procedure accurately accounts for all expenses, but might require considerable explanation to government income tax authorities. Hence, interest is not used when sinking fund deposits are not made to an outside source.

Example 5.4

Rework Example 5.3 to compare S.L.D. and S.F.D. (Sum of Digits).

SOLUTION

Table 5.2 illustrates depreciation over 10 years for the investment in Example 5.4 as calculated by both the sinking-fund and straight-line methods. Figure 5.4 compares the book values obtained by the two methods as a line graph. As Table 5.2 and Figure 5.5 show, at the end of the second year the depreciation deposit into the sinking fund is \$34,440, but interest on the previous deposit is 0.15 × \$34,440 (deposit for the first year), or \$5,160. This is repeated for the third year with 15% interest on \$39,600 (\$34,440 + \$5,160), and so on for each year. Before

TABLE 5.2
Solution of Example 5.4

End of year	Total in sinking fund depreciation reserve	Annual interest, 15% of column 2	Annual deposit	Annual charge	Book value at end of year	Annual charge	Book value
1	2	3	4	5	6	7	8
Start	0	0	0	0	\$700,000	0	\$700,000
1	\$34,440	0	\$34,440	\$34,440	665,560	\$70,000	630,000
2	74,040	\$51,600 (15% of 74,040)	34,440	39,600	625,960	70,000	560,000
3	119,580	11,100	34,440	45,540	580,420	70,000	490,000
4	171,960	17,940	34,440	52,380	528,040	70,000	420,000
5	232,200	25,800	34,440	60,240	467,800	70,000	350,000
6	301,540	34,900	34,440	69,340	398,460	70,000	280,000
7	381,280	45,300	34,440	79,740	318,720	70,000	210,000

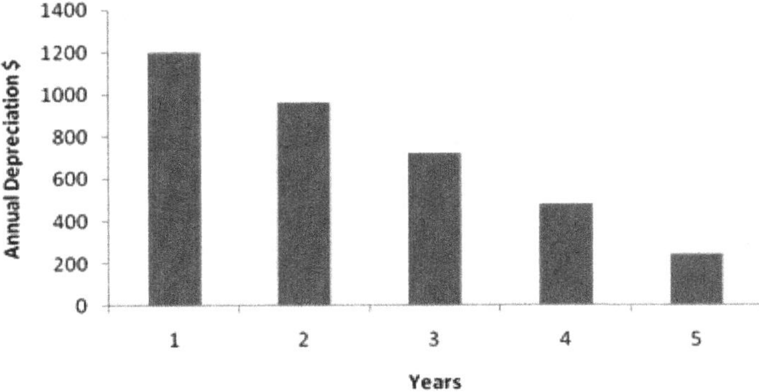

FIGURE 5.5 Individual annual depreciation.

the petroleum company can earn interest of $150,000 for the second year it must deposit $34,440 in the sinking fund and pay $5,160 interest on a total of $39,600 to the sinking-fund depreciation reserve.

Figure 5.6 shows how the straight-line and sinking-fund methods differ. The curve of the sinking fund bulges from the straight-line method curve, yet both eventually meet at the end of the 10th year.

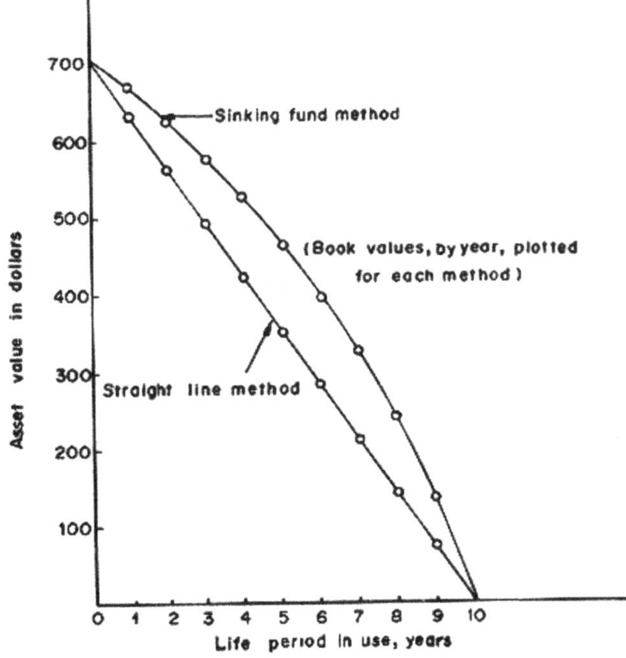

FIGURE 5.6 Comparison of straight-line and sinking-fund methods of calculating depreciation.

5.4 COMPARISON BETWEEN THE DEPRECIATION METHODS

The choice of the best depreciation is not a straightforward task. It is not our purpose to explore here the details of depreciation accounting methods. Suffice it to say that the following factors are important in choosing one method of depreciation and not the other:

1. Type and function of property: lifetime, salvage value
2. Time value of money (interest)
3. Simplicity
4. Choose the one for which the present worth of all depreciation charges is a maximum

In the absence of guidelines and for quick results, the following rules are recommended:

1. Use straight-line depreciation (simple)
2. Take the useful lifetime of the asset = 10 years
3. Assume salvage value = zero

Now, we can make the following specific comparison:

Straight-line and sinking fund versus	Declining balance and sum of digits
• Annual depreciations costs are *constant*. • Asset value is higher for (S.F.D.) because of the effect of *i*, as compared to (S.L.D.). • S.L.D. is simple and widely used. • S.F.D. is seldom used. It is applicable for assets that are sound in performance and stand little chance of becoming obsolete.	• Annual depreciation costs are *changing*, greater in early life than in later years. • Used for equipment where the greater proportion of production occurs in the early part of life, or when operating costs increase with age. • Both methods are classified as "accelerated depreciation" type. They provide higher financial protection. • For D.D.B.D., the annual fixed percentage factor is constant, while for S.D.D. it is changing.

Berg et al worked on a model in selecting best method for calculation between the straight-line depreciation method and an accelerated depreciation method like sum of the digits and double declining method. They found that straight-line depreciation can be better than other depreciation methods as the other method is usually considered in empirical literature on accounting method choice. They also concluded that while making a selection between straight-line and accelerated method, it is very necessary to consider the uncertainty in future cash flows and the structure of tax system.

No land states that declining balance method is the most prominent type of accelerated depreciation used in financial reporting. However, he adds drawback that at the end of the asset's useful life this method depreciate the asset to its salvage value. Different companies use various ways to adjust this problem.

5.5 METHODS FOR DETERMINING DEPLETION

5.5.1 Background

When limited natural resources, such as crude oil and natural gas, are consumed, the term "depletion" is used to indicate the decrease in value, which has occurred. As some of the oil is pumped up and sold, the reserve of oil shrinks and the value of the oil property normally depreciation-like charge applied to account for the exhaustion of natural resources.

5.5.2 Methods Used for Depletion

If a depletion allowance is to be used, there are two possible methods of calculating its value:

1. Fixed percentage method
2. Cost-per-unit basis

For the fixed percentage method, the percentage depletion is usually set by government ruling (in the United States it has been 22% of net sales), but in no case can the fixed percentage exceed 50% of net income before deduction of depletion.

In the cost-per-unit method, the amount of depletion charged to each barrel, or ton, of crude produced is determined by the ratio of intangible development cost plus the depletable costs divided by the estimated total units potentially recoverable. This then gives a cost per unit, which is in either barrels or tons depending on how the estimated total units potentially recoverable are given.

The total units recoverable may be estimated if the number of years of production and the production rates can be estimated. For oil and gas wells the calculations vary with the nature of the production curve and the allowable flow permitted by conservation authorities of the government of the oil-producing country. A mathematical analysis is used for estimating the total barrels of oil potentially recoverable under certain assumed conditions.

Example 5.5

Given the following:

The intangible development costs, excluding a $1,000,000 bonus to land owner, all occur in the first year = $8,000,000.

Depreciable capital such as: casing, machinery, derricks, rigs, etc. = $45,000,000.

Estimated life of equipment = 9 years.

Assume that 1,500,000 bbl of crude oil are produced and sold the first year at $100.0/bbl.

Assume the annual operating expenses (and others) = $2,500,000.

Estimate the depletion charge using a fixed percentage rate of 27.5% of net sales.

SOLUTION

The depletion charge is based on a 3-year period.

Cost items for the first year ($):

Net sales for 1,500,000 bbl at $100.0/bbl	= 150,000,000
Annual depreciation ($45,000,000/g)	= 5,000,000
First-year expenses	= 2,500,000
The depletion allowance = (0.275) (150,000,000) = 412,250,000	

In order to check on the criterion that the depletion allowance ($412,250,000) does not exceed 50% of the net income (before allowing for depletion), the following calculations are carried out (in $):

Net sales (revenue)		150,000,000
First-year expenses	2,500,000	
Development expenses	8,000,000	
Annual depreciation charges	5,000,000	
Total expenses		155,000,000
Total net income (profit)		134,500,000
50% of net income		672,500,000

Thus, the maximum allowable depletion will be $67.25 million and not $41.25 million. The $1,000,000 bonus in this problem is recovered as part of the depletion charge.

Example 5.6

Solve Example 5.5 using the cost-per-unit method and then compare the two methods used in calculating the depletion allowance.

SOLUTION

The depletion charge using cost-per-unit method:

$$= \frac{\text{Sum of development and bonus costs}}{\text{recoverable oil reserves}}$$

$$= \frac{8000000 + 100000}{720000}$$

$$= \$12.5/\text{bbl}$$

The allowable depletion based on the cost-per-unit method for the first year:

$$= (12.5)(1500,000) = \$187,50000$$

This amount of $1,875,000 would be allowed even if it exceeded the value permitted by the fixed-percentage method. However, the cost-per-unit method must be used *each year* once it has been adapted.

5.5.3 Summary and Comparison

1st: The allowable first-year charges for capital recovery:

Basis	Percentage Depletion ($)	Cost Per Unit Depletion ($)
Annual depreciation	5,000,000	5,000,000
Development expenses	8,000,000	included (5.10)
First-year depletion	67,250,000	18,750,000
Total first-year charges for capital recovery	80,250,000	23,750,000

2nd: Net income for first-year (net revenue – total costs):

Net revenue (sales)	**150,000,000**			**150,000,000**
Operating expenses	2,500,000		2,500,000	
Development expenses (100% incurred in first year)	8,000,000		(included)	
Depreciation expenses	5,000,000		5,000,000	
Depletion expenses	67,250,000		18,750,000	
Total costs		82,750,000		26,250,000
Net income		67,250,000		123,750,000

3rd: Net income for the second year*:

Net revenue (sales)		**150,000,000**		**150,000,000**
Operating expenses	2,000,000†		2,000,000	
Depreciation expenses	5,000,000		5,000,000	
Depletion	41,250,000‡		18,750,000	
Total costs		48,250,000		25,750,000
Net income		101,750,000		124,250,000

* Assume the same sales as in the first year.

† Operating expenses for the second year are assumed to be less than for the first year.

‡ For this case $41,250,000 calculated by the fixed percentage method (27.5%) is less than the 50% criterion: 150,000,000 – 7,000,000 (0.5) = $146,500,000.

One can conclude from the above calculations that the percentage depletion method promotes the recovery of a greater amount of oil-reserve depleted value, or $67,250,000 to $18,750,000 for the cost-per-unit method. But there are no more development costs incurred after the first year's $8,000,000. The $800,000 was a large factor in determining the amount of depletion. Of course, new development charges could be incurred in other years, and would then be included in determining the amount of depletion.

Also, any additional development costs or any changes in estimated recoverable oil will require a recalculation of the cost-per-unit depletion rate, which is then used to determine depletion in subsequent years.

Although the net income by the cost-per-unit depletion method is greater in both the first and second years for the example given, the total net income plus capital recovery for 2 years added together by the percentage depletion is equal to the cost-per-unit depletion method total.

However, the percentage method has an advantage in a lower profits tax over the cost-per-unit method with reported lower net incomes for each year (in the first year, $67,250,000 to $123,750,000, and in the second year, $101,750,000 to $124,250,000). But the cost-per-unit method does have an economic advantage where rights to oil resources are purchased outright or leased at a relatively higher price to the seller because the net income figures are greater with this method.

The comparison of both the methods. In these figures, bar charts for both the years using both methods are shown.

Accounting for depletion can be complicated because of the uncertainties of future development costs, uncertainties about actual recoveries of oil from proven reserves, uncertainties about future value of oil reserves as selling prices go down or up and uncertainties about the scale of operations, that is, the magnitude of production of oil. Variations in all or any of these factors may result in changes in the depletable value, necessitating separate calculations each year for the depletion charges.

When there is an increase in value of oil reserves as opposed to an increase in amount of proven oil reserves, or a big increase in selling prices, accretion rather than depletion is practiced to show the increase or "growth" in the oil reserve. When such increase in value results an allowance for it must be made in the accounts of the oil company. Life time of Assets are documented in Table 5.3

TABLE 5.3
Estimated Service Life of Some Assets

	Life (years)
Group I: General business assets	
1. Office furniture, fixtures, machines, equipment	10
2. Transportation	
a. Aircraft	6
b. Automobile	3
c. Buses	9
d. General-purpose trucks	4–6
e. Railroad cars (except for railroad companies)	15
f. Tractor units	4
g. Trailers	6
h. Water transportation equipment	18
3. Land and site improvements (not otherwise covered)	20
4. Buildings (apartments, banks, factories, hotels, stores, warehouses)	40–60
Group II: Nonmanufacturing activities (excluding transportation, communications, and public utilities)	
1. Agriculture	
Group III: Manufacturing	
1. Aerospace industry	8
2. Apparel and textile products	9
3. Cement (excluding concrete products)	20

TABLE 5.3
Estimated Service Life of Some Assets *(Continued)*

	Life (years)
4. Chemicals and allied products	11
5. Electrical equipment	
a. Electrical equipment in general	12
b. Electronic equipment	8
6. Fabricated metal products	12
7. Food products, except grains, sugar, and vegetable oil products	12
8. Glass products	14
9. Grain and grain-mill products	17
10. Knitwear and knit products	9
11. Leather products	11
12. Lumber, wood products, and furniture	10
13. Machinery not otherwise listed	12
14. Metalworking machinery	12
15. Motor vehicles and parts	12
16. Paper and allied products	
a. Pulp and paper	16
b. Paper conversion	12
17. Petroleum and natural gas	
a. Contract drilling and field service	6
b. Company exploration, drilling, and production	14
c. Petroleum refining	16
d. Marketing	
18. Plastic products	11
19. Primary metals	
a. Ferrous metals	18
b. Nonferrous metals	14
20. Printing and publishing	11
21. Scientific instruments, optical and clock manufacturing	12
22. Railroad transportation equipment	12
23. Rubber products	14
24. Ship and boat building	12
25. Stone and clay products	15
26. Sugar products	18
27. Textile mill products	12–14
28. Tobacco products	15
29. Vegetable oil products	18
30. Other manufacturing in general	12
Group IV: Transportation, communication, and public utilities	
1. Air transport	6
2. Central steam production and distribution	28
3. Electric utilities	
a. Hydraulic	50
b. Nuclear	20

(Continued)

TABLE 5.3
Estimated Service Life of Some Assets *(Continued)*

	Life (years)
c. Steam	28
d. Transmission and distribution	30
4. Gas utilities	
a. Distribution	35
b. Manufacture	30
c. Natural-gas production	14
d. Trunk pipelines and storage	22
5. Motor transport (freight)	8
6. Motor transport (passengers)	8
7. Pipeline transportation	22
8. Radio and television broadcasting	6
9. Railroads	
a. Machinery and equipment	14
b. Structures and similar improvements	30
c. Grading and other right of way improvements	variable
d. Wharves and docks	20
10. Telephone and telegraph communications	variable
11. Water transportation	20
12. Water utilities	50

Source: Peters and Timmerhaus (1981).

Source: Max Peters, Klaus Timmerhaus and Ronald West, *Plant Design and Economics for Chemical Engineers*, fifth edition, McGraw-Hill Education, 2003.

NOTATION

A_d, A_r Annual depreciation and annual capital recovery defined by Equations (4.12) a (4.15), respectively

a Designates a specific year in the useful lifetime (n).

d Annual depreciation rate ($/yr).

d_a Annual depreciation rate for the year (a).

f Fixed percentage factor defined by Equation (5.3)

$\overline{f}$ Accelerated depreciation factor defined by Equation (5.4)

n Number of useful (service) years of life.

V_a Value of an asset at year a

V_o Original value of an asset ($).

V_s Salvage value of an asset ($), also referred to as V_n

Section II.II: Profitability Analysis and Evaluation

INTRODUCTION

The basic aim of financial measures and profitability analysis is to provide some economic-decision yardsticks for the attractiveness of a venture or a project, where the expected benefits (revenues) must exceed the total production costs.

Capital expenditure proposals must be sufficiently specific to permit their justification for either: exploration and production operations, surface petroleum operations, petroleum refining, and expansion purposes or for cost reduction improvements and/or necessary replacements. In reality, an evaluation of capital expenditure proposals is both technical and economic in nature. First, there are the technical feasibilities and validities associated with a project, next comes the economic evaluation and viability.

In the economic phase of evaluation, oil management may find that it has more investment opportunities than capital to invest, or more capital to invest than investment opportunities. Whichever situation exists, oil management needs to resort to some economic criteria for selecting or rejecting investment proposals. Management's decision is, in either case, likely to be based largely on the measures of financial return on the investment.

The most common measures, methods, and economic indicators of economically evaluating the return on capital investment are presented in this section using the following chapters:

Chapter 6: Annual Rate of Return (A.R.R.) & Payout Period (P.P.)
Chapter 7: Discounted Cash Flow Rate of Return (D.C.F.R.) & Present Value Index (N.P.I.)
Chapter 8: Analysis of Alternatives, Selections, and Replacements
Chapter 9: Feasibility Study
Chapter 10: Model Solved Examples

6 Annual Rate of Return (A.R.R.) & Payout Period (P.P.)

Effective Profitability Indicators.

6.1 OVERVIEW

Classification of these methods, in chapter, into two groups is considered, where the time value of the cash flow received from a project is the criterion used in this classification:

- Time value of money is neglected. Two methods fall in this group. They are known as the annual rate of return (R.O.I.) and the payment period (P.P.).
- Time value of money is considered. Two methods represent this group. They are known as the discounted cash flow of return (D.C.F.R.) and the net present value (N.P.V.).

In general, no one method is by itself a sufficient basis for judgment. A combination of more than one profitability standard is needed to approve or recommend a venture. In addition, it must be recognized that such a quantified profitability measure would serve as a guide. Many unpredictable factors and uncertainties cannot be accounted for, specifically those in exploration and production operations.

Based on this classification, the R.O.I. and P.P. are described as "rough" or "crude" quick methods, while the D.C.F.R. and N.P.V. are known to be accurate realistic and time-demanding indicators.

6.2 THE ANNUAL RATE OF RETURN (R.O.I.) IS DEFINED BY THE EQUATION

$$\text{R.O.I.} = (\text{annual profit/capital investment})(100)$$

Consideration of the income taxes is provided in calculating the R.O.I. by using either "net" profit or "gross" profit.

TABLE 6.1
Average Return on Investment for Crude Oil Desalting (Solution of Example 6.1)

		Project 1			Project 2		
		Income before Depreciation	Depreciation Allowance	Net earning after Depreciation	Income before Depreciation	Depreciation Allowance	Net earning after Depreciation
Year	1	$400,000	$250,000	$150,000	$75,000	$200,000	-$125,000
	2	$350,000	$250,000	$100,000	$180,000	$200,000	-$20,000
	3	$300,000	$250,000	$50,000	$300,000	$200,000	$100,000
	4	$275,000	$250,000	$25,000	$400,000	$200,000	$200,000
	5				$600,000	$200,000	$400,000
Sum				$325,000			$555,000
Average Investment				$500,000			$500,000
Average Earning				$81,250			$111,000
Average Rate of Return			16.25%			22.20%	

For oil ventures, where the cash flow extends over a number of years, the average rate of return is calculated using an average value for the profit, by dividing the sum of the annual profits by the useful lifetime:

$$\text{R.O.I.} = \left[\frac{\sum_{1=y}^{n} \text{annual profits}}{n} \right] / (\text{capital investment})(100)$$

The main drawback of this method is the fact that money received in the future (cash flow) is treated as money of present value (which is less, of course).

Example 6.1

It is necessary to calculate the R.O.I. for two projects involving the desalting of crude oil; each has an initial investment of $1 million. The useful life of project 1 is 4 years and of project 2, it is 5 years. The earnings pattern is given next in Table 6.1.

SOLUTION

The average R.O.I. is calculated for both projects as shown in Table 6.1. The final answers are:

R.O.I. for project 1 = 16.25%
R.O.I. for project 2 = 22.2%

6.3 PAYOUT PERIOD (P.P.), PAYBACK TIME, OR CASH RECOVERY PERIOD

P.P. is defined as the time required for the recovery of the depreciable capital investment in the form of cash flow to the project. Cash flow would imply the total income minus all costs except depreciation.

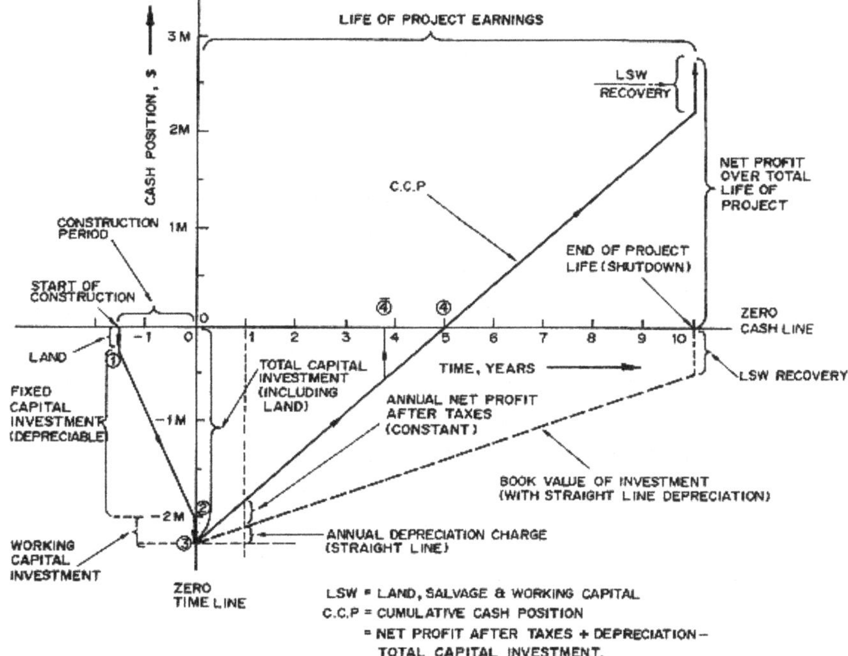

FIGURE 6.1 Cumulative cash flow diagram.

Mathematically, this is given by Equation (6.1), where the interest charge on capital investment is neglected:

$$\text{Payout period (years) (P.P.)} = \frac{\text{depreciable capital investment}}{\text{average annual cash flow}}$$

A cumulative cash flow diagram, shown in Figure 6.1, illustrates some of the basic concepts, including the P.P. It is briefly explained as follows:

- Investment for land (if needed) comes first followed by
- Investment for the depreciable asset, throughout the construction period (points 1–2).
- The need for the working capital comes next for startup and actual production (points 2–3).
- Production starts now at point 3 (zero time) and goes all the way profitably to cross the zero cash line at point 4. This point corresponds to the time spent to recover the cumulative expenditure, which consists of: capital of land + capital cost of depreciable assets + working capital.
- The payout period will accordingly be defined by point 4, that is, the time required to recover the depreciable capital only. Point 4, could be considered

as an alternative way (but different in value) to define payout period as the time needed for the cumulative expenditure to balance the cumulative cash flow exactly.

Usually oil companies seek to recover most of their capital investments in a short payback period, mostly because of uncertainty about the future and the need to have funds available for later investments. This becomes especially important when the company is short of cash—emphasis on rapid recovery of cash invested in capital projects may be a necessity.

The payback period is used by oil companies in ascertaining the desirability of capital expenditures, because it is a means of rating capital proposals. It is particularly good as a "screening" means relative to various capital proposals. For example, expenditures for units may not be made by an oil refinery unless the payback period is no longer than 3 years. On the other hand, the proposed purchase of a subsidiary may not be considered further unless the payback period is 5 years or less. But payback has its drawbacks. For example, payback ignores the actual useful length of life of a project. Also, no calculation of income beyond the payback period is made. Payback is not a direct measure of earning power, so the payback method can lead to decisions that are really not in the best interests of an oil company.

7 Discounted Cash Flow Rate of Return (D.C.F.R.) and Present Value Index (P.V.I.)

A continuation of Economic Indicators

7.1 OVERVIEW

Discounted cash flow (D.C.F.R.) is a valuation method used to estimate the value of an investment based on its future *cash flows*. D.C.F. analysis attempts to figure out the value of an investment today, based on projections of how much money it will generate in the future.

Now, if we have an oil asset (oil well, surface treatment facilities, a refining unit, etc.) with an initial capital investment "*P*," generating annual cash flow over a lifetime *n*, then the D.C.F.R. is defined as the rate of return, or interest rate that can be applied to yearly cash flow, so that the sum of their present value equals *P*.

7.2 DISCOUNTED CASH FLOW RATE OF RETURN (D.C.F.R.)

From the computational point of view, D.C.F.R. cannot be expressed by an equation or formula, similar to the previous methods. A three-step procedure involving trial and error is required to solve such problems. The following example 7.1 illustrates the basic concepts:

Example 7.1

Solved Example (7.1): Assume an oil company is offered a lease of oil wells which would require a total capital investment of $110,000 for equipment used for production. This capital includes $10,000 working money, $90,000 depreciable investment and $10,000 salvage value for a lifetime of 5 years.

SOLUTION

Assume an oil company is offered a lease of oil wells, which would require a total capital investment of $110,000 for equipment used for production. This capital includes $10,000 working money, $90,000 depreciable investment, and $10,000 salvage value for a lifetime of 5 years.

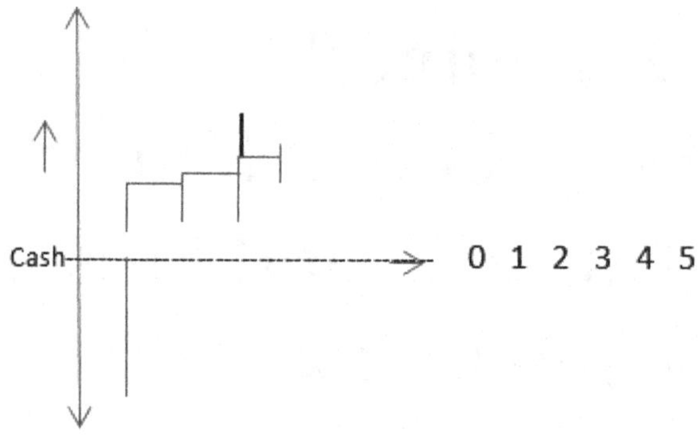

Year	Cash flow ($10³)
0	-110
1	30
2	31
3	36
4	40
5	43

Cash flow to project (after taxes) gained by selling the oil is as given in Figure 6.2. Based **on calculating the D.C.F.R., a decision has to be made: should this project be accepted?**

Two approaches are presented to handle the D.C.F.R.

7.2.1 1st Approach: Using the Future Worth

Our target is to set the following equity:

By the end of 5 years, the future worth of the cash flow recovered from oil sales) and should break even with the future worth of the capital investment, had it been deposited for compound interest in a bank at an interest rate *i*.

This amounts to:

$$F_o = F_B$$

where $F_B = 110{,}000(1+i)^5$, for banking and $F_o = \Sigma_{i=1}^{5} F_i$, for oil investment, which represent the cash flow to the project, compounded on the basis of end-of-year income.

Hence,

$$F_o = 30,000(1+i)^4 + 31,000(1+i)^3 + 36,000(1+i)^2 + 40,000(1+i) + 43,000 + 20,000$$

Notice that the $20,000 represents the sum of working capital and salvage value; both are released by the end of the 5th year.

Setting up $F_B = F_O$, we have one equation involving i as the only unknown, which could be calculated by trial and error. The value of i is found to be 0.207, that is, the D.C.F.R. = 20.7%.

7.2.2 2nd Approach: Using the Discounting Technique

Our objective here is to discount the annual cash flow to present values using an assumed value of i. The correct i is the one that makes the sum of the discounted cash flow equals to the present value of capital investment, P. The solution involves using the following equation:

$$P = \sum_{y=1}^{5} p_y$$

where:

$$p_y = (\text{annual cash flow})_y \; d_y = (\text{A.C.F.})_y \left(\frac{1}{1+i}\right)^y$$

for the year y, between 1 and 5.

Another important criterion that can be used in order to arrive at the correct value of i in the discounting of the cash flow is given by the following relationship:

D.C.F.R. is the value that makes P.V.I. = 1.0, where P.V.I. is defined by

$$\text{P.V.I.} = \frac{\text{sum of discounted cash flow (present value)}}{\text{initial capital investment}}$$

The solution of this example applying the discount factor is illustrated in Table 7.1.

If the annual cash flow has been constant from year to year, say A $/yr, then the following can be applied:

$$A\left[\frac{1}{(1+i)} + \frac{1}{(1+i)^2} + \cdots + \frac{1}{(1+i)^n}\right] = P$$

TABLE 7.1
D.C.F.R. for Investment of Lease of Oil Wells

		i = 15%		i = 20%		i = 25%		i = 20.7%	
Year (y)	Cash flow	dy	Persent Value ($)	dy	Persent Value ($)	dy	Persent Value ($)	dy	Persent Value ($)
0	110,000								
1	30,000	0.8696	26,088	0.8333	24,999	0.8000	24,000	0.8290	24,870
2	31,000	0.7561	23,439	0.6944	21,526	0.6400	19,840	0.6870	21,297
3	36,000	0.6575	23,670	0.5787	20,833	0.5120	18,432	0.5700	20,520
4	40,000	0.5718	22,872	0.4823	19,292	0.4096	16,384	0.4720	18,880
5	43,000	0.4971	21,375	0.4019	17,282	0.3277	14,091	0.3910	16,813
	20,000								
	Total		117,444		103,932		92,747		102,380
	P.V.I		1.07		0.94		0.84		0.93

Multiplying both sides of the above equation by $(1 + i)^n$, we get:

$$A[(1+i)^{n-1} + (1+i)^{n-2} + \cdots + 1] = P(1+i)^n$$

The sum of the geometric series in the left-hand side is given by:

$$\frac{(1+i)^n - 1}{i}$$

Hence the equation can be rewritten in the form:

$$P(1+i)^n = A\frac{(1+i)^n - 1}{i}$$

The future worth of P, if invested in the bank, is given by:

$$F_B = P(1+i)^n$$

The future worth of the annual cash flow received from oil investment (A), if compounded in a sinking-fund deposit, is given by:

$$F_o = A\frac{(1+i)^n - 1}{i}$$

Now, this equation can be used to calculate directly the D.C.F.R. by trial and error knowing the values of A, P, and n.

The D.C.F.R. thus represents the maximum interest rate at which money could be borrowed to finance an oil project.

7.3 NET PRESENT VALUE (N.P.V.)

7.3.1 Introduction

Net present value (N.P.V.) is the difference between the present value of cash inflows and the present value of cash outflows over a period of time. N.P.V. is used in capital budgeting and investment planning to analyze the profitability of a projected investment or project.

The D.C.F.R. method is based on finding the interest rate that satisfied the conditions implied by the method. Here we provide a value for "i" that is an acceptable rate of return on the investment and then calculate the discounted value (present value) of the cash flow using this i. The net present value is then given by:

$$\text{N.P.V.} = (\text{present value of cash flow discounted at a given i}) - \text{capital investment}$$

Example 7.2

Calculate the N.P.V. of the cash flow for the oil lease described above if money is worth 15%.

SOLUTION

At $i = 0.15$, the annual cash flow is discounted. The present value of the sum of the cash flows = \$127,000. The N.P.V. is directly calculated using the following equation:

$$\begin{aligned}\text{N.P.V.} &= 127{,}000 - 110{,}000\\ &= \underline{\$17{,}000}\end{aligned}$$

That is, the oil lease can generate \$17,000 (evaluated at today's dollar value) over and above the totally recovered capital investment. Solution is illustrated as given in Table 7.2

TABLE 7.2
D.C.F.R. for Investment in a Lease of Oil Wells

		i= 5%		i= 20%		i= 25%		i= 20.7%	
Year (y)	Cash flow	dy	Persent Value ($)	dy	Persent Value ($)	dy	Persent Value ($)	dy	Persent Value ($)
0	110,000								
1	30,000	0.8696	26,088	0.8333	24,999	0.8000	24,000	0.8290	24,870
2	31,000	0.7561	23,439	0.6944	21,526	0.6400	19,840	0.6870	21,297
3	36,000	0.6575	23,670	0.5787	20,833	0.5120	18,432	0.5700	20,520
4	40,000	0.5718	22,872	0.4823	19,292	0.4096	16,384	0.4720	18,880
5	43,000	0.4971	21,375	0.4019	17,282	0.3277	14,091	0.3910	16,813
	20,000								
	Total		117,444		103,932		92,747		102,380
	P.V.I		1.07		0.94		0.84		0.93
	N.P.V		$7,444.40		-$6,067.70		-$17,252.90		-$7,620.00

7.4 COMMENTS ON THE TECHNIQUES OF ECONOMIC ANALYSIS

All four methods described earlier determine in one way or the other, the return on investment or the attractiveness of a project.

To evaluate whether a project, or a proposal on a project for the future, is yielding, or will yield, a good or bad return, the return on investment must be compared to a standard acceptable level of profit, which the oil company wishes to maintain. The internal cutoff rate (or breakeven point for return) is the cost of capital, which is the rate of borrowing money at the time of use of these measures for calculating return on investment. There is no precise agreement on how oil management calculates cost of capital, but it should include both the cost of borrowed funds and the cost of equity financing (when applicable). The following are some important observations to take care of:

1. As mentioned before, the R.O.I. and P.P. are economic indicators to be used for rough and quick preliminary analyses. The R.O.I. method does not include the time value of money and involves some approximation for estimating average income or cash flow. The P.P., on the other hand, ignores the useful life of an asset (later years of project life) and does not consider the working capital.
2. The D.C.F.R. and N.P.V. are regarded as the most generally acceptable economic indexes to be used in the oil industry. They take into account the following factors:
   - Cash flows and their magnitude
   - Lifetime of project
   - Time value of money

Although the D.C.F.R. involves a trial-and-error calculation, computers can be easily used in this regard. The D.C.F.R. is characterized by the following:

- It gives no indication of the cash value.
- It measures the efficiency of utilizing a capital investment.
- It does not indicate the magnitude of the profits.

It is recommended for projects where the supply of capital is restricted and capital funds must be rationed to selected projects.

The N.P.V. method, on the other hand, is considered to measure "profit". The values reported by the N.P.V. yield the direct cash measure of the success of a project; hence, they are additive (compare with D.C.F.R.).

3. Any of the methods described in this chapter and proposed for economic evaluation in oil projects should be used with discretion and with due regard for its merits and demerits. Each index provides limited knowledge which is helpful in making project decisions. No major investment decision should

be totally based on a single criterion. A more careful study should be considered for oil projects ending in different conclusions as a result of using different economic indicators.

4. Other important factors to consider in economic evaluation are discussed next. Every oil company has to consider that certain investments will not yield a "measurable profit" because some investments may be needed to improve employee or community goodwill, or to meet legal requirements of the government under which the oil operations are located. For example, investments in equipment to reduce air or water pollutants and investments in the social well-being of the community may not contribute dollars to equity of a company. These are examples of those investments which will not yield a measurable profit. And oil companies must face some of these "opportunities", especially when their operations are in countries other than that in which their main administrative offices are located.

An efficient oil company is aware of this type of investment and makes plans in advance for these expenditures. Oil management must, therefore, increase its return on those investments yielding measurable profits accordingly, so that the portfolios of profit and nonprofit investments taken together yield a sufficient overall return. For example, suppose that an oil company has calculated its "needed" return (or cost of capital for owners of the oil company) to be approximately 15%. But 25% of its investments are nonprofit, or "necessity", projects. This means that 75% of its investments are "profitable" ones. To cover the 25% that are nonprofit investments, the returns on the 75% that are profitable will have to be approximately 20% (15% ÷ 75%). Thus, oil companies need not only to appraise all potential investments individually, but also to constantly view the position of their portfolios of profit and nonprofit investments taken together.

7.5 MODEL EXAMPLES

Example 7.3

If an oil company expects a cash flow of $800,000 by the end of 10 years, and 10% is the current interest rate on money, calculate the N.P.V. of this venture.

SOLUTION

No capital investment is involved here, so the problem is simply a discounting procedure.

The present value of the cash flow

$$= 800{,}000(1+0.1)-10$$
$$= \$308{,}000$$

Example 7.4

Assume that a distillation unit with an initial cost of $200,000 is expected to have a useful life of 10 years, with a salvage value of $10,000 at the end of its life. Also, it is expected to generate a net cash flow above maintenance and expenses amounting to $50,000 each year. Assuming a discount rate 10% calculate the N.P.V.

SOLUTION

The present value of the annual cash flow can be found using the following equation:

$$P = A\frac{(1+i)^n - 1}{i(1+i)^n}$$

$$= 50{,}000\frac{(1.1)^{10} - 1}{0.1(1.1)^{10}}$$

$$= 50{,}000(6.144) = \underline{\$307{,}25}$$

where this factor is readily obtained from tables found in Appendix A. Calculations are given in the following tabulated form (Table 7.3).

TABLE 7.3
N.P.V. of Distillation Column

Distillation Unit Initial Cost	$200,000
Useful Life / Year	10
Salvage Value	$10,000
Net Cash Flow each Year	$50,000
D.C.F.R	$10
Present value of cash flow of 50000 annally, for 10 years at 10 %	$307,250
The Peresnt Value of Cash Flows for 10 years, Minus original investment of $200,000	$107,250
Persent Value of $10,000 salavage value to be recived at the end of years at 10%	$3,860
Total value of net cash receipts plus persent value	$111,110

NOTATION AND NOMENCLATURE

D.C.F.R Discounted cash flow rate of return (percentage), defined by Equation

d_y Discount factor, $(1 + i)^{-n}$, for the year y

F_B Future worth of an investment if deposited in the bank

F_o Future worth of compounded cash flows, generated from oil project

N.P.V. Net present value ($)

P	Present value of an asset equals sum of discounted cash flows and is given by: P = sum of p_y over the years, $y = 1$ to $y = n$
P.P.	Payout period (years), defined by Equation
P.V.I.	Present value index (dimensionless), defined by Equation
R.O.I.	Return on Investment, defined as the annual rate of return (percentage) by Equations
y	Designates a year within the lifetime *n*

8 Analysis of Alternatives, Selections, and Replacements

A decision is made between alternatives or the selection is based in this chapter on applying the economic indicators presented before in Chapters 6 and 7.

8.1 INTRODUCTION

This technique of analysis of alternatives and selection can assume many different aspects.

Typical example are: the choice among alternative processes proposed for enhanced oil recovery in oil fields, the choice among alternative methods of cooling process streams in gas plants or the choice among alternative designs of heat exchangers, waste-heat boilers, pumps, or any piece of equipment.

A SELECTED CASE (EXAMPLE 8.1)

An oil company is offered a lease of a group of oil wells in which primary production is getting to an end, and the major condition of this offer is to undertake a secondary recovery project (water injection) by the end of the 5th year. The capital investment of this project is estimated to be $650,000. In return, the revenue in the form of cash flow realized from this lease is as follows:

- $50,000/year for the 1st 4 years
- $100,000/year for the 1st 4 years from the 6th to 20th years

A comparison has to be made between the two alternatives: To invest or not to invest? In other words, should the project be accepted or not?

SOLUTION

Such a situation could be handled by using the annual cost/present worth economic approach as will be explained later.

Economic alternatives can be classified into two main categories:

- To choose among different ways to invest money not necessarily to accomplish the same job, in which case the decision is influenced by management rather than by technical people;

- To choose among alternative assets or equipment doing the same job, where mutually exclusive choices are considered and decision is made by technical people. Mutually exclusive projects imply that when two alternatives are compared, one project or the other is selected (but not both).

Many cases of alternative analysis can be handled with the "differential technique" or finding the "rate of return on the extra investment", for the difference between two alternate investments. The following methods are presented:

1. Differential approach (Δ approach) or return on extra investment (R.O.E.I.)
2. Total equivalent annual cost (T.E.A.C.)/present value method
3. Total capitalized method

To identify the problem at hand is of prime importance to identify the problem at hand to be either income expansion, or cost reduction:

- *Profit or income expansion*: where revenues (cash flows) are generated, and maximization of the profit is required, or
- *Cost reduction*: where no cash flows are given; instead expenses are known and reduction in costs is the criterion.

Replacement analysis, on the other hand, can be considered some sort of alternative analysis for investment tied up with an old asset versus an additional or (a replacement) investment. This situation is encountered to replace worn, inadequate or obsolete equipment, and physical assets.

8.2 DIFFERENTIAL APPROACH (Δ APPROACH), OR RETURN ON EXTRA INVESTMENT (R.O.E.I.)

The differential approach is a concept, which could be applied for selection among alternatives for a group of equipment, plants, processes, or oil-related venture projects. The principle of minimum capital investment is applied in this method in the following sense: For a set of alternatives needed for a given job and doing the same function, choose the minimum investment as the base one on *base plan*.

The differential approach to be used as a criterion for selecting alternatives is summarized by the following procedure:

1. Select the minimum capital investment (C.I.) as our base plan, compute ΔC.I. (difference in capital investment) for the alternatives.
2. Compute Δprofit (difference in cash income) for the alternatives, for the income-expansion problem, and Δsaving (difference in annual costs) for the alternatives, for the cost-reduction problem.

3. Calculate the rate of R.O.E.I. as follows:

$$\Delta\text{C.I.} \rightarrow \text{profit; R.O.E.I.} = \left(\frac{\Delta\text{profit}}{\Delta\text{C.I.}}\right)100$$

$$\Delta\text{C.I.} \rightarrow \text{saving; R.O.E.I.} = \left(\frac{\Delta\text{saving}}{\Delta\text{C.I.}}\right)100$$

- Check to see that the preferred choice should have an R.O.E.I. greater than a minimum value prescribed by management.

For alternatives involving small increments in capital investment, the best (most economical) alternative is arrived at by either graphic or analytical solutions.

The following solved examples illustrates these principles.

Example 8.2

In the alkanolamine sweetening process of natural gas, two types of coolers have been suggested for the amine solvent: type A and type B. Using the data given next (Table 8.1) recommend which alternative should be used if both types are acceptable technically. The minimum rate of return on money invested is 15% and the economic lifetime is 10 years for the coolers.

SOLUTION

Use straight-line depreciation of 10% of C.I.

The problem is a cost-reduction type.

TABLE 8.1
Data for Example 8.2

	Type A	Type B	
Capital Investment (C.I.)	10,000	15,000	
n, years	10	10	
A. dep. =C.I./n	1,000	1,500	
A. oper. cost	3,000	1,500	
Total annual cost = A. dep. + A. oper. cost	4,000	3,000	
A. rate of return (given)	0.15	0.15	
Diff. in C.I.			5,000
Diff. in annual cost (saving)			1,000
Annual percentage saving = diff. in annual cost (saving)/diff. in C.I.			20%
Rate of return on the extra capital is greater than 15%.			
Therefore, Type B is recommended.			
Assuming that n for type B changes between 5 and 15 years.			
Plot the annual % saving versus n, as shown next in Figure 8.1.			

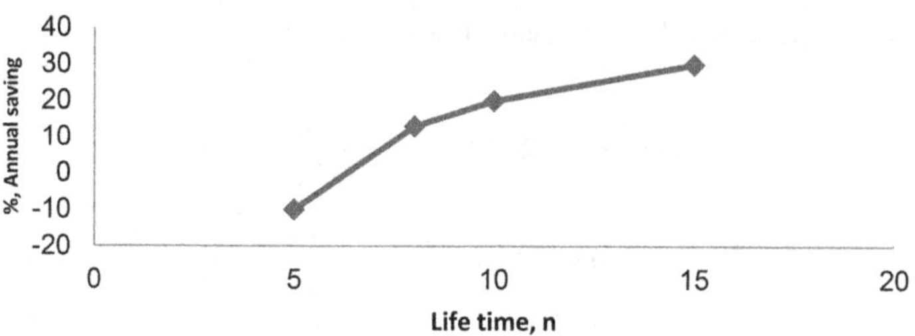

FIGURE 8.1 Change of A.% saving versus lifetime of type B.

Example 8.3

Insulation thickness is important for heat exchangers in the oil industry. One situation was encountered in the sulfur recovery plant from hydrogen sulfide gas (H_2S) (which has to be removed from natural gas). A heat exchanger was designed and recommendation was made for ***four*** possible thicknesses of insulation. The costs and savings related to these cases are as follows. Which one is recommended for 15% minimum R.O.I.?

ILLUSTRATION OF SOLUTION IS GIVEN IN TABLE 8.2

For 15% minimum R.O.I., calculations indicate that all four proposals are acceptable, since they generate R.O.I. > 15%, each. Now, we can apply the differential approach as indicated above. However, let us use the graphic analysis technique, since the problem involves small-investment increments. Referring to Figure 8.2, the annual savings/C.I. curve is drawn as shown using the above data. As can be seen, by increasing the C.I., the annual savings are increased until we hit the optimum point, *M*, which represents the maximum savings. Then, by drawing

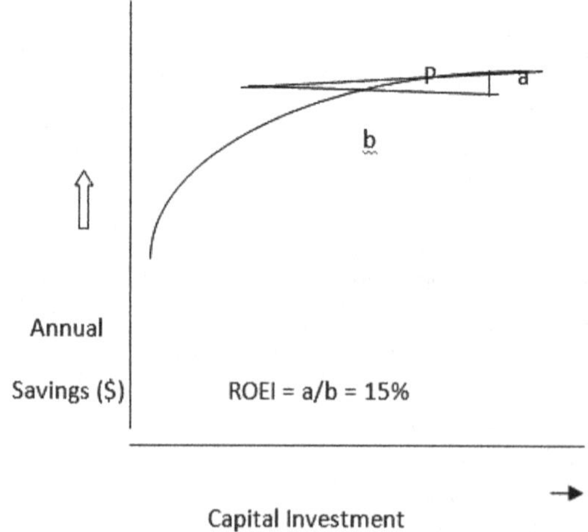

FIGURE 8.2 Graphical presentation of differential solution (Example 8.3).

TABLE 8.2
Data for Example 8.3

Parameter	1 inch Insulation	2 inch Insulation	3 inch Insulation	4 inch Insulation
Cost of insulation ($)	1,200	1,600	1,800	1,870
Savings (Btu/hr)	300,000	350,000	370,000	380,000
Value of savings ($/yr)[a]	648	756	799	821
Annual depreciation cost ($/yr)[b]	120	160	180	187
Annual profit ($)	528	596	619	634
R.O.I.	44.0%	37.3%	34.4%	33.9%

[a] Based on $0.3 per million Btu of the heat recovered and 300 working days per year.
[b] Based on 10-year lifetime.

our tangent line at *P*, we can achieve an R.O.E.I. of about 17% when using C.I. of nearly $1,600, or an insulation of 2-inch thickness.

The R.O.E.I. or differential method has one big drawback if applied to alternatives with different economic lifetimes. This puts a constraint on using it for these situations, which can be handled by other methods to be discussed next. Figure 8.2 illustrates this case.

	1	2	3	4
1st comparing 1 to 2	acceptable as a basis	17.0%	—	—
2nd comparing 2 to 3	—	-basis-	11.5%	—
3rd comparing 2 to 4	—	-basis-	—	14.1%

Conclusion: Design 2 is recommended; it gives more profit than design 1.

While, R.O.E.I. is 17%, which is >15% (minimum).

Figures 8.3 and 8.4 are graphical plots to illustrate the results obtained in solving this example.

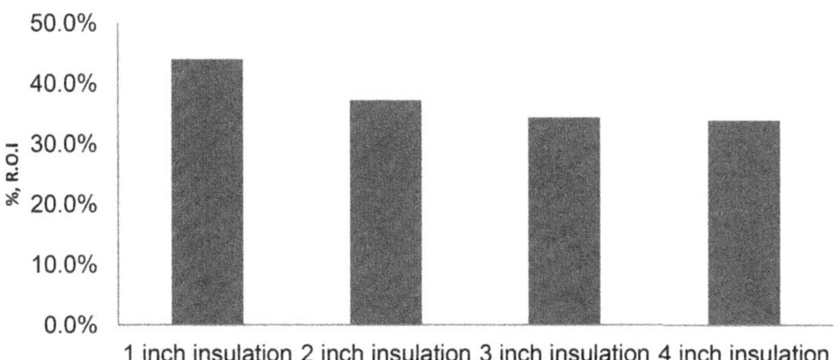

FIGURE 8.3 Change of R.O.I. savings versus types.

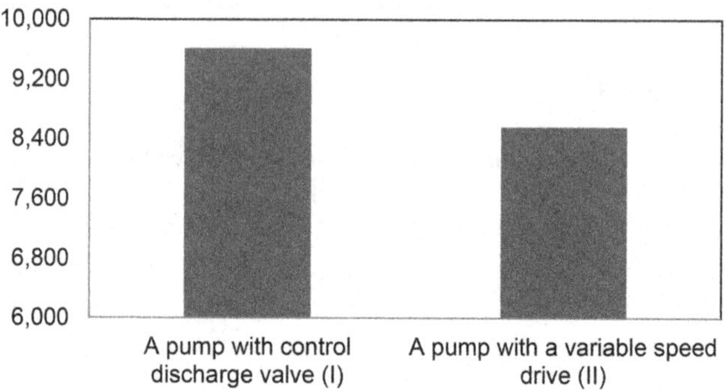

FIGURE 8.4 Comparison between types of pumps.

8.3 TOTAL EQUIVALENT ANNUAL COST (T.E.A.C.)/PRESENT VALUE METHOD

In this method, all costs incurred in buying, installing, operating, and maintaining an asset are put on the some *datum*; that is, on *annual basis*. Generally, the annual equivalent costs are brought to the present value for all alternatives.

Specifically, the T.E.A.C. is the sum of the annual cost of capital recovery (initial capital plus interest on it) and other annual operating costs. (Remember that depreciation costs cannot be included with the annual operating costs. They are taken care of in the cost of capital recovery.)

$$\text{T.E.A.C.} = A_r + \text{other annual operating costs}$$

Where:

$$A_r = P\left(\frac{i(1+i)^n}{(1+i)^n - 1}\right)$$

Example 8.4

Recommend which arrangement to select out of the following two cases, where energy saving is required by using higher capital investment:

	A pump with control discharge valve (I)	A pump with a variable speed drive (II)
C.I. ($)	13,000	17,000
Annual cost of energy for pumping ($)	6,000	2,800
Annual maintenance costs ($)	1,500	3,000
Lifetime (yr)	10	10

Assume $i = 10\%$ and the salvage value is negligible.

TABLE 8.3
Data for Example 8.4

	A Pump with Control Discharge Valve (I)	A Pump with a Variable Speed Drive (II)
Capital Investment (C.I.)	13,000	17,000
Lifetime (yr)	10	10
Annual maintenance costs ($)	1,500	3,000
Annual cost of energy for pumping ($)	6,000	2,800
Ar	2,116	2,767
T.E.A.C.	9,616	8,567

Assume $i = 10\%$ and the salvage value is negligible. Which design is to be recommended?
System II is recommended, since T.E.A.C. is less than for system I. Figure 8.4 represents the results.

Figure 8.4 is a graphical plot to illustrate such results.

SOLUTION IS GIVEN IN TABLE 8.3

$$\text{For system I: } A_r = 13{,}000\left[\frac{0.1(1.1)^{10}}{(1.1)^{10}-1)}\right]$$

$$= \underline{\$2{,}116}$$

$$\text{For system II: } A_r = 17{,}000\left[\frac{0.1(1.1)^{10}}{(1.1)^{10}-1)}\right]$$

$$= \$2{,}767$$

$$\text{T.E.A.C. for I:} = 2{,}116 + 6{,}000 + 1{,}500$$

$$= 9{,}616$$

$$\text{T.E.A.C. for II:} = 2{,}767 + 2{,}800 + 3{,}000$$

$$= 8{,}567$$

Example 8.5

Given: Assume the same two heat exchangers given in Example 7.2 with the same annual costs, economic lives, salvage values, and investments, and with the cost of capital once again 8%.

Wanted: Compare the two alternatives using the present worth values for each of the possibilities, as well as total equivalent capital at the "present" time of consideration of purchase of heat exchangers.

SOLUTION

Using the present value method, a series of known uniform annual costs are reduced to an equivalent present value. This allows one to estimate the dollar value at the present time that is equivalent to the amount of annual costs for some fixed years of service by two alternatives. But uniform annual costs must first be determined, and this is what the present value method does. (The annual cost method does not determine uniform annual costs.)

TABLE 8.4
Data for Example 8.6

	Purchase Possibility A	Purchase Possibility B
1. Present value of original (initial) costs	\$15,000	\$40,000
2. Present value of salvage value; formula "Find P, Given F", or factor	\$500 × 0.4632 = 232	\$1,000 × 0.4632 = 463
3. Present value of annual costs: (total costs) × factor of formula "Find P, Given A"	\$12,100 × 6.710 = \$81,191	\$6,800 × 6.710 = \$45,628
4. (1) – (2) + (3)	\$95,959	\$85,165

Now, for each of the possibilities, the present values of installations and the salvage values must be added and deducted, respectively, to current value of annual costs for 10 years in order to get total equivalent capital requirements.

The following calculations are carried out to find the the equivalent capital at 8% Table 8.4 illustrates the solution.

A comparison of the calculations for equivalent capital involved, \$95,959 for possibility A and \$85,165 for possibility B for the present time on an economic basis, indicates \$10,794 less favoring possibility B. In other words, a savings of \$1,610 in annual costs, as given by the annual cost method, favoring possibility B is reflected in a \$10,794 reduction in equivalent present value of possibility B when annual costs are uniform and determined with the use Total "present" equivalent capital at 8%: of the present value method.

Under conditions of low interest rates, the present or current value of possibility B, which is \$40,000, can still be less than the current worth of \$15,000 for possibility A. Thus, the interest rate is important in order to determine the present value. Lower interest rates, such as say 5% instead of 8%, favor even more the use of higher initial investments, in this case the \$40,000 stainless steel heat exchanger rather than the \$15,000 steel-copper exchanger, because the relative cost for the use of money is lower. Example 8.7 confirms these results. A summary of the solution using excel is given next, followed by a graphical chart as shown in Figure 8.5.

Example 8.6: (summary)

	A: (Steel and Copper) Heat Exchanger	B: Stainless Steel Heat Exchanger
Capital Investment cost(C.I.)	15,000	40,000
Lifetime (yr)	10	10
Salvage value*0.4632	232	463
Annual labor, maintenance, repairs, and operational expenses costs (\$)	11,500	4,000
Annual direct costs (\$)	600	2,800
Total annual costs	12,100	6,800
Present value of annual costs	81,191	45,628
Total "present" equivalent capital at 8%	95,959	85,165

FIGURE 8.5 Change of total annual costs for different types of H.E.

Example 8.7

Compare the relative annual costs and current present values of the two alternatives in Examples 8.5 and 8.6 for 10 years of service if money is worth 5% instead of 8%.

SOLUTION

a. *For the annual cost method at 5%:*

	Purchase Possibility A	**Purchase Possibility B**
Annual costs:		
Capital recovery = $14,500 × 0.1295 ("Find A, Given P") + (0.05)($500) =	$1,903	$39,000 × 0.1295 + (0.05)($1,000) = 5,101
Capital recovery = $14,500 × 0.1295 ("Find A, Given P") + (0.05)($500) =	$1,903	$39,000 × 0.1295 + (0.05)($1,000) = 5,101
Labor, maintenance, etc.	11,500	4,000
Other direct costs	600	2,800
Total annual costs	$13,903	$11,901

Compared to the costs when the interest rate is 8% (see Example 7.2, total annual costs for each possibility are lower when the interest rate is 5%. But the *difference* in annual costs is greater when the interest rate is lower. At 8%, the difference is $1,610 less in favor of possibility B, whereas at 5% it is $2,002 in favor of possibility B. Thus, lower costs of borrowing favor alternatives with large investment amounts more than alternatives with lower investment amounts.

b. *For the present value (present worth) at 5%:*

	Purchase Possibility A	**Purchase Possibility B**
Present worth of original costs (initial)	$15,000	$40,000
Present worth of salvage value ("Find P, Given F")	$500 × 0.6139 = 307	$1,000 × 0.6139 = 614
Present worth of annual costs ("Find P, Given A")	$12,100 × 7.722 = 93,436	$6,800 × 7.722 = 52,509
Total "present" equivalent capital at 5%	$108,129	$91,895

At 5%, the equivalent capital for purchase possibility B is $16,234 ($108,129 – $91,895) less than purchase possibility A. With lower interest rates, differences in equivalent capital are greater: $16,234 between alternatives at 5% and $10,794 between alternates at 8%. However, total present equivalent capital amounts are greater with lower interest rates: Totals at 5% are $108,129 and $91,895 for possibilities A and B, respectively, and at 8% are $95,959 and $85,165.

A comparison of these results and those obtained when money is worth 8% shows that (a) the time-money series is equivalent to larger capital requirements, and (b) the difference in equivalent present value is greater in favor of purchase possibility B than it is for A when money is worth only 5%.

Example 8.8

The overhead condenser in a stabilization unit of a natural gasoline plant has to be made of corrosion-resistant material. Two types are offered; both have the same capacity (surface area); however, the costs are different because of different alloying materials:

	Condenser A	**Condenser B**
C.I. ($)	23,000	39,000
n (years)	4	7

If money can be invested at 8%, which condenser would you recommend based on the T.C.C.?

SOLUTION

$$K_A = 23{,}000\left(\frac{(1.08)^4}{(1.08)^4 - 1}\right)$$

$$= \$86{,}000$$

$$K_B = 39{,}000\left(\frac{(1.08)^7}{(1.08)^7 - 1}\right)$$

$$= \$93{,}000$$

Therefore, condenser type A is selected (lower K).

Example 8.9

Solve Example 8.5 using the capitalized cost technique for 8% and 5% annual interest rates.

SOLUTION

Two methods are presented:

1. *Using the relationship givenbefore, (direct application)*:
   a. For $i = 8\%$:

	Purchase Possibility A	**Purchase Possibility B**
n (year)	10	10
C_R ($)	14,500	39,000
V_s ($)	500	1,000
Total operating cost ($/yr)	12,100	6,800

$$K_A(\$) = 500 + 14{,}500(1.8629) + \frac{12{,}100}{0.08} = \$178{,}762$$

$$K_B(\$) = 1000 + 39{,}000(1.8629) + \frac{6800}{0.08} = \$158{,}653$$

b. For $i = 5\%$:

$$K_A = 500 + 14{,}500(2.59) + \frac{12{,}100}{0.05} = \$280{,}055$$

$$K_b = 100 + 39{,}000(2.59) + \frac{6800}{0.05} = \$238{,}010$$

2. *Using step-by-step procedure (detailed)*:
   a. Capital requirements through capitalization, *with interest at 8%*, are as follows:

	Purchase Possibility A	Purchase Possibility B
Total annual costs:		
Net capital invested factor of formula ("Find A, Given F")	\$14,500 × 0.06903 = \$1,001	\$39,000 × 0.06903 = \$2,692
Annual labor, maintenance, operational costs, etc.	11,500	4,000
Other direct annual costs	600	2,800
Total annual costs to be capitalized	\$13,101	9,492
Capitalization of annual costs	\$13,101/0.08 = 163,763	\$9,492/0.08 = 118,650
Initial costs of annual costs	15,000	40,000
Total capitalized cost when money is worth 8%	\$178,763	\$158,650

   b. Capital requirements through capitalization, *with interest at 5%*, are as follows:

	Purchase Possibility A	Purchase Possibility B
Total annual costs:		
Net capital invested factor of formula ("Find A, Given F")	\$14,500 × 0.0795 = \$1,153	\$39,000 × 0.0795 = \$3,100
Annual labor, maintenance, operational costs, etc.	1,500	4,000
Other direct annual costs	600	2,800
Total annual cost to be capitalized	\$13,253	\$9,900
Capitalization of annual costs	\$13,253/0.05 = 265,060	\$9,900/0.05 = 198,000
Initial costs of investment	15,000	40,000
Total capitalized cost when money is worth 5%	280,060	\$238,000

It is clear that both the direct and detailed methods give the same final answer; however, one would be reluctant to use the latter approach.

At the lower interest rate of 5%, the capitalized cost is $42,060 less for possibility B ($280,060 – $238,000). The results illustrate the peculiar effect of the interest rate and emphasize the potential difficulties in comparing alternates on either a present value or a capitalized cost basis. When cost of capital is high, total capitalized costs become lower, but differences between capitalized costs of higher and lower investment amounts favor higher investments more when cost of capital (interest rate) is lower.

The interest rate is the determining factor, although the relative size of such individual items as initial costs, annual labor costs, annual material, repairs, maintenance, and other costs, when compared to capital recovery costs, can affect total equivalent capital involved.

The important point is that the interest based on the going value of money is always lower than the rate for a venture involving a risk. The engineer using the going rate for interest will bias his comparisons in favor of the alternative equivalent to oil capital requirements. Because of this, the annual cost method is preferred, but the service lives of the alternatives should be equal and annual costs of alternatives should be uniform. When different service lives are involved, or where nonuniform annual expenditures must be compared for alternatives, it is better to use the present value method and put all costs on a comparable basis in order to get accurate results and avoid "distortions" of costs.

8.4 REPLACEMENT ANALYSIS

In the oil industry, the usual experience is that assets are retired while they are still physically capable of continuing to render their service either in the oil field, in transportation systems or in the refining operations. The question is: how can we make the decision to replace an asset?

The decision to make such replacement should generally be made on the grounds of economy along with engineering fundamentals applicable to oil operations. That is, replacing a worn, obsolete, or inadequate asset can be translated into the language of economics.

Reasons behind a replacement can be defined either as "*a must*", that is, we have to replace, otherwise the operation will come to a halt, or "*optional*" in which case the asset is functioning, but there is a need for a more efficient or modern type. Such a classification is illustrated in Figure 8.6.

The principles governing replacement are best explained by using the word "defender" to stand for the old asset and the word "challenger" to identify the possible new candidate that will make the replacement. In order to utilize the challenger/defender analogy for replacement comparison, the following factors must be considered (Valle-Riestra, 1983):

All input/output of cash flows associated with the asset have to be known or estimated. This applies in particular to maintenance and operating costs of both defender and challenger.

Cost estimation of the value of the defender (market value/book value) must be made.

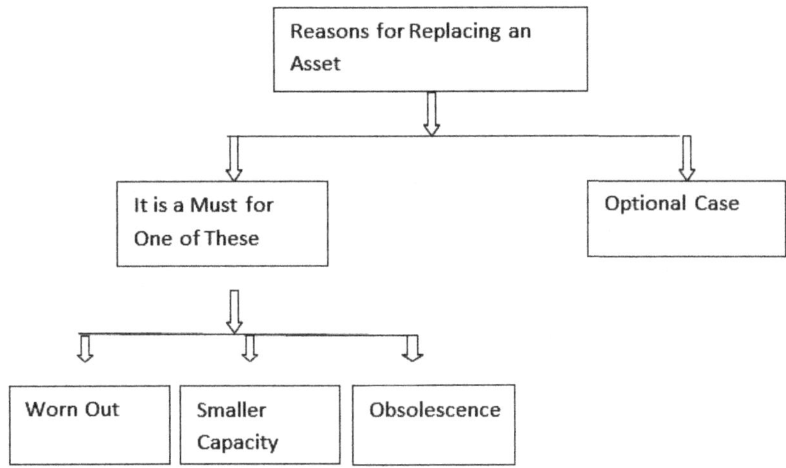

FIGURE 8.6 Replacement analysis.

Methods recommended earlier for the comparison of alternatives such as T.E.A.C., present worth or Δ approach could be applied. In other words, no new techniques are provided. Tax obligations or credits should be considered.

Example 8.10

A tank farm is receiving crude oil through a pipeline. Periodic measurements of the crude oil level are made. The annual labor cost for the manual operation is estimated to be \$50,000. However, if an automated level-measuring system is installed, it will cost \$150,000. Maintenance and operating expenses of the system are \$15,000 and \$5,000, respectively. The system will be operated for 5 years.

Should the automated level-measuring system be installed? Assume interest rate is 10%.

SOLUTION

Two alternatives must be compared:

Alternative 1: Manual operation

$$\text{Annual cost} = \$50,000$$

Alternative 2: Automated level-measuring system

$$\begin{aligned}\text{Annual cost} &= \text{capital recovery cost} + \text{operating maintenance}\\ &= 150,000\left[\frac{0.1(1.1)^5}{(1.1)^5}\right] + 15,000 + 5,000\\ &= 39,570 + 20,000 = \$59,570\end{aligned}$$

The manual operation, alternative 1, is less expensive.

Example 8.11

An oil company has an existing steam-generation unit. Its cost when new is $30,000, its lifetime is 10 years and it has a salvage value of zero. The annual operating cost is $22,000. After it has been in use for 5 years, the estimated book value of the unit is found to be $6,000. The remaining lifetime now is only 3 years.

It has been proposed to replace this unit by another new one. Its cost is $40,000, lifetime 10 years, operating costs $15,000/yr and zero salvage value. Should we continue using this unit or go for the replacement?

The company requires 10% R.O.I.

SOLUTION

Old Unit	**Replacement**
V_o($): 30,000	$40,000
n: 10 years	n: 10 years
Operating: $22,000/yr After 5 years of use	Operating: $15,000/yr
V_5 = $6,000	
3 years are left only	
$V_s = 0$	

Now, take these 3 years for comparison:

$d = \frac{6000}{3} = \$2{,}000/\text{yr}$	$d = \frac{40{,}000}{10} = 4{,}000$
Operating costs = $22,000/yr	Operating costs = 15,000
Total cost = $24,000/yr	Total cost = $19,000

Therefore, savings = 24,000 – 19,000 = $5,000/yr.

If replacement takes place, R.O.E.I. = (5,000/$34,000)100 = 14.7%.

Example 8.12

Consider a control valve that becomes obsolete 3 years before it has been fully depreciated. When fully depreciated, the valve will have a salvage value of $400, but at this time (3 years before), it has a trade-in (or resale) value of $1,000. If the book value (original cost – total depreciation to date) is $760, there is a favorable "bonus" to management of $240 in trade-in.

But the bonus of $240 is irrelevant as a sunk cost. If a minimum rate of return is assumed as 10% before taxes, the question is whether the obsolete control valve with 3 years to go before being fully depreciated should be replaced now by a new valve. Calculations are needed to compare the old valve with a new valve, which would cost $5,000 and have an eventual salvage value of $500 and a service life of 10 years.

SOLUTION

Annual cost of *old* valve:	
Capital recovery costs (760)(0.40211) 10% for 3 years + 0.10 × $400	= $346.00
Operating and maintenance costs (estimated)	= $1,820.00
Total annual cost of old valve	= $2,166.00
Then, tentative annual cost of *new* valve:	
Capital recovery costs ($4,500)(0.16275) 10% for 10 years + 0.10 × $500	782.00
Operating and maintenance cost (estimated)	= $1,000.00
Total annual cost of new valve	= $1,782.00

By comparing the old control valve with the new valve, we can see that purchasing the new valve now would mean an annual savings of $384 or ($2,166 – $1,782). If the old valve is depreciated out, only the salvage value of $400 could be allowed on capital recovery.

9 Feasibility Study

9.1 INTRODUCTION

A feasibility study is an analysis to determine whether the project is technically and financially feasible that takes all of a project's relevant factors into account—including economic, technical, legal, and scheduling considerations. This is done in order to determine if the project under consideration would be completed successfully.

In other words, the study tries to determine whether the project is technically and financially feasible. Financially feasible means whether the project is feasible within the estimated cost. A feasibility study also determines whether a project makes good business sense, i.e., will it be profitable?

9.2 PROPOSED METHODOLOGY

The following is a proposed outline to follow for solving a feasibility study aiming for financial viability of a project:

9.3 APPLICATIONS

SELECTED CASE NO 1

The following example illustrates the application of a feasibility study in order to provide an assessment of the practicality of a proposed project. The project involves the installment of a waste-heat-recovery system.

A waste heat recovery unit (WHRU) is an energy recovery heat exchanger that transfers heat from process outputs at high temperature to another part of the process for some purpose, usually increased efficiency. Waste heat found in the exhaust gas of various processes or even from the exhaust stream of a conditioning unit can be used to preheat the incoming gas. This is one of the basic methods for recovery of waste heat, as shown in Figure 9.1.

Consider the unit shown in the diagram, Figure 9.2, a two-pass water tube waste heat boiler. It was suggested to buy such unit and install it in an oil field.

The following information is provided in order to carry out a feasibility study:

- Cost of unit = $250,000 with lifetime 10 years
- Value of waste-heat to be recovered by the unit is anticipated to be = $40,000 annually
- Maintenance fees for the unit = $10,000 annually
- Money could be invested by the management at 10% annually

What is your recommendation using the feasibility study?

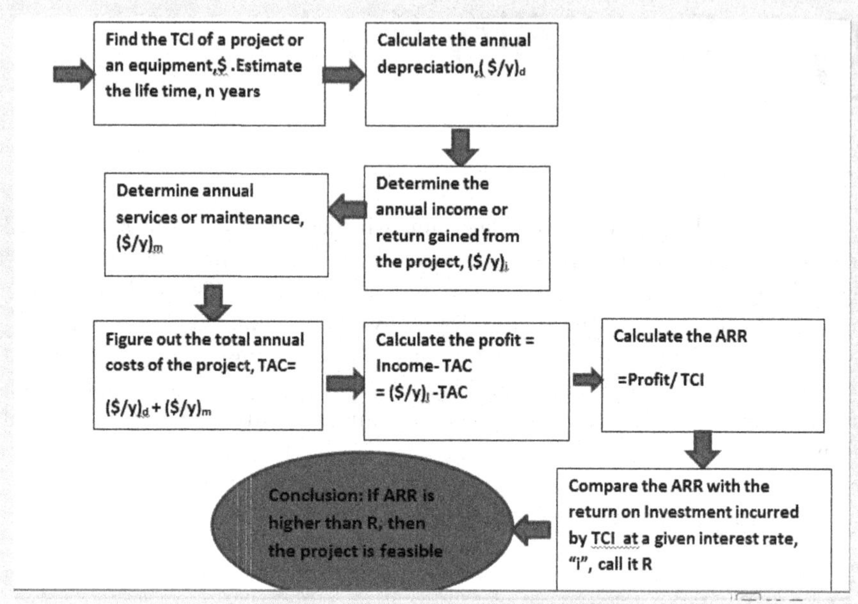

FIGURE 9.1 Proposed outline to solve feasibility study problems.

SOLUTION

$$\text{The annual depreciation of the unit over the lifetime} = 250{,}000/10$$

$$= \$25{,}000$$

$$\text{The total annual costs TAC of operating the WHB} = 25{,}000 + 10{,}000$$

$$= \$35{,}000$$

$$\text{Income} - \text{TAC} = 40{,}000 - 35{,}000 = \$5{,}000$$

The feasibility analysis indicates a profit of $5,000. This is equivalent to ARR $=[5{,}000/250{,}000] \times 100 = 2\%$.

If money would be invested by the management at 10% annual rate of return:

$$\text{ARR will be} = 250{,}000 \times 0.10$$

$$= \$25{,}000$$

CONCLUSION

The feasibility study shows that under the given conditions, it is not feasible to install the waste heat boiler (WHB).

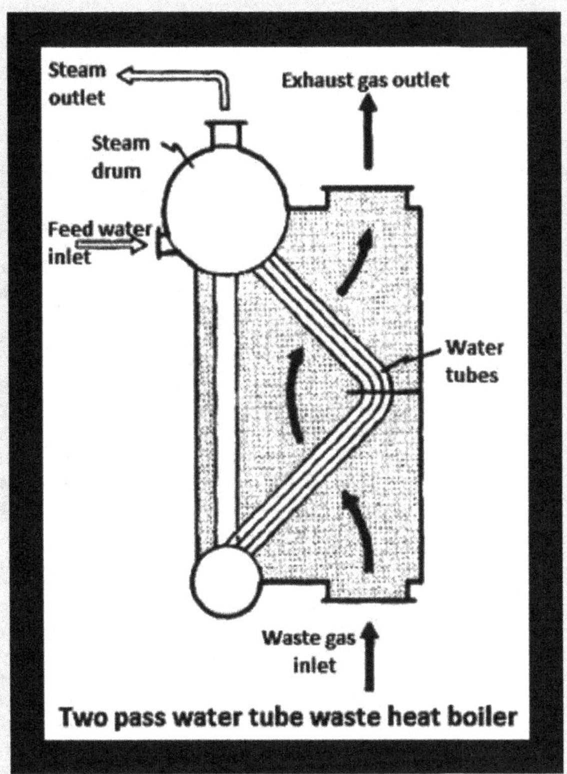

FIGURE 9.2 A two-pass waste heat boiler.

SELECTED CASE NO 2

A feasibility study carried out for an oil company indicated that it is possible to invest $1 million in either one of two projects. Anticipated cash flows generated by the two projects over the useful lifetime are given next: Table 9.1.

Year	Project 1	Project 2
1	400,000	100,000
2	320,000	200,000
3	200,000	300,000
4	300,000	400,000
5	100,000	500,000
Total anticipated cash Flow	1,320,000	1,500,000

At 8% Discount

Year	Discount factor for 8 %	Project 1 Cash Flow	Project 1 Discounted Value (col. 1*col. 2)	Project 2 Cash Flow	Project 2 Discounted Value (col. 1*col. 4)
	1	2	3	4	5
1	0.926	400,000	370,400	100,000	92,600
2	0.856	320,000	273,920	200,000	171,200
3	0.794	200,000	158,800	300,000	238,200
4	0.735	300,000	220,500	400,000	294,000
5	0.681	100,000	68,100	500,000	340,500
			$1,091,723		$1,136,505

At 10% Discount

Year	Discount factor for 10 %	Project 1 Cash Flow	Project 1 Discounted Value (col. 1*col. 2)	Project 2 Cash Flow	Project 2 Discounted Value (col. 1*col. 4)
	1	2	3	4	5
1	0.909	400,000	363,600	100,000	90,900
2	0.826	320,000	264,320	200,000	165,200
3	0.751	200,000	150,200	300,000	225,300
4	0.683	300,000	204,900	400,000	273,200
5	0.621	100,000	62,100	500,000	310,500
			$1,045,123		$1,065,105

At 12% Discount

Year	Discount factor for 12 %	Project 1 Cash Flow	Project 1 Discounted Value (col. 1*col. 2)	Project 2 Cash Flow	Project 2 Discounted Value (col. 1*col. 4)
	1	2	3	4	5
1	0.893	400,000	357,200	100,000	89,300
2	0.797	320,000	255,040	200,000	159,400
3	0.712	200,000	142,400	300,000	213,600
4	0.636	300,000	190,800	400,000	254,400
5	0.567	100,000	56,700	500,000	283,500
			$1,002,143		$1,000,205

Give your recommendations of which project you choose based on the net present value (N.P.V.). Use selected values for the discount interest rate (more than one).

1. Compute the discounted cash flow rate of return (D.C.F.R.) for each project.

SOLUTION

For (1), calculation is done for three different discount interest rates: 8%, 10%, and 12%. In addition, a graphic plot is presented (Figure 9.1 for the change of the discounted value (present value) of the cash flows for both projects with the discount rate.

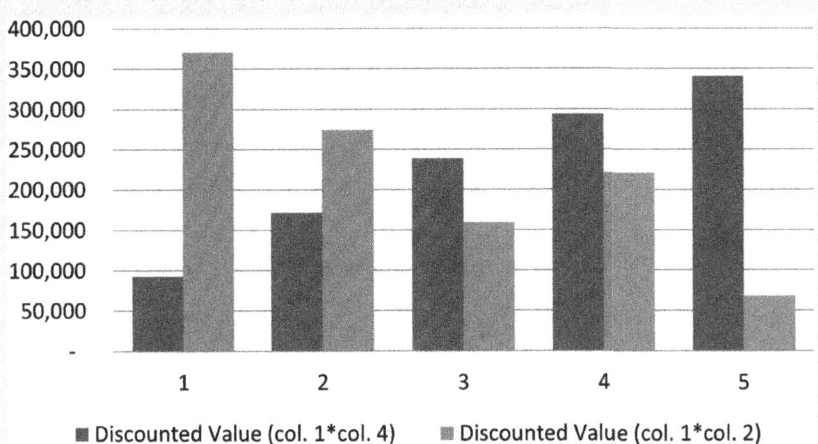

Next example is fordistillaion column, shown in Table 9.2 (Table 6.7)

TABLE 9.2
N.P.V. of a Distillation

Distillation Unit Initial Cost	$200,000
Useful Life / Year	10
Salvage Value	$10,000
Net Cash Flow each Year	$50,000
D.C.F.R	$10
Present value of cash flow of 50000 annally, for 10 years at 10 %	$307,250
The Peresnt Value of Cash Flows for 10 years, Minus original investment of $200,000	$107,250
Persent Value of $10,000 salavage value to be recived at the end of years at 10%	$3,860
Total value of net cash receipts plus persent value	$111,110

Therefore, as the example shows, the choice between the two projects depends on the discount rate used. Usually, the oil company's cost of capital for investing in the project will determine which project is selected.

From the data, it can be seen that at a discount of 12%, the present value of cash flow from project 2 is $1 million; and at a discount of over 12%, the present value of cash flow gives us the discount rated amount of under $1 million. This analysis of the present value of cash flow gives us the discount rate at which anticipated cash flow equals the initial investment, which is the D.C.F.R. For project 2, it is about 12%; for project 1, it is about 13%.

10 Model Solved Examples

PROBLEM 10.1

A sum of \$10,000 is borrowed by a refining oil company. Propose four different equivalent plans of money payments for this capital over a period of 10 years assuming the interest rate is 6%.

SOLUTION

As shown in Table 10.1, plan I involves the annual payment of the interest only (\$600) until the end. Plans II and III involve systematic reduction of the principal of the debt (\$10,000). For plan II this is done by uniform repayment of principal (\$1,000/yr) along with diminishing interest, while for plan III a scheme is devised to allow for uniform annual payment for both capital and interest all the way through until the end (\$1,359). For plan IV, on the other hand, payment is done only once at the end of the 10th year. The equivalence of the four payments is further illustrated in Figure 10.1

PROBLEM 10.2

Assume a petroleum company investment of \$10 million for an expansion to a current refinery, allocated \$1,000,000 for land and \$7,000,000 for fixed and other physical properties subject to depreciation. Additional capital of \$2,000,000 is available for operation purposes, but this sum is not subject to depreciation. Investors want a 15% interest rate (or earning rate to investors) on their money for a 10-year period. The sinking-fund method will be used, with depreciation figured at 15% per year. No income taxes are involved in order to simplify the example.

SOLUTION

First-year profit before deducting the sinking-fund depreciation charge made at the earning rate of 15% interest, and assuming no salvage value for the physical properties, is 0.15 × \$1,000,000, or \$150,000 per year.

But the oil company must earn enough additional money annually to pay for the depreciation occurring on the depreciable capital of \$700,000.

Using sinking-fund depreciation and a 15% interest rate for the sinking fund, the annual deposit in the fund is given by:

$$A_d = \frac{\$700,000 \times 0.15}{(1.15)^{10} - 1} = \$34,440$$

TABLE 10.1
Summary for the Four Plans for Solving Example 10.1

Year	Investment	I ($)	II ($)	III ($)	IV ($)
0	$ 10,000				
1		600	1,600	1,359	
2		600	1,540	1,359	
3		600	1,480	1,359	
4		600	1,420	1,359	
5		600	1,360	1,359	
6		600	1,300	1,359	
7		600	1,240	1,369	
8		600	1,180	1,359	
9		600	1,120	1,359	
10		10,600	1,060	1,359	17,908

Thus, company profits before depreciation must total $184,440 ($150,000 + $34,440) and not merely $150,000 in the first year. Actually, the $184,440 in the first year represents:

$34,440 = the sum of annual depreciation charge
$105,000 = the 15% interest on the un-depreciated part of the depreciable capital which is, in the first year or before any deductions, 0.15 × $700,000
$45,000 = the 15% interest on the non depreciable capital, or 0.15 × $300,000
$184,440 = the total for the first year

Thus $139,440 ($105,000 + $34,440) is needed to cover (1) the depreciation deposit in the sinking fund, and (2) the interest on the depreciable capital for that year. This is also calculated by using:

$$A_r = \$700{,}000 \frac{0.15(1.15)^{10}}{(1.15)^{10} - 1} = \$139{,}440$$

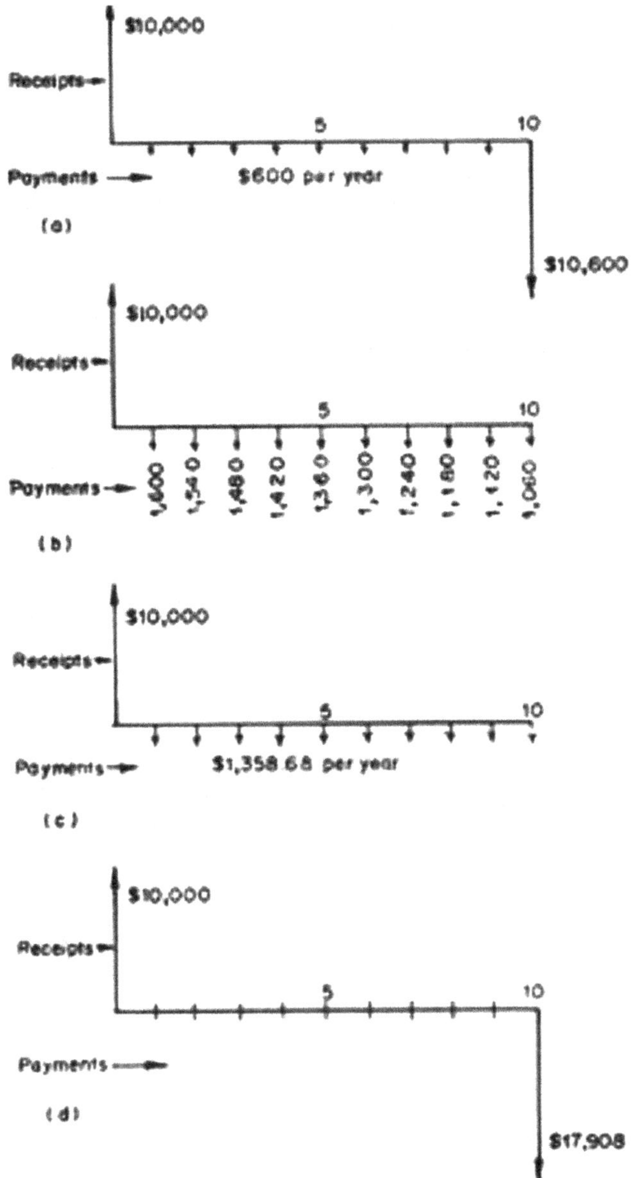

FIGURE 10.1 Equivalence of the four payments.

In each succeeding year, the book value of the depreciable capital decreases, but the depreciation reserve increases in such a manner that the sum of the two always equals to $700,000 and the total annual interest remains constant at $105,000 even though the interest charges on each component vary.

The biggest drawback to the actual use of the sinking-fund method in business is the fact that businesses rarely maintain an actual depreciation sinking fund. The interest rate which could be obtained on such deposits would be small, probably not over 6% in the petroleum business, according to financial experts of the oil industry. An active business, such as an oil company operation, is constantly in need of working capital. This capital will usually earn much more than 6%.

A reasonable rule is that all values should be kept invested in the oil business and not remain idle. As a result, a fictitious depreciation fund is often used: The amounts which have been charged to.

PROBLEM 10.3

With reference to the investment made to procure boilers for surface facilities in an oil field, as shown in Table 10.2, calculate the payback period (P.P.) for each alternative and give reasons for selecting one and not the other.

SOLUTION

P.P. is readily calculated as follows:

$$\text{P.P.} = 50{,}000(\$) \,/\, \frac{50{,}000(\$/\text{yr})}{4}$$
$$= 4 \text{ years, for both cases}$$

As far as the P.P. as a criterion for choice, the number of years to recover the depreciable capital is the same for both types of boilers. However, the recovery of investment for boiler 1 is faster than for boiler 2 (e.g.,, compare \$20,000 to \$5,000 for the 1st year), as shown in Figure 10.2. Therefore, from the standpoint of cost of money (time value of money), investment in boiler 1 is preferable to investment in boiler 2.

TABLE 10.2
Comparison of Two Boiler Investment (Example 10.3)

	Cash Flow	
Year	Boiler 1	Boiler 2
0	50,000	50,000
1	20,000	5,000
2	15,000	10,000
3	10,000	15,000
4	5,000	20,000
Total Cash Flow	50,000	50,000
Payback period	P.P 1 =	4 Years
	P.P 2 =	4 Years

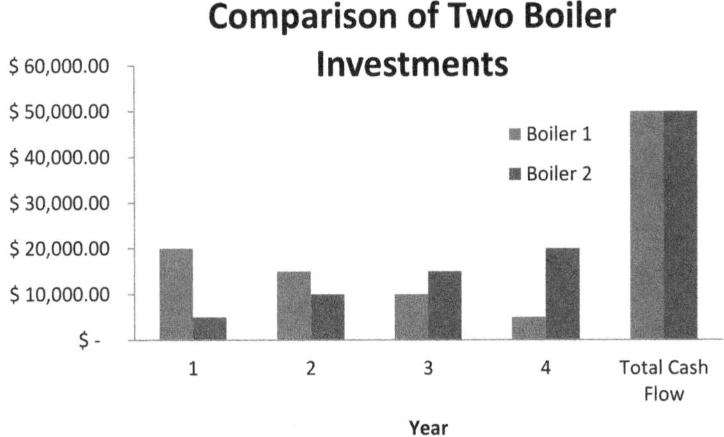

FIGURE 10.2 Solution of Problem 10.3.

This example points out that, when using the payout period method, oil management should also observe the rapidity of cash flows between alternatives. The alternatives may have the same number of years-to-pay-back, as they do here, but one may be more favorable than the other because the largest amount of cash flow comes in the first few years. This could be an excellent point in favor of investment in one alternative over another when both have approximately the same payout periods. It could be a strong factor in selection of one especially if a greater amount of cash "back" is needed early in the investment (Figure 10.3).

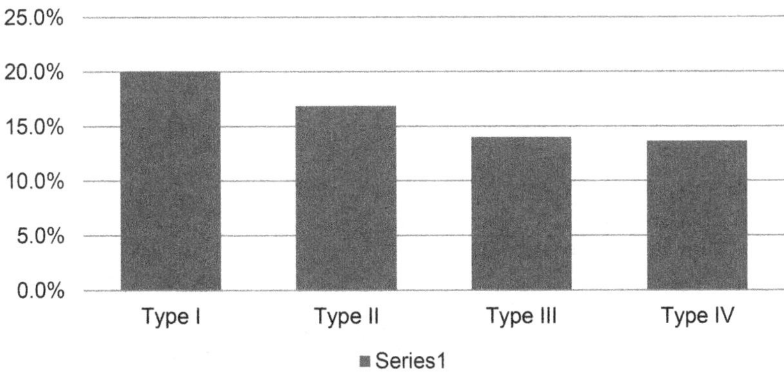

FIGURE 10.3 Change of return on investment (R.O.I.) percentage saving versus different types.

10.1 MODEL EXAMPLES

Example 10.1

If an oil company expects a cash flow of \$800,000 by the end of 10 years, and 10% is the current interest rate on money, calculate the net present value (N.P.V.) of this venture.

SOLUTION

No capital investment is involved here, so the problem is simply a discounting procedure.

The present value of the cash flow:

$$= 800,000(1+0.1)-10$$

$$= \$308,000$$

Example 10.2

Assume that a distillation unit with an initial cost of \$200,000 is expected to have a useful life of 10 years, with a salvage value of \$10,000 at the end of its life. Also, it is expected to generate a net cash flow above maintenance and expenses amounting to \$50,000 each year. Assuming a selected discount rate of 10%, calculate the N.P.V.

$$P = A\frac{(1+i)^n - 1}{i(1+i)^n}$$

$$= 50,000\frac{(1.1)^{10} - 1}{0.1(1.1)^{10}}$$

$$= 50,000(6.144) = \underline{\$307,25}$$

where this factor is readily obtained from tables found in Appendix A. Calculations are given in the Table 10.3.

TABLE 10.3
N.P.V. for Distillation Column

Distillation Unit Initial Cost	\$200,000
Useful Life / Year	10
Salvage Value	\$10,000
Net Cash Flow each Year	\$50,000
D.C.F.R	\$10
Present value of cash flow of 50000 annally, for 10 years at 10 %	\$307,250
The Peresnt Value of Cash Flows for 10 years, Minus original investment of \$200,000	\$107,250
Persent Value of \$10,000 salavage value to be recived at the end of years at 10%	\$3,860
Total value of net cash receipts plus persent value	\$111,110

TABLE 10.4
Summary for the Five Types

	Type I	Type II	Type III	Type IV
Capital Investment (C.I.)	10,000	16,000	20,000	26,000
n, years	5	5	5	5
A. dep. = C.I./n	2,000	3,200	4,000	5,200
A. oper. cost	100	100	100	100
Total annual cost = A. dep. + A. oper. cost	2,100	3,300	4,100	5,300
Revenue (income) $/yr	4,100	6,000	6,900	8,850
Annual profit	2,000	2,700	2,800	3,550
R.O.I.	20.0%	16.9%	14.0%	13.7%

Example 10.3

Instead of flaring the associated natural gas separated along with crude oil, it was decided to recover the lost heat by using the waste-heat recovery system (W.H.R.S.). For *pilot test runs,* four designs were offered; each has a lifetime of 5 years. The savings and costs associated with each are as follows (Table 10.4).

All four designs seem to be acceptable as far as the minimum annual R.O.I., exceeding 10% (required by management). Which design is to be recommended?

SOLUTION

Using incremental comparison:

	1	**2**	**3**	**4**
1st comparing 1 to 2	acceptable as a basis	11.7%	—	—
2nd comparing 2 to 3	—	-basis-	2.5%	—
3rd comparing 2 to 4	—	-basis-	—	8.5%

Conclusion: Design 2 is recommended; it gives more profit than design 1, while return on extra investment (R.O.E.I.) is 11.7%, which is >10% (minimum).

Figures 10.3 and 10.4 are bar charts to illustrate the solution of the problem.

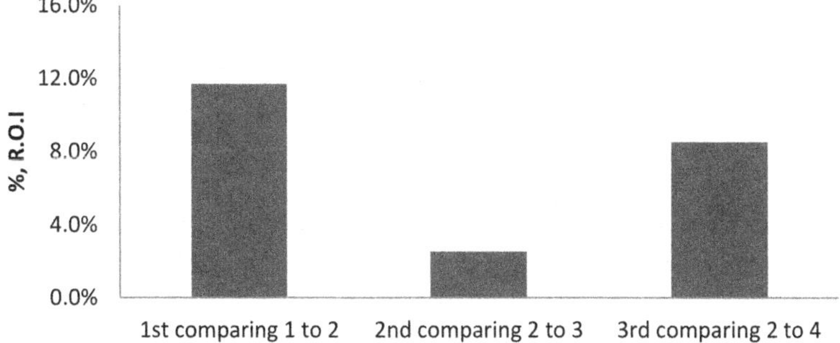

FIGURE 10.4 Incremental comparison versus different types.

Example 10.4

Given: Consider two possibilities related to the purchase of a heat exchanger for an oil refinery to replace an older model for which annual costs are running around $20,950. Other details are as follows:

Purchase possibility A is a heat exchanger constructed with materials of steel and copper. Its investment cost is $15,000. Its economic service life is estimated to be 10 years, and salvage value at the end of the 10th year is estimated at $500. Annual labor, maintenance, repairs, and operational expenses are estimated at $11,500; other annual direct costs are 4% of the investment cost of $15,000, or $600, when operating under optimum conditions.

Purchase possibility B is a stainless steel heat exchanger with an investment value of $40,000. Its economic life is also regarded as 10 years, with a scrap value of $1,000 at end of the 10th year. Annual labor, maintenance, repairs, and operational expenses are estimated at $4,000; other annual direct costs are 7% of the investment cost of $40,000, or $2,800, when operating under optimum conditions.

The current cost of capital is 8%.

Find: Using the annual cost method, determine which purchase possibility would be more economical with respect to annual costs.

SOLUTION

Purchase possibility A with capital recovery, formula "Find A, Given P"		Purchase possibility B with capital recovery, formula "Find A, Given P"	
(Original cost – salvage value) (recovery factor) + (salvage value) (interest rate)		(Original cost – salvage value) (recovery factor) + (salvage value) (interest rate)	
($15,000 – $500) (0.1490) + ($500) (0.08) = $2,201 capital recovery of original cost and salvage value		($40,000 – $1,000) (0.1490) + ($1,000) (0.08) = $5,891 capital recovery of original cost and salvage value	
Summary of annual costs with capital recovery		Summary of annual costs with capital recovery	
Recovery of capital	$2,201	Recovery of capital	$5,891
Annual costs, maint., repairs	11,500	Annual costs, maint., repairs	4,000
Annual costs, optimum conditions	600	Annual costs, optimum conditions	2,800
Total annual costs	$14,301	Total annual costs	$12,691

With a potential savings in annual cost of $1,610 ($14,301 – $12,691) in favor of the stainless steel heat exchanger, purchase possibility B appears to be the more feasible "buy" according to the annual cost method. (Only differences in costs, with cost items common to both purchase possibilities, were used.) Furthermore, the salvage value of the stainless steel exchanger ($1,000) is $500 more than for the steel-copper exchanger (Figure 10.5).

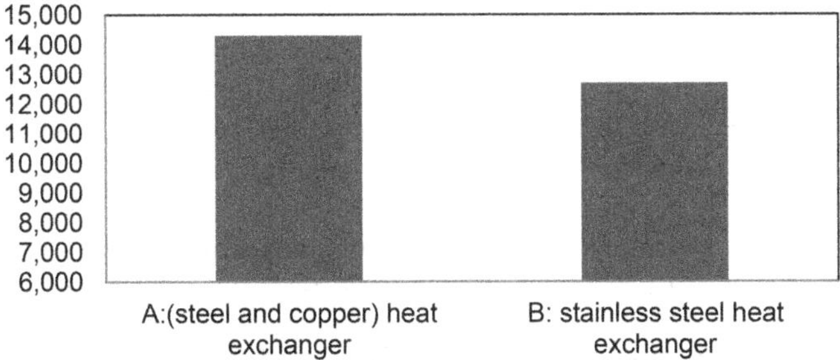

FIGURE 10.5 Comparison between types A and B.

The annual cost method is used where the same costs for each alternative recur annually almost in the same manner. For a series of costs which are nonuniform, an average annual cost equivalent might be calculated. For alternatives with different lifetimes, the time period for comparison might be that of the alternative with the shortest life.

Whereas the annual cost method does not give the relative amounts of capital, but the present value method does. The present value method reduces all costs to equivalent capital at a given date.

Example 10.4 (summary):

	A: (Steel and Copper) Heat Exchanger	B: (Stainless Steel) Heat Exchanger
Capital Investment cost (C.I.)	15,000	40,000
Lifetime (yr)	10	10
Salvage value	500	1,000
Annual labor, maintenance, repairs, and operational expenses costs ($)	11,500	4,000
Annual direct costs ($)	600	2,800
The current cost of capital is 8%	1,200	3,200
Capital recovery	2,201	5,891
Total annual costs	14,301	12,691

Purchase possibility B appears to be the more feasible "buy" according to the annual cost method. (Only differences in costs, with cost items common to both purchase possibilities, were used.) Furthermore, the salvage value of the stainless steel exchanger ($1,000) is $500 more than for the steel-copper exchanger.

Example 10.5

A feasibility study carried out for an oil company indicated that it is possible to invest $1 million in either one of two projects. Anticipated cash flows generated by the two projects over the useful lifetime are given in Table 10.4.

1. Give your recommendations of which project you choose based on the N.P.V. Use selected values for the discount interest rate (more than one).
2. Compute the discounted cash flow rate of return (D.C.F.R.) for each project.

SOLUTION

For (1), calculation is done for three different discount interest rates: 8%, 10%, and 12%, as shown in Table 10.5. In addition, a graphic plot is presented (Figure 6.8) for the change of the discounted value (present value) of the cash flows for both projects with the discount rate.

In summarizing the results of Table 10.5, if the cash flows of project 1 and project 2 are discounted at 8%, project 2 is preferable; if the cash flows are discounted at 10%, project 2 is preferred to project 1 because the present value of project 2 is almost $14,000 more; and if the cash flows are discounted at 12%, project 1 is slightly preferable to project 2, and will continue to be preferable to project 2 as discount rates go higher than 12%.

Therefore, as the example shows, the choice between the two projects depends on the discount rate used. Usually, the oil company's cost of capital for investing in the project will determine which project is selected.

From the data, it can be seen that at a discount of 12%, the present value of cash flow from project 2 is $1 million; and at a discount of over 12%, the present value of cash flow gives us the discount rated amount of under $1 million. This analysis of the present value of cash flow gives us the discount rate at which anticipated cash flow equals the initial investment, which is the D.C.F.R. For project 2, it is about 12%; for project 1, it is about 13%.

Example 10.6

During field operations, the manager in charge is considering the purchase and the installation of a new pump that will deliver crude oil at a faster rate than the existing one.

The purchase and the installation of the new pump will require an immediate layout of $15,000. This pump, however, will recover the costs by the end of 1 year.

The relevant cash flows for the case.

TABLE 10.5
Solution of Example 10.6

Year	Project 1	Project 2
1	400,000	100,000
2	320,000	200,000
3	200,000	300,000
4	300,000	400,000
5	100,000	500,000
Total anticipated cash Flow	1,320,000	1,500,000

At 8% Discount					
Year		Project 1		Project 2	
	Discount factor for 8 %	Cash Flow	Discounted Value (col. 1*col. 2)	Cash Flow	Discounted Value (col. 1*col. 4)
	1	2	3	4	5
1	0.926	400,000	370,400	100,000	92,600
2	0.856	320,000	273,920	200,000	171,200
3	0.794	200,000	158,800	300,000	238,200
4	0.735	300,000	220,500	400,000	294,000
5	0.681	100,000	68,100	500,000	340,500
			$1,091,723		$1,136,505

At 10% Discount					
Year		Project 1		Project 2	
	Discount factor for 10 %	Cash Flow	Discounted Value (col. 1*col. 2)	Cash Flow	Discounted Value (col. 1*col. 4)
	1	2	3	4	5
1	0.909	400,000	363,600	100,000	90,900
2	0.826	320,000	264,320	200,000	165,200
3	0.751	200,000	150,200	300,000	225,300
4	0.683	300,000	204,900	400,000	273,200
5	0.621	100,000	62,100	500,000	310,500
			$1,045,123		$1,065,105

At 12% Discount					
Year		Project 1		Project 2	
	Discount factor for 12 %	Cash Flow	Discounted Value (col. 1*col. 2)	Cash Flow	Discounted Value (col. 1*col. 4)
	1	2	3	4	5
1	0.893	400,000	357,200	100,000	89,300
2	0.797	320,000	255,040	200,000	159,400
3	0.712	200,000	142,400	300,000	213,600
4	0.636	300,000	190,800	400,000	254,400
5	0.567	100,000	56,700	500,000	283,500
			$1,002,143		$1,000,205

Data for Example

	Year		
	0	**1**	**2**
Install new (larger pump)	−15,000	19,000	0
Operate existing (old pump)	0	95,000	95,000

If the oil company requires 10% minimum annual rate of return on money invested, which alternative should be chosen?

SOLUTION

The present worth method is applied in solving this problem (see Chapter 6).

Calculate the present worth for both alternatives, where:

Present worth = Present values of cash flows, discounted at 10% – Initial capital investment

a. For the new pump: P.V. = (190,000)/1.1 = $172,727

$$\text{Present W} = 172{,}727 - 15{,}000$$

$$= \underline{\$157{,}727}$$

b. For the old pump: P.V. = (95,000)/1.1 + (95,000)/$(1.1)^2$

$$= 78{,}512 + 86{,}363$$

$$= \$164{,}875$$

Based on the above results, keep the old pump. It gives higher present value.

Example 10.7

The XYZ oil production company was offered a lease deal for oil wells on which the primary reserves are close to exhaustion. The major condition of the deal is to carry out secondary recovery operation using water-flood at the end of the 5 years. No immediate payment by the XYZ Company is required. The relevant cash flows are estimated as given in Table 10.6.

TABLE 10.6
Data for Example

Year				Net Present Worth @ 10%
0	1–4	5	6–20	
0	$50,000	–$650,000	$100,000	$227,000

The decision to be made: should the lease and the secondary flood proposal be accepted?

Justify your answer, and check the present worth (P.W.) value.

SOLUTION

The fact that the proposal at hand gives a **positive** P.W., makes it a viable one. The project should be undertaken.

Next, calculation is carried out to check the P.W. reported in Table 10.2.

The cash flows are discounted to present values, at 10%. Using the compound interest factors listed in Appendix B, the following results are obtained:

$$\text{The discounted values} = 50{,}000(3.1698) - 650{,}000(0.5645) + 100{,}000(4.7227)$$

$$= 158{,}490 - 403{,}585 + 472{,}270$$

$$= \mathbf{\$227{,}175}$$

Section II.III: Mathematical Approaches

INTRODUCTION: SESSION II.III

Chapters 11 and 12 share one common identity. These two chapters are classified as mathematical tools to evaluate economic profitability of oil projects. They involve mathematical manipulations in order to arrive to an economic decision, that is why are grouped as "mathematical approach" as shown in the graph.

Fundamentals of economic engineering economic analysis is presented next

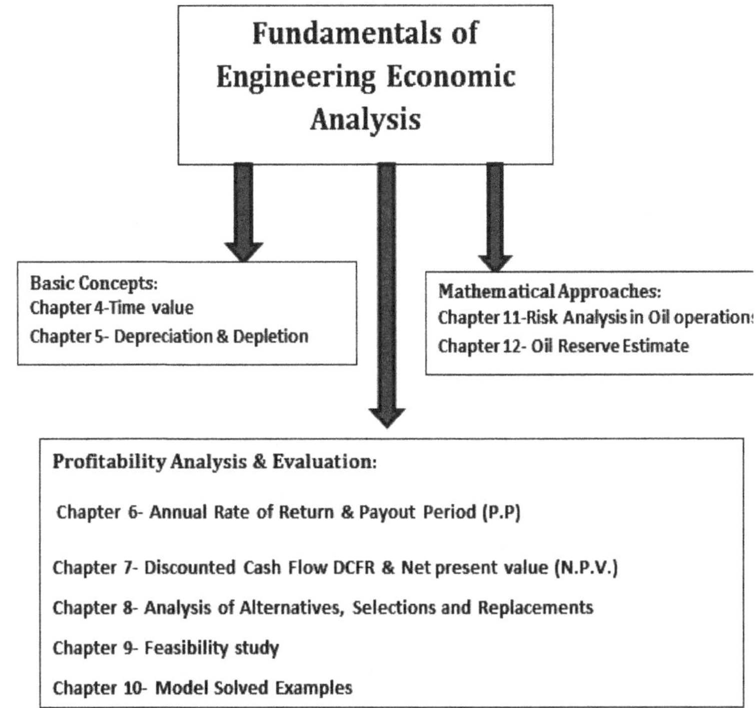

11 Risk and Decision Analysis in Oil Operations

11.1 INTRODUCTION

The ability to evaluate critical risk factors for oil and gas projects is crucial to optimizing outcomes and planning for effective and cost-efficient risk mitigation programs. Substantial investments are required for gas and oil exploration and production projects.

This chapter highlights methods, tools, and techniques used to study risk, uncertainty, and decision analysis for oil operations. It does not give details to any of these methods.

Some types of risks are briefly described in Figure 11.1.

Why oil projects? The *oil and gas* industry invests significant money and other resources in projects with highly *uncertain* outcomes. Risks and uncertainties are everywhere in many oil projects. Risk and uncertainty impact decision-making by the projects chosen. In the first place how these projects are developed, and what is their economic performance. In simple words, a risk assessment is a careful examination of anything that may cause harm during the course of a given work. Once this is done, you will then be able to decide upon the most appropriate action to take to minimize the likelihood of expected hurt by someone.

On the other hand, decision-making is defined as the selection of a course of action from among alternatives or the selection of a logical choice from available options. Risk is a prevalent issue in investment decisions because it could not be avoided, but can be managed. Investment without risk element might not be a worth-while investment because overcoming risk could launch the business into unprecedented success.

Some important applications involving risk and economic analysis in oil operations may include:

- Reserve Estimate
- Exploration and Production (E&P)
- Recovery Factors
- Expected Production Profile

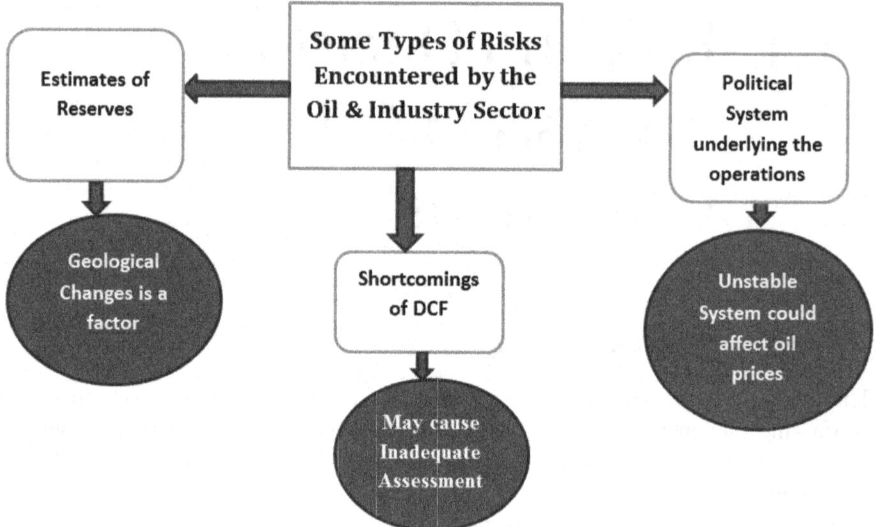

FIGURE 11.1 Representation of risks encountered by the oil industry.

11.2 THEORY AND METHODS

In the study of risk analysis and uncertainties in this field, probabilistic concepts and tools are commonly used to describe projects under risk and uncertainty, the following methods are used:

- Monte Carlo Simulation
- Decision Trees
- Commercial Software
- Engineering Economy
- Economic Indicators
- Database

Risk and decision analysis software is as diverse as the analysis methods themselves. There are programs to do Monte Carlo simulation and decision tree analysis. Analytic models to do economics can be linked to both Monte Carlo simulation and decision trees. Monte Carlo simulation (also known as the Monte Carlo method) lets you see all the possible outcomes of your decisions and assess the impact of risk, allowing for better decision-making under uncertainty.

A decision situation is called a *decision under risk* when the decision maker considers several states of nature, and the probabilities of their occurrence are explicitly stated. A decision situation where several states are possible and sufficient information is not available to assign probability values to their occurrence is termed a *decision under uncertainty*.

Decision situations can be classified as follows:

- Decision-making under certainty, where complete information is assumed or available
- Decision-making under risk, where partial information is known
- Decision-making under uncertainty, where limited information is available

Because of space limitation in this volume, detailed description of the methods is not included.

11.3 APPLICATIONS

CASE 1

Suppose an oil company would like to assign three drilling rigs to drill oil wells at three different stratigraphic locations in a manner that will minimize total drilling time. The drilling times in days are presented in Table 11.1

TABLE 11.1
Drilling Times in Days

Drilling Times in Days for Three Different Oil Wells

Well Number	1	2	3
Rig Number			
A	30	70	40
B	40	60	60
C	30	80	50

This example illustrates the concept of complete enumeration. This means examining every payoff, one at a time, comparing the payoffs to each other, and discarding inferior solutions. The process continues until all payoffs are examined. By complete enumeration as shown in the solution given in Table 11.2, all the alternatives are listed. It is clear that alternative number 5 is the best choice since the total drilling time is the minimum.

TABLE 11.2
Complete Enumeration Solution

Alternative	Assignment	Total Drilling Time
1	A-1, B-2, C-3	30 + 60 + 50 = 140
2	A-1, B-3, C-2	30 + 60 + 80 = 170
3	A-2, B-1, C-3	70 + 40 + 50 = 160
4	A-2, B-3, C-1	70 + 60 + 30 = 160
5	A-3, B-2, C-1	40 + 60 + 30 = 130 ←
6	A-3, B-1, C-2	40 + 40 + 80 = 160

CASE 2

Consider an investment of a company, engaged in oil field services, of \$10,000 over a 4-year period that returns *Rt* at the end of year *t*, with *Rt* being a statistically independent random variable. The following probability distribution is assumed for *Rt*.

R_t	Probability
\$2,000	0.10
\$3,000	0.20
\$4,000	0.30
\$5,000	0.40

SOLUTION

The expected value of the return in a given year is given by:

$$\text{EV(return)} = 2000(0.10) + 3000(0.20) + 4000(0.30)5000(0.40)$$

$$= 4000$$

The variance of an annual return is determined as follows:

$$\text{Variance (return)} = \{(2000 - 4000)\}^2(0.10) + \{(3000 - 4000)\}^2(0.2) + \ldots.$$

$$= 1{,}000{,}000$$

CASE 3

An oil firm has four alternatives from which one is to be selected. The probability distributions describing the likelihood of occurrence of the present worth of cash flow amounts, expected values, and variance for each alternative are given in Table 11.3.

TABLE 11.3
Comparison of Four Alternatives of an Oil Firm

	Present Worth of Cash Flow (\$1,000)						
Alternatives	–\$40	10	60	110	160	EV	Var
A1	0.2	0.2	0.2	0.2	0.2	60	$5 * 10^9$
A2	0.1	0.2	0.4	0.2	0.1	60	$3 * 10^9$
A3	0.0	0.4	0.3	0.2	0.1	60	$2.5 * 10^9 \leftarrow$
A4	0.1	0.2	0.3	0.3	0.1	65	$3.85 * 10^9$

SOLUTION

For any given alternative, the decision maker wishes to maximize the expected value and at the same time to minimize the variance of the present worth of the cash flow. If equal weights to the expected value and variance are given, then the values of the expected value-variance criterion will be as computed in the last column of Table 11.3.

Based on the expected value-variance criterion, alternative A3 should be selected.

CASE 4

An oil-drilling company is considering bidding on a $110 million contract for drilling oil wells. The company estimates that it has a 60% chance of winning the contract at this bid. If the company wins the contract, it will have three alternatives: (1) to drill the oil wells using the company's existing facilities, (2) to drill the oil wells using new facilities, and (3) to subcontract the drilling to a number of smaller companies. The results from these alternatives are given as follows:

Outcomes	Probability	Profit ($ million)
1. Using existing facilities:		
Success	0.30	600
Moderate	0.60	300
Failure	0.10	–100
2. Using new facilities:		
Success	0.50	300
Moderate	0.30	200
Failure	0.20	–40
3. Subcontract:		
Moderate	1.00	250

The cost of preparing the contract proposal is $2 million. If the company does not make a bid, it will invest in an alternative venture with a guaranteed profit of $30 million. Construct a sequential decision tree for this decision situation and determine if the company should make a bid.

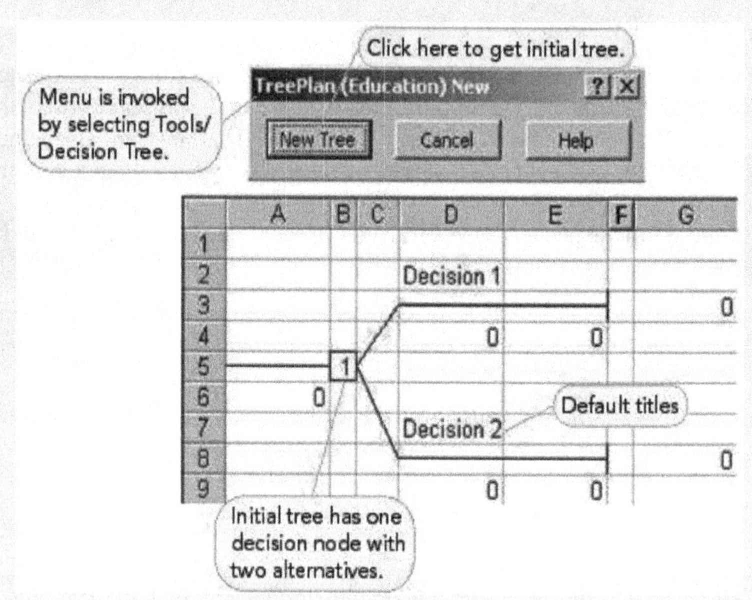

FIGURE 11.2 Starting the TreePlan program and select a tree.

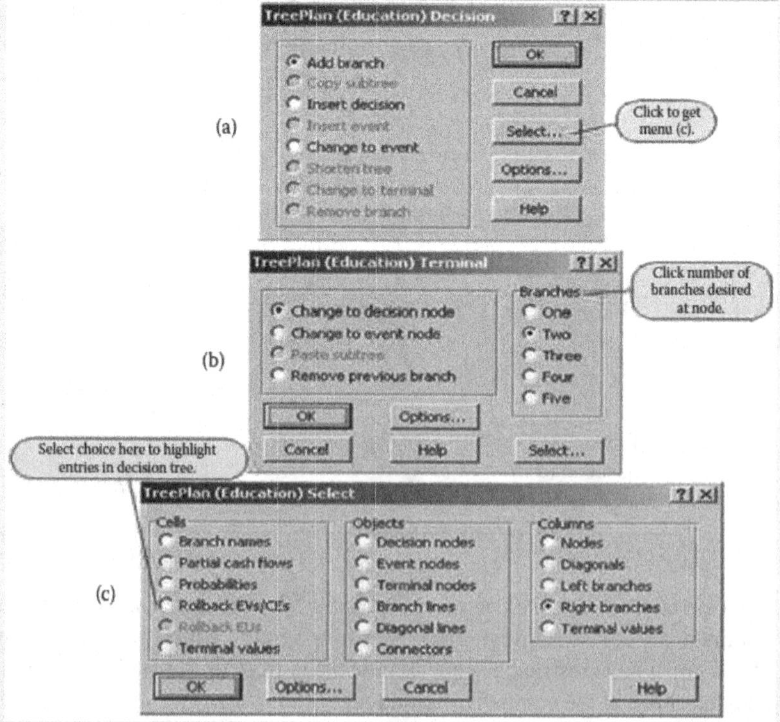

FIGURE 11.3 TreePlan dialogue boxes.

SOLUTION

The oil company should make the bid because this will result in an expected payoff value of $143.2 million. The problem is solved using an academic version of Microsoft Excel Add-in, TreePlan Software. To construct a decision tree with TreePlan, go to Tools menu and choose Decision Tree, which brings up the TreePlan as shown in Figure 11.2.

The dialogue boxes used by TreePlan for constructing a decision tree are shown in Figure 11.3. The dialogue boxes enable us to adding decision nodes, state of nature nodes, decision alternative branches, state of nature branches, probabilities, payoffs, and all other tree parameters.

12 Oil Reserves and Reserve Estimate

12.1 INTRODUCTION

This chapter consists of two sections: Section 1 deals with reserves and reserve estimate. Section 2 is devoted to economic evaluation and applications, with solved examples.

Evaluation of an oil property depends on the development of the underground accumulation of hydrocarbons and on the amount of money that will be received from selling the produced hydrocarbons. Such evaluation includes estimate of reserves, estimate of gross income, estimate of net income after all types of taxes and production costs, and calculations of present worth value of the property.

Development of an oil and/or gas reservoir depends on the producible amount of hydrocarbons. This amount is called "reserves". The proved reserve is that form of reserve which is recoverable by the force of natural energy existing in the reservoir or by secondary processes. The probable reserve is the reserve which has not been proved by production at a commercial flow rate.

Reserve estimation is one of the most essential tasks in the petroleum industry. The total estimated amount of oil in an oil reservoir, including both producible and non-producible oil, is called "oil in place". Practically speaking, because of reservoir characteristics and limitations in petroleum extraction technologies, only a fraction of this oil can be brought to the surface, and it is only this producible fraction that is considered to be reserve.

An oil evaluation study has as its primary purpose the determination of the value of oil in place. Such evaluation includes estimates of reserves. Methods most commonly used to estimate the reserve of recoverable hydrocarbons are presented first in this chapter. These include: volumetric, material balance, and decline curve methods. The role of economic evaluation for oil properties is illustrated by many case studies.

SECTION 1: RESERVES AND RESERVE ESTIMATE

12.2 OVERVIEW

Many factors impact the demand for and supply of oil and natural gas, influence how and where energy suppliers invest their capital, and determine the manner in which countries compete to attract foreign investment. Globally, we currently consume the equivalent of over 11 billion tonnes of oil from fossil fuels every year. Crude oil reserves are vanishing at a rate of more than 4 billion tonnes per year—so if we carry on, as we are, our known oil deposits could run out in just over 53 years.

World oil supply derives from the following factors:

- The investment decisions of individual companies
- The political decisions of countries in regard to licensing
- The degree of foreign investment and a multitude of other variables that influence system dynamics, including price, inventory levels, geopolitics, market psychology and manipulation, organization of the petroleum exporting countries (OPEC) policy, exchange rates, unexpected events, and resource availability

In its latest Statistical Review of World Energy, BP estimated the world had 1.7297 trillion barrels of crude oil remaining at the end of 2018. That was up from 1.7275 trillion barrels a year earlier and 1.4938 trillion barrels in 2008 (Nov 11, 2019).

Oil and gas assets represent the majority of value of an E&P company.

The *Oil and Gas Financial Journal* describes reserves as "a measurable value of a company's worth and a basic measure of its life span".

Interest in the determinants of investment in crude oil and natural gas reserves derives from three basic sources:

Oil and gas assets represent the majority of value of an E&P company.

The *Oil and Gas Financial Journal* describes reserves as "a measurable value of a company's worth and a basic measure of its life span".

Interest in the determinants of investment in crude oil and natural gas reserves derives from three basic sources:

- First, it is always interesting to find a satisfactory explanation of investment behavior in any industry
- Second, an aspect of the current concern with the "energy crisis" is the domestic crude petroleum industry's productive capacity, which is an increasing function of the stock of proved oil and gas reserves
- Third, there is a decades-old controversy over the special provisions of the federal corporation income tax law

The American Society of Appraisers defines the fair market value as reserves as:

The price, expressed in terms of cash equivalents, at which property would change hands between a hypothetical willing and able buyer and a hypothetical willing and able seller, acting at arm's length in an open and unrestricted market, when neither is under compulsion to buy nor sell and when both have reasonable knowledge of the relevant facts.

The American Society of Appraisers recognizes three general approaches to valuation:

1. The Cost Approach,
2. The Income Approach, and
3. The Market Approach.

When valuing acreage rights comparable transactions do provide the best indication of value. However, when valuing reserves, a D.C.F. (discounted cash flow) is often the best way to allocate value to different reserve categories because comparable transactions are very rare as the details needed to compare these specific characteristics of reserves are rarely disclosed.

The total present worth of future income is then discounted further, a percentage based on market conditions, to determine the fair market value. The costs of any expected additional equipment necessary to realize the profits are included in the annual expense, and the proceeds of any expected salvaged of equipment is included in the appropriate annual income.

12.3 METHODS OF RESERVE ESTIMATE

The methods most commonly used to estimate the reserve of recoverable hydrocarbons are:

1. Volumetric methods
2. Material balance methods
3. Decline curve methods

Each of these methods will be discussed separately.

12.3.1 Volumetric Methods

The estimation of reserve is done on the basis of an equation which is not complicated to use provided the required data are available. The data include the area of the production zone (A), the formation thickness (h), the porosity (ϕ), and the initial water saturation (S_{wi}). The equation has the form:

$$N = \frac{7758\ Ah\phi(1 - S_{wi})}{B_{oi}} \tag{12.1}$$

where:

N = bbls of initial oil in place at surface temperature and pressure Condition, which is called stock tank

B_{oj} = initial oil formation volume factor, which is defined as bbl at reservoir condition (rb), divided by bbl at surface condition (STB)

Once the recovery factor is known, then the recoverable oil can be known. The bulk volume of the reservoir can be calculated using subsurface and isopachous maps. The isopachous map consists of isopach lines that connect points of formations having equal thickness. The areas lying between the isopach lines of the entire reservoir under consideration are used to calculate the volume contained in it.

Simpson's rule, trapezoidal rule and pyramidal rule are normally used to determine the reservoir bulk volume (V_B). Simpson's rule provides the following equation:

$$V_B = h/3(A_o + 4A_1 + 2A_2 + 4A_3 + 2A_{n-2} + 4A_{n-1} + 4A_{n-1} + A_n) + t_n A_n \tag{12.2}$$

where:

h = interval between the isopach lines in ft
B_o = area in acres enclosed by successive isopach lines in acres
A_1, A_2, A_3, A_n = areas enclosed by successive isopach lines in acres
t_n = average thickness above the top

Trapezoidal rule provides the following equation:

$$V_B = h/2(A_o + 2A_1 + 2A_2 + \cdots 2A_{n-1} + A_n) + T_n A_n \tag{12.3}$$

Pyramidal rule has the form:

$$V_B = h/3(A_n + A_{n+1} + \sqrt{A_n A_{n+1}}) \tag{12.4}$$

This equation calculates the reservoir bulk volume between any two successive areas (ΔV_B), and the total reservoir bulk volume is the summation of all the calculated bulk volumes.

The accuracy of trapezoidal rule and pyramidal rule depends on the ratio of the successive areas. If the ratio of the areas is smaller than 0.5, the pyramidal rule is used; otherwise the trapezoidal rule is used.

The formula as provided in Equation (12.1) can be applied to calculate free gas in a gas reservoir as given below:

$$G = 43560\ V_B\ \phi(1 - S_w)/B_g \tag{12.5}$$

where:

G = gas in place
B_g = gas formation volume factor
V_B = reservoir bulk volume
S_w = connate water

12.3.2 Material Balance Equation

Material balance equation accounts for the fluids that leave, enter, or accumulate in the reservoir at any time. The oil reservoir is classified as an undersaturated or saturated reservoir based on the reservoir pressure. A reservoir with pressure higher than the bubble point pressure is considered to be an *under-saturated* reservoir. The material balance for such reservoir, with the assumption that the oil is produced by the fluid expansion only and the reservoir is constant, is derived below:

Assume that the initial production, P_i, dropped to P due to N_p STB produced. Then,
Initial volume = NB_{oi} bbl at the reservoir condition, rb
Final volume = $(N - N_p)B_o$ bbl at the reservoir condition, rb

Since the reservoir volume is constant, then:

$$\begin{aligned} \text{Initial volume} &= \text{Final volume} \\ NB_{oi} &= (N - N_p)B_{oi} \\ N &= N_pB_o/(B_o - B_{oi}) \end{aligned} \tag{12.6}$$

A reservoir with pressure lower than the bubble point pressure will cause gas to form, resulting in a free gas phase. Such a reservoir is called a *saturated* reservoir. The derivation of material balance equation for this case is given next:

$$\begin{aligned} \text{The initial volume} &= NB_{oi} \\ \text{Final volume} &= \text{remaining oil} + \text{free gas} \\ &= (N - N_p)B_o + G_fB_g \\ G_f &= \text{initial gas} - \text{remaining gas} - \text{produced gas} \\ &= NRsi - (N - N_p)R_s - N_pR_p \end{aligned}$$

Assume the reservoir volume is constant, then:

$$\begin{aligned} \text{Initial volume} &= \text{Final volume} \\ NB_{oi} &= (N - N_p)B_o + (NR_{si} - (N - N_p)R_s \\ &\quad - N_pR_p)B_p \\ &= N(B_o + B_g(R_{si} - R_s) \\ &\quad - N_pR_p)B_p \\ &= N(B_o + B_g(R_{si} - R_s) \\ &\quad - N_p(B_o + B_g(R_p - R_s)) \\ N_p(B_o + B_g(R_p - R_s)) &= N(B_o + B_{oi} + B_g(R_{si} - R_s)) \\ N &= N_p(B_o + B_g(R_p - R_s))/B_o - B_{oi} \\ &\quad + B_g(R_{si} - R_s) \end{aligned} \tag{12.7}$$

where:

N = oil in place, rb
N_P = oil produced, STB
B_o = formation volume factor, rb/STB
B_{oi} = initial formation volume factor, rb/STB
B_g = gas formation volume factor, rb/STB
R_{Si} = initial gas in solution, SCF/STB
R_s = gas in solution at a pressure lower than P_i
R_p = cumulative gas-oil ratio

If the reservoir has a *gas cap* at the time of discovery, then the material balance equation will have the form:

$$\begin{aligned} N = N_p(B_o + B_g(R_p - R_s)/B_o - B_{oi} \\ + B_g(R_{si} - R_s) + mB_{oi}(B_g/B_{gi} - 1) \end{aligned} \tag{12.8}$$

where:

m = volume of free gas/oil volume
$= G_f B_{gi}/NB_{oi}$

If the reservoir is *under water drive*, the water influx as well as the water production needs to be added to the material balance. Then, Equations (12.7) and (12.8) become:

$$\begin{aligned} N = N_p(B_o + B_g(R_p - R_s)) - W_e \\ + B_w W_p/B_o - B_{oi} + B_g(R_{si} - R_s) \end{aligned} \tag{12.9}$$

$$\begin{aligned} N = N_p(B_o + B_g(R_p - R_s)) - W_e + B_w W_p/B_o \\ - B_{oi} + B_g(R_{si} - R_s) + mB_{oi}(B_g/B_{gi} - 1) \end{aligned} \tag{12.10}$$

All these terms, except N_p, R_p, W_e, and W_p, are functions of pressure and also are properties of the fluids. These data should be measured in the laboratory. R_p depends on the production history. It is the quotient of both the gas produced (G_p) and the oil produced (N_p). A water influx can be calculated by using different methods depending on the flowing conditions. The boundary pressure as well as the time are used to calculate the water influx. The value of m is determined from the log data, which provides the gas-oil and oil-water contacts and also from the core data. Therefore, the accuracy of the calculated oil in place depends upon how accrately we take these measurements for such calculations.

12.3.3 Material Balance Equation for Gas Reservoir

a. *No water drive*: If the reservoir volume stays constant and G_p, gas produced during a time *t*, and B_{gi} drop to B_g; then material balance is given by Equation (12.11) as follows:

$$\begin{aligned} \text{Initial volume} &= \text{Final volume} \\ GB_{gi} &= B_g(G - G_p) \\ &= B_g g - G_g G_p \\ G_p B_g &= G(B_g - B_{gi}) \\ G &= G_p B_g / B_g - B_{gi} \end{aligned} \tag{12.11}$$

b. *With water drive:* The material balance:

$$G = G_p B_g - W_{cd} + W_p B_w / B_g - B_{gi} \tag{12.12}$$

If the measured data are accurate, the calculated gas in place will always be accurate. In Equation (12.12), the water influx can be found using the pressure drop during the production history with other parameters.

12.3.4 Material Balance Equation, Straight-line Concept

The material balance equation given in Equation (12.10) may be expressed as a straight-line equation which will have the form:

$$F = NE_o + N_m E_g + W_e \tag{12.13}$$

where:

$$\begin{aligned} F &= N_p(B_o + B_g(R_p - R_s)) + W_p + W_w,\ \text{rb} \\ E_o &= [B_o - B_{oi} + B_g(R_{si} - R_s)]\text{rb/STB} \\ E_g &= B_{oi}[(B_g / B_{gi}) - 1)]\text{rd/TB} \end{aligned}$$

F represents the total underground withdrawal while E_o denotes the oil expansion and the expansion of associated gas, while E_g represents the gas cap expansion.

Equation (12.13) includes all the drive mechanisms. If any one of these mechanisms is not acting in the reservoir, then the term representing such a mechanism must be deleted from the equation.

a. *No water drive, no original gas cap*:

$$\begin{aligned} W_c &= O \text{ and } m = O \\ F &= NE_o \end{aligned} \tag{12.14}$$

A plot of F versus E_o gives a straight line passing through the origin with a slope of N (initial oil in place).

b. *No water drive ($W_c = 0$)*. Equation (12.13) will be reduced to:

$$F = N(E_o + mE_g) \tag{12.15}$$

Again plotting F versus $(E_o + mE_g)$ yields a straight line passing through the origin with a slope of N.

c. *No water drive and m is not known*. Equation (12.15) can be written differently:

$$F/E_o = N + mNE_g/E_o \tag{12.16}$$

A plot of F/E_o versus E_g/E_o should result in a straight line with the intercept of N with Y-axis. The value of m can be known from the slope.

d. *For water drive reservoir, $m = 0$*, Equation (12.13) will have the form:

$$F = NE_o + W_e$$

Divide by E_o:

$$F/E_o = N + W_e/E_o \tag{12.17}$$

A plot of F/E_o versus W_e/E_o should give a straight line with N being the Y intercept provide the calculated water influx is correct.

The same concept can be applied to the gas reservoir to express the gas material balance equation as a straight line. Equation (12.12) becomes:

$$\begin{aligned} G_pB_g &= GE_g \\ \text{where } E_g &= B_g - B_{gi} \end{aligned} \tag{12.18}$$

Plotting G_pB_g versus E_g should give a straight line with G being the slope. If the reservoir is under water drive, Equation (12.12) can be written as:

$$\begin{aligned} GE_g &= G_pB_g - W_e + W_p \\ GE_g &= G_pB_g + W_p - W_e \\ W_e + GE_g &= G_pB_g + W_p \end{aligned}$$

Divide by E_g:

$$W_e/E_g + G = G_pB_g + W_p/E_g \tag{12.19}$$

A plot of $G_pB_g + W_p/E_g$ versus W_e/E_g should result in a straight line with G being the Y intercept.

Using the straight-line technique to estimate oil or gas reserves will minimize the error in the calculated reserve because a number of data will be used for the reserve estimation and the error in the data will be averaged.

The gas in place can be estimated by another approach which requires plotting P/z versus cumulative gas production for a volumetric reservoir. Such a plot results in a straight line with G being the X-axis intercept. Estimation of gas reserve using early production data may result in error by as much as a factor of 2. Therefore, this method should be used only when the cumulative gas production reaches a stage of about 20% of the gas in place.

12.3.5 Decline Curve Methods

Predicting the reserve using decline curve methods requires production rate of all the wells. The production rate generally declines with time, reaching an end point, which is referred to as the economic limit. The economic limit is a production rate at which the income will just meet the direct operating cost of a well or a certain field. Typical decline curve analysis consists of plotting production rate versus time and trying to fit the obtained data into a straight line or other forms which can be extrapolated up to the economic limit to estimate the reserve on the assumption that all the factors affecting the well performance have exactly the same effect in the future as they had in the past.

The commonly used decline curve methods are:

1. Constant percentage decline
2. Hyperbolic decline
3. Harmonic decline

12.3.5.1 Constant Percentage Decline

The constant percentage decline is known as the exponential decline and is used widely more than the other forms of decline due to its simplicity. In this case, the decline rate is assumed to be constant during the production time. The decline rate in production rate with time is:

$$D = -\Delta q / (q / \Delta t) \tag{12.20}$$

where:

D = decline rate
$\Delta q = q_i - q$; q_i is initial production rate and q is production at a time (t)
Δt = time t required for q_i to decline to q

Integrating Equation (12.20) to get rate-time relation:

$$-\int_0^t D\,dt = \int_{q_i}^{q} \frac{dq}{q} \qquad (12.21)$$

$$q = q_i e^{-Dt}$$

Integrating Equation (12.21) with respect to time:

$$-\int_0^t q\,dt = q_i \int_o^t e^{-Dt}$$

or

$$N_p = -q_i / D(1 - e^{-Dt}) \qquad (12.22)$$

From Equation (12.22):

$$q/q_i = e^{-Dt}$$

Substitute in Equation (12.22); then:

$$N_p = q_i - q / D \qquad (12.23)$$

Equation (12.23) can be rearranged as follows:

$$q = q_i - N_p D \qquad (12.24)$$

A plot of q versus N_p will result in a straight line. The slope of the line is D and q_i is the intercept of the Y-axis. Equation (12.21) also yields a straight line if q is plotted against t on semilog paper. The slope of such a plot is D and the intercept is q_i. The N_p is the cumulative production between any two production rates.

$$q = q_i e^{-Dt}$$
$$-\ln q / q_i = Dt$$

When the decline rate is not constant, then the hyperbolic decline can be assumed and the decline rate varies according to the following equation:

$$D = D_i (q / q_i)^n \qquad (12.25)$$

where:

n = decline constant between zero and 1
D_i = initial decline rate

The general equation for hyperbolic rate decline can be obtained by substituting Equation (12.17) into Equation (12.21) and then integrating the resulting equation. The equation thus finally derived will have the form:

$$q = \frac{q_i}{(1 + Dnt)^{1/n}} \tag{12.26}$$

The cumulative production rate obtained from the hyperbolic decline can be derived as follows:

$$N_p = \int_0^t q\,dt$$

Equation (12.20) can be written as:

$$D = -\frac{\frac{dq}{q}}{dt}$$

Substitute D value from Equation (12.25) in the above equation, and then substitute q value in the equation to calculate N_p:

$$\begin{aligned} N_p &= \int_{q_1}^{q_2} \frac{dq}{D_i \left(\frac{q}{q_i}\right)^n} \\ &= \frac{q_i^n}{D_i} \int_{q_1}^{q_2} \frac{dq}{q^n} \\ &= \frac{q_i^n}{D_i(1-n)} \left(q_1^{1-n} - q_2^{1-n}\right) \end{aligned} \tag{12.27}$$

The values of q_i, D_i, and n are assumed to be known and are constant, and thereafter Equation (12.27) can be used without any difficulty. The values of q_i, D_i, and n can be obtained by comparing the actual decline data with a series of curves of hyperbolic type. A plot of q/q_i versus time may fit in one of the curves which gives the values of q_i D_i, and n.

12.3.5.2 Harmonic Decline

In this curve, when the decline rate is not constant, it decreases as the production rate increases. Such a varying rate in decline is called a harmonic decline. It also occurs if the decline constant n of Equation (12.27) is 1. An equation derived for such decline is:

$$q = q_i / 1 + a_i t \tag{12.28}$$

This type of decline may take place in reservoirs where gravity drainage controls the production. Gravity drainage exists in tilted reservoirs where oil production is affected by drainage of oil from upstructure to downstructure which causes segregation of gas and oil in the reservoir. The cumulative production can be obtained by integrating Equation (12.24) with respect to time:

$$N_p = \int_0^t q\,dt$$

$$N_p = q_i \int_0^t \frac{dt}{1 + a_i t} \tag{12.29}$$

$$= q_i / a_i \ \ln(1 + a_i t)$$

But from Equation (12.25):

$$(1 + a_i t) = q / q_i$$

Substitute in Equation (12.25)

$$N_p = (q_i / q_i) \ln(q / q_i) \tag{12.30}$$

A graphic harmonic decline analysis can be obtained by writing Equation (12.24) as:

$$\begin{gathered} 1 / q_i (1 + a_i t) = 1 / q \\ 1 / q_i + (a_i t) 1 / q_i = 1 / q \end{gathered} \tag{12.31}$$

Plotting $1/q$ versus t on Cartesian coordinates should result in a straight line, with a_i/q_i being the slope and $1/q_i$ the intercept with $1/q$-axis. From the slope a_i can be known. Also Equation (12.30) can be rewritten in a different form:

$$\begin{aligned} N_p &= q_i / a_i (\ln q - \ln q_i) \\ (a_i / q_i) N_p &= \ln q - \ln q_i \\ \ln q_i + (a_i / q_i) N_p &= \ln q \end{aligned} \tag{12.32}$$

A plot of q versus N_p on a semilog paper will result in a straight line with slope being a_i/q_i and intercept q_i. This straight line can be extrapolated into the economic limit to calculate the reserve.

12.4 COMPARISON OF THE METHODS

Comparison of all the predictive methods depends on the data available and on the accuracy of these data. Volumetric methods are usually used in the early life of the reservoir while the material balance equations or the decline curve methods can be used when enough data are collected. However, material balance equation techniques depend on many measurements, such as B_o, B_g, R_s, R_p, and total production; hence more error is anticipated in the calculated reserves. The error in the calculated reserve by the decline curve is less than with other methods.

SECTION 2: ECONOMIC EVALUATION AND APPLICATIONS

Evaluation of an oil property is concerned with its money value, which measures the profitability of such an oil property. The profitability depends on the *development* of underground accumulations of hydrocarbons and on the sale value of the hydrocarbons, which helps to estimate the present worth value of such property at any time under certain specified conditions. The gross income of hydrocarbon sales depends on the current prices of oil and gas and on the predicted economic conditions. The net profit is related to all the expenses that are deducted from the gross income, such as operating cost, which includes the expenses required to produce the hydrocarbon and to maintain the reservoir, taxes, and royalty when applicable.

The following applications and case studies illustrate the role of economic evaluation for an oil property.

Example 12.1

Given the following data:

Area = 1,200 acres
Formation thickness = 20 ft
Average porosity = 20%
Connate water = 25%
Formation volume factor = 1.3 rb/STB
Initial gas in solution (Rsi) = 650 SCF/STB

a. Calculate the oil in place.
b. Calculate the total gas in solution.

SOLUTION

Part (a):

$$N = 7758 = Ah\phi(1-S_w)B_{oi}$$
$$= 7758 \times 1200 \times 20 \times .2(1-.25)/1.2$$
$$= 24,274,000 \text{ STB}$$

Then, oil in place = **24,274,000 STB**

Part (b):

$$\text{Total gas in solution} = (\text{oil in-place})(\text{initial gas in solution})$$
$$= (N)(Rsi)$$
$$= 24,274,000 \times 650$$
$$= 15.78 \times 10^9 \text{SCF}$$

Example 12.2

An oil reservoir has a gas cap at the time of discovery. The size of this gas cap is not known. The production data and the fluid properties are given as a function of pressure in Table 12.1.

a. Calculate the oil in place using the material balance equation as a straight line.
b. Use the material balance equation itself.

SOLUTION

Since the production was due to gas cap expansion and the gas cap size is not known, the following equation can be used:

$$F/E_0 = N + mN\, E_g/E_0 \tag{12.33}$$

All the calculations are given in Table 12.2.

TABLE 12.1
Data for Example 12.2

P, psi	N_p, STB	B_o, rb/STB	R_s, SCF/STB	B_g, rb/SCF	R_p, SCF/STB
3,200	0	1.35	520	0.000932	0
2,950	2.50×10^8	1.345	444	0.00095	950
1,800	3.37×10^8	1.34	435	0.000995	1,000
2,765	4.95×10^8	1.32	410	0.0011	1,150
2,500	6.62×10^8	1.308	395	0.00123	1,280

TABLE 12.2
Solution for Example 12.2

P, psi	F	E_o	E_g	F/E_o	E_g/E_o
2,950	4.57×10^8	0.0672	0.0255	6.8×10^9	0.379
2,800	$6\ 407 \times 10^8$	0.0745	0.09125	$8\ 6 \times 10^9$	1.22
2,650	10.56×10^8	0.091	0.238	11.6×10^9	2.615
2,500	15.87×10^8	0.1118	0.4182	14.1×10^9	3.743

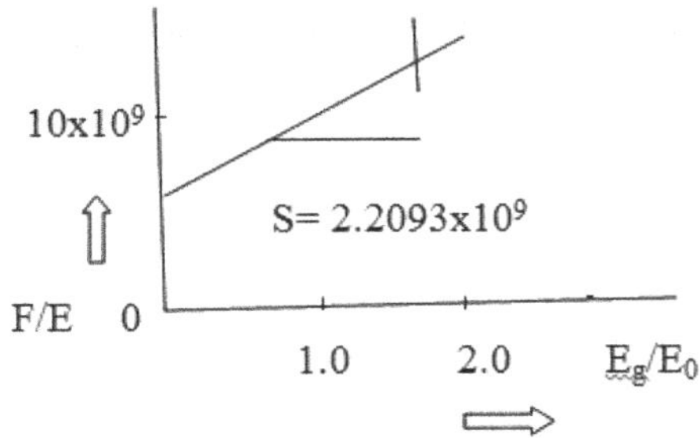

FIGURE 12.1 Solution of Example 12.2.

Plotting F/E_o against E_g/E_o as shown in Figure 12.1 yields a straight line. The values of the intercept and the slope are given as follows:

$$\text{Y intercept} = 5.9 \times 10^9$$
$$\text{Slope} = 2.2093 \times 10^9$$

From Equation (12.19), the Y intercept is N and the slope is mN, then:

$$m = \text{slope/N} = 2.2093 \times 10^9 / 5.9 \times 10^9$$
$$= 0.3745$$

Now, m is known, the material balance equation can be used to calculate initial oil in place. Equation (12.10) will be used:

$$N(\text{At, P} = 2950\,\text{psi}) = 1.084 \times 10^{10}\,\text{STB}$$
$$N(\text{At, P} = 2500\,\text{psi}) = 5.803 \times 10^9\,\text{STB}$$

Since m is known, Equation (12.15) can be used to determine the oil in place N. The calculation is shown in Table 12.3.

TABLE 12.3
Data to Determine Oil in Place, *N*

P	F	$E_o + E_g$
2,950	4.57×10^8	0.07675
2,800	6.407×10^8	0.1087
2,650	10.56×10^8	0.180
2,500	15.87×10^8	0.2684

Example 12.3

For application of the constant decline curve, the following production history for a well is given:

Year	B/day
1	9,600
2	7,200
3	6,700
4	5,700
5	5,200
6	4,650
7	4,300
8	3,800

a. Estimate the remaining life of this field if the economic limit is 800 B/D.
b. What is the recoverable oil as of year 8?
c. What is the net income if the price of oil is assumed to be \$85/bbl?

SOLUTION

Since the decline rate follows the constant percentage decline, then a plot of q versus time on semi-log is recommended gives a straight line. The slope of the line represents the decline rate, D.

$$D = -(2.3)\frac{\log 3{,}800 - \log 7{,}300}{(8-2)\times 12} = 0.02086/\text{month}$$

a. Using Equation (12.21), the revising number of years can be calculated as follows:

$$q = q_i e^{-Dt}$$
$$-\ln q/q_i = Dt$$

or

$$t = \frac{-\ln q/q_i}{D} = \frac{-\ln(800/3{,}800)}{0.02086}$$
$$= 74.695\,\text{month}$$
$$= 6.225\,\text{year}$$

b.
$$\text{Recoverable oil} = q_1 - q_2/D$$
$$= \frac{3{,}800 - 800}{0.02086}$$
$$= 143{,}816 \text{ bbl}$$

c. Total income = 143,816 × 85 = $12,224,360

If the operating expenses is taken to be $38/bbl, then the gross income =$6,759,352
If this gross income is to be taxed at 46%, the net profit =$**3,650,000**

Example 12.4

Use the calculated oil in place in Example 12.1 assuming the following values:

Sale value of the oil = $85/bbl
Operating costs = $47/bbl
The calculated oil in place = 24,274,000 STB
Gross income = (Oil in place) × Price
= 24,274,000 bbl × 85 $/bbl = $20.6 × 10^8
Production taxes = 20.6 × 10^8 × 0.046 = $0.9476 × 10^8
Operating Costs = 24,274,000 × 47 = $11.4 × 10^8
Net income = Gross income – (Production costs + Operating costs)
= 20.6 × 10^8 – (0.9476 + 11.4) × 10^8
= $**8.2524 × 10^8**

This calculation excludes any capital expenditure that may be justified in the future. Also, the calculation is based on today's oil price, which may change in the future.

Example 12.5

A similar calculation can be done for Example 12.2 assuming the oil price, operating cost, and production taxes are the same as used in the previous calculations.

SOLUTION

Gross income = oil in place × price
= 5.9 × 10^9 × 85
= $500 × 10^9
Production taxes = $500 × 10^9 × 0.046
= $23 × 10^9
Operating costs = 5.9 × 10^9 × 47
= $277 × 10^9
Net Income = Gross income – (Production taxes + Operating costs)
= $500 × 10^9 – 300 x 10^9
= $**200 × 10^9**

Again, this net income excludes any capital expenditures that may be needed in the future. Other taxes that may be applicable are not combined.

Section III

Engineering Decisions through Economic Impact Analysis: Applications and Real World Examples

This section could be considered the backbone of our text. It represents the application of economic analysis to many engineering problems encountered in various sectors of petroleum operations. It illustrates how economic analysis is applied to solve engineering problems in different facets of the oil industry. Addressing relevant problems involving oil-engineering decisions is our main target in this section. Feasibility studies are presented for a number of cases. Three main operations that underlie the oil and gas industry, from prospects to finished products are as shown next:

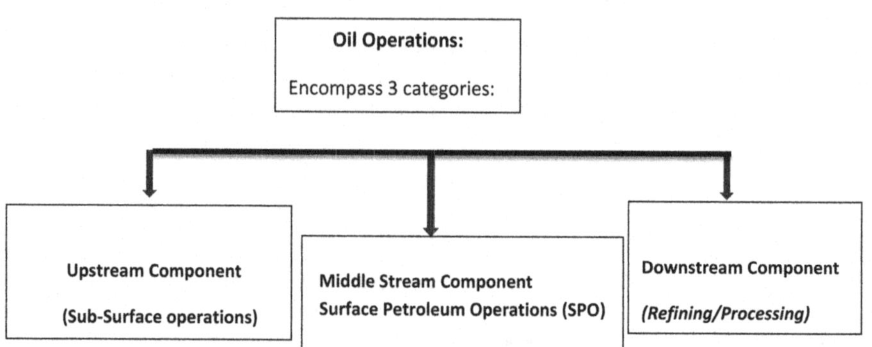
Oil Operations:
Encompass 3 categories:
Upstream Component
(Sub-Surface operations)
Middle Stream Component
Surface Petroleum Operations (SPO)
Downstream Component
(Refining/Processing)

Section III.I: Upstream Operations (Subsurface Operations)

Three main operations that underlie the oil and gas industry, from prospects to finished products are as shown in the text.

Section III.II: Upstream Operations (Subsurface Operations): Exploration and Production (E&P)

13 (E&P) Exploration, Drilling, and Oil Production—Part 1

13.1 INTRODUCTION

Exploration includes prospecting, seismic, and drilling activities that take place before the development of a field is finally decided. Symbolically the two operations of exploration plus drilling leading to oil production are drawn as shown Figure 13.1.

Exploration and production (E&P) is known as the *upstream* segment of the oil and gas industry. Exploration involves two distinctive stages: Exploration surveying and Exploration drilling. The contents of the chapter include both types of exploration. In practice, the process of oil and gas exploration and production typically involves four stages:

- Exploration
- Well Development
- Production
- Abandonment

Exploration and drilling is presented first as an integrated topic because of the relationship of the two processes. This is followed by presenting oil production in Chapter 15.

13.2 EXPLORATION AND DRILLING

Hydrocarbon exploration is a high-risk investment and risk assessment is paramount for successful exploration portfolio management. Virtually every oilfield decision is founded on profitability. With no control of oil and gas prices, and facing steadily rising costs and declining reserves, companies' basic decisions are based on constantly moving targets. Simply we can say that a producing oil and gas property is a series of cash payments projected in the future.

Technology aspects covered in this chapter deal with the very first activity in finding oil. It highlights, first, methods used for search of oil or oil exploration. Types of drilled wells, their numbers, and spacing are discussed next; and the use of economic balance and binomial expansion is proposed to solve relevant problems. The cost of finding oil and the size of capital expenditures in oil fields are considered in the chapter as well.

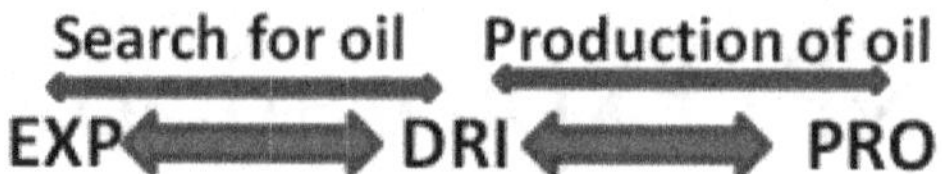

FIGURE 13.1 A schematic representation for the interaction between exploration, drilling, and production.

Applications and case studies include problems on optimization of the number of wells to drill, the cumulative binomial probability of success in drilling wells, and many other operations.

Knowledge of the basic principles as well as some of the common terms and concepts encountered in the oil fields is desirable for complete understanding of the subject. Geological formations, origin and accumulation of petroleum, oil reservoirs and their classification, petroleum prospecting practices, drilling and development operations, and many others are important in our engineering economics discussion.

Since our purpose here is not an explanation of the technical operations in petroleum production, we intend to highlight only the topics pertinent to the economic appraisal or valuation of "an oil property". The "oil property" as defined is meant to include any property with underground accumulations of liquid and/or gaseous hydrocarbons that might be produced at a profit.

Additional background materials on oil production methods and the estimation of recoverable oil reserves are given in the next chapter.

13.3 THE SEARCH FOR OIL: EXPLORATION

The first prerequisite to satisfying man's requirements for refined petroleum products is to find crude oil. Oil searchers, like farmers and fishermen, are actually in a contest with nature to provide the products to meet human needs. They are all trying to harvest a crop.

But the oil searcher has one problem that the farmer does not have. Before the oil man can harvest his crop, he has to find it. Even the fisherman's problem is not as difficult, since locating a school of fish is simple compared to finding an oil field. The oil searcher is really a kind of detective. His hunt for new fields is a search that never ends; the needle in the haystack could not be harder to find than oil in previously untested territories.

Today, petroleum prospecting and hence its discovery is credited to what is called "*subsurface studies*". This includes:

- The use of geophysical instruments
- Cuttings made by the bit as the well is drilled
- Core samples collected from the well
- Special graphs called "*logs*," generated by running some tools into the oil wells during the drilling operations

The net result of these studies is the preparation of different kinds of geological maps that show the changes in the shape of subsurface structures with depth.

The *geophysical techniques* encompass three methods:

1. Seismic method
2. Magnetic method
3. Gravitational method

Each of these techniques utilizes the principles of physical forces and the properties of the earth. For example, in the seismic method, creation of artificial earthquake waves is established by firing high explosives into holes. The rates of travel of these waves are analyzed by echo sounding techniques. The most recently invented instruments are: reflection seismographs, gravimeters, and airborne magnetrometers. Such devices enable geophysicists to explore not only the surface and the subsurface conditions of the earth searching for oil, but the lunar surface and depths as well. These sophisticated lunar experiments monitor the earth's magnetic and gravitational properties from space.

Stratigraphy, on the other hand, involves drilling a well basically to obtain stratigraphic correlation and information. Complete sections of the well formations are exposed and rock samples are taken while drilling operation is in progress. The success in finding oil will depend to a large degree on the accuracy of *well logging*. Several kinds of well logs are known; the most commonly accepted ones are:

1. Drillers logs
2. Sample logs
3. Electric logs
4. Radioactivity logs
5. Acoustic logs

Once the data are collected using core samples and wire-line logs of various kinds, *contour maps* are prepared. Generally a contour map consists of a number of contours, or lines, on which every point is at the same elevation above or below sea level of a given area. These lines must be at regular depth intervals to enable geologists to depict three-dimensional shapes.

Other means of exploring for oil include detailed ground geological surveys aided by preliminary results of aerial photography and photo-geological work.

13.4 OIL RESERVOIRS AND CLASSIFICATION

The two most important prerequisites for an oil accumulation to occur are:

1. A trap that acts as a barrier to fluid flow
2. A porous and permeable bed or reservoir rock

Thus, each geological formation, irrespective of age or composition, must process these physical properties of *porosity* and *permeability* in order to be described as a "reservoir rock".

Some of the reservoir-rock characteristics are as follows:

1. Although porosity and permeability are important as individual parameters, neither of them is of value in the absence of the other.
2. The reservoir is judged by its thickness and porosity, that is, by the abundance of interconnected voids, which provide passages for the fluids to flow.

Flow capacity or permeability depends on porosity to some extent, but porosity does not depend on permeability. In other words, reservoir rocks of high porosity are not necessarily of high permeability, and those of low porosity are not necessarily of low permeability. Generally speaking, *sandstone* reservoirs are more porous than *limestone*.

A *reservoir* may be defined as anybody of underground rocks with a continuously connected system of void spaces filled with hydrocarbon fluids, which can move toward wells—drilled into the rocks—under the influence of either natural or artificial driving forces. If the volume of the hydrocarbons produced by the wells is sufficient to permit an "*economic recovery*", then the accumulation is known as a commercial reservoir and usually referred to as a *proven reserve*.

Reservoirs, on the other hand, could be described as a "resource base", which is the sum total of crude oil, natural gas, and natural gas liquids in the ground within an identified geographic area. The reservoir thus includes all stocks, including some stocks which are unrecoverable and therefore *not* included in "proven reserves".

Proven reserves refer to the reserve stocks of immediate or short-term economic feasibility of extraction; therefore, stocks which are known to exist but cannot profitably be extracted are excluded from reserves. The cost limits, or as far as one can go on profitably employing these reserves, are those costs consistent with the taking of "normal" risk and commercial production.

The void spaces of proven reservoirs normally contain some *interstitial water* (or connate water) along with the hydrocarbons. Since most of this water is held in space by some sort of capillary forces, reservoir rocks turn out to be saturated with the *three* reservoir fluids: oil (liquid), gas, and water.

An *oil field* consists of all "pools" or reservoirs underlying a continuous geographic area, with no large enclosed subareas being considered unproductive.

13.5 THE ROLE OF DRILLED WELLS IN DEVELOPMENT

All of the activities described earlier for oil exploration lead only to an evaluation of the probability that oil is in a particular location. Once it seems probable that there really is oil, wells must be drilled. Reservoir and oil fields are discovered only by drilling to sufficient depths to verify what was recommended by an exploration team. The following stages in well drilling are identified:

1. Wildcat wells, exploratory wells, or test wells are drilled first for such probing purposes.
2. An unsuccessful wildcat well is called a *dry hole*.

3. A successful wildcat well is called a *discovery well.*
4. Subsequent wells drilled into proven reservoirs for production purposes are called *development wells.*

Such a stage-wise classification is illustrated in Figure 13.2

As far as the "*test wells*" are concerned, the following should be noted:

a. Drilling of test wells is the most costly single operation in oil exploration (this will be discussed further).
b. One exploratory well alone does not indicate extensive oil accumulation. Other wells, carefully located near the well where oil has been discovered, are drilled to discover if there is a reservoir in the area and approximately how much is available and can be recovered. Thus, it is dcsirable:

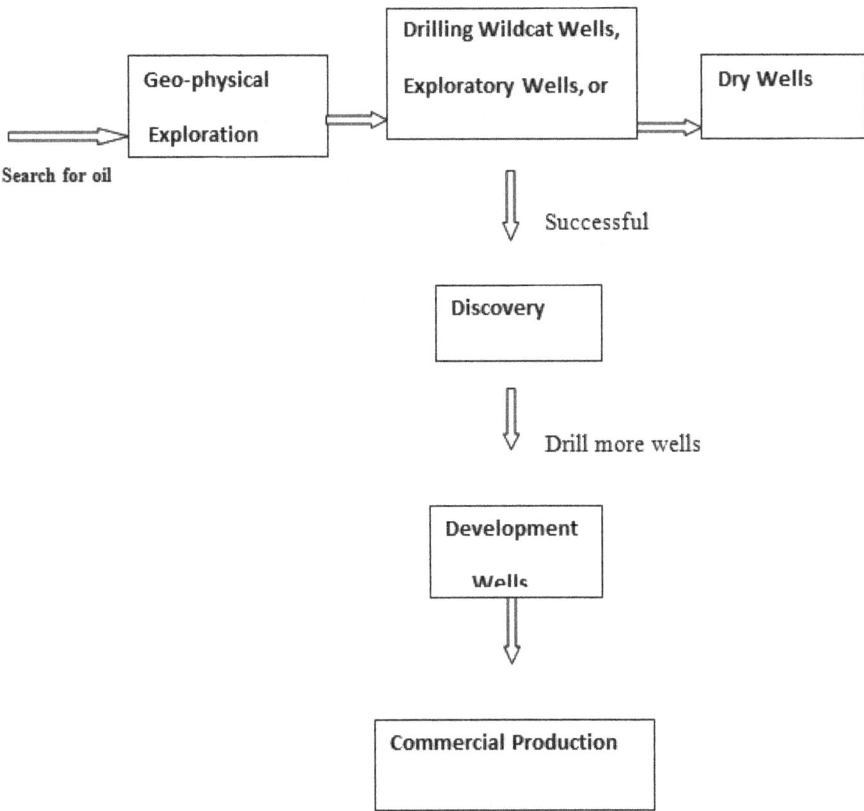

FIGURE 13.2 Different stages in well drilling.

1st to obtain reliable information as to the quantity of oil (and gas) which is recoverable, so an economic and proper size and type of surface crude oil production plant can be setup, and

2nd to determine from the samples of the reservoir the characteristics of the oil itself, the nature and amount of oil in the reservoir. The raising of oil to the ground surface and then the handling of the oil at ground surface will depend to a great extent on the nature of the oil itself and its associated gas. Crude oil can range from very heavy viscous oil, almost a tar, with little or no gas dissolved in it and under very low pressure, down to an extremely light, straw-colored oil with a considerable volume of gas, known as a condensate-type crude. The condensate-type crude is more likely to be found at great depths. Under conditions of high pressure and temperature which exist at deep levels, the crude is usually in the gaseous stage. Between the extremes of a heavy viscous oil and a very light oil, there is an infinite variety of crude oil. The manner of producing these crudes is decided only after examining samples, which show their characteristics and physical attributes.

"Intelligent wells" are increasing in popularity. These contain permanent monitoring sensors that measure pressure, temperature, and flow and telemeter these data to surface. More importantly, these wells contain surface-adjustable downhole flow-control devices, so, based on the dynamic production information from all the wells in the reservoir, flow rates can be optimized without having to perform a costly intervention.

13.6 NUMBER OF WELLS AND WELL SPACING

The location as well as the number of wells drilled into a proven reservoir raises many questions:

"How many wells should we drill in the reservoir?"
"How close should the wells be?"
"How many wells do we need before we can lay pipelines economically?"

Usually, use of the "economic balance" will provide answers to this type of question.

13.7 DRILLING OPERATIONS

There are two methods of drilling a well, the cable tool and the rotary methods. No matter which method is used, a derrick is necessary to support the drilling equipment.

Cable tool drilling is the older method of drilling. In this method a hole is punched into the earth by repeatedly lifting and dropping a heavy cutting tool, a bit, hung from a cable. Today, however, practically all wells are drilled by the rotary method.

Rotary drilling bores a hole into the earth much as a carpenter bores a hole into a piece of wood with a brace and bit. In the middle of the derrick floor there is a horizontal steel turntable, which is rotated by machinery. This rotary table grips and turns a pipe extending through it downward into the earth. At the lower end of the pipe, a bit is fastened to it.

As the drill chews its way farther and farther down, more drill pipe is attached to it at the upper end. As section after section of drill is added, the drill pipe becomes almost as flexible as a thin steel rod. Controlling the drill pipe under such conditions, and keeping the hole straight as well, is very difficult and requires great skill in drilling.

During the drilling, a mixture of water, special clays, and chemicals, known as drilling mud, is pumped down through the hollow drill pipe and circulated back to the surface in the space between the outside of the pipe and the walls of the pipe. This drilling mud serves several purposes, including lubricating and cooling the bit and flushing rock cuttings to the surface.

As the drilling hole is deepened, it is lined with successive lengths of steel pipe, called casings. Each string of casing slides down inside the previous one and extends all the ways to the surface. Cement is pumped between these successive strings of casing, and seals against any leakage of oil, gas, or water.

To achieve large annual additions to reserves and to output, the rate of drilling must be stepped up sharply. Barrels added per foot drilled are one of the best indicators of the results of drilling effort. This measure should not show a decline. A projection of the trend of barrels added per foot of drilling should be established for oil companies engaged in production.

13.8 FACTORS AFFECTING PENETRATION IN DRILLING

Studies made by experts from drilling and equivalent companies indicate that there is a positive effect of weight and speed of rotation on penetration rate, or feet per hour of drilling. This is true whether toothed or carbide-studded bits are used.

Past experience has shown that proper penetration rate of weight on bit rotary speed and hydraulic horsepower can be plotted on a graph to determine optimum drilling at minimum drilling cost. Thus, the penetration rate of a bit varies with weight on bit, rate of rotation, and hydraulic horsepower.

13.9 COSTS OF DRILLING

An increase in depth increases drilling costs. Actually, costs increase exponentially with depth, even for a "normal" trouble-free well. Also, an increase in depth can increase the chances of mechanical problems. This adds to the cost of drilling.

Increased depth also reduces available information about potential reservoirs, as to quality of crude oil and quantity available (proven reserves). Risks increase with uncertainties as to reservoir quantity and quality available.

Costs of drilling depend on:

- The kind of oil and what potential energy the oil possesses by virtue of its initial pressure in its reservoir.
- By the amount of dissolved gas it may contain. In many cases the crude may have enough potential energy to permit a well to flow large quantities of oil to the surface without any artificial assistance, such as use of gas or water injection.

This is quite prevalent in oil wells in the Middle East. But when oil cannot flow unaided, or when the pressure in the reservoir has decreased to a pressure that is too

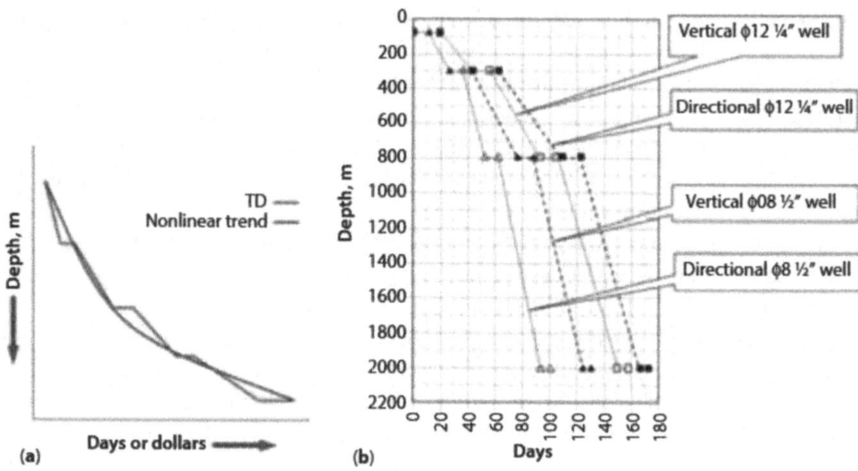

FIGURE 13.3 Drilling cost as a function of well depth (Hossain and Al-Majed, 2015).

low to be economical, costly mechanisms which lift oil to the ground surface must be employed. Furthermore, low pressure in the reservoir and low gas content generally go together. This kind of crude, therefore, must be handled in a different manner.

Land rig operating rates vary between $8k/day and $45k/day, largely depending on the region and rig type. The North American market is the most important indicator of land rig rates. US High Spec Land Rigs Average Day Rates. Data is based on multiple companies reports, averaged and rounded up.

The cost of drilling has plummeted from $4.5 million to $2.6 million. The average drill cost per foot was lowered from $245 to $143. That means drilling costs have been reduced for Chesapeake by about 42%, which offsets the drop in the commodity cost (for natural gas).

Drilling costs will depend on the depth of the well and the daily rig rate. The rig daily rate will vary according to the rig type, water depth, distance from shore and *drilling* depth. One development well will *cost* 55–88 MM$ plus completion *costs* (+80%) totaling 99–158.4 MM$. Drilling cost as a function of well depth is given in Figure 13.3, by Hossain and Al-Majed, 2015.

13.9.1 Statistical Solution for Cost Estimation in Oil Well Drilling

A group at Mexico proposed Bit Program for an onshore well in Mexico. Traditionally, the poor bit performances are discharged, understood as being the result of bad practices that shall not be repeated. A handful of good runs are elected as results to be repeated or improved.

Some wells are drilled with the casing itself replacing the conventional drill string, disclosed as economic by saving time circulating, running casing, and reducing non-productive time (Patel, D. et al, 2018).

Intervals of the planned well are divided by the expected rate of penetration (ROP) by depth, generating the Drilling Hours seen in a Bit Program (Table 13.1).

TABLE 13.1
Statistical Program for Cost Estimation of Oil Drilling

Bit	Section	Bit	Bit	Depth	Length	Drilling	ROP	Acc.	WOB	RPM	Mud	Remarks
n°	n°	size	type	Out	Drilled	Hours	-	Hours	-	-	Weight	-
-	-	(in)	-	(m)	(m)	(h)	(m/h)	(h)	(ton)	-	(sg)	-
1	1	26	115	1,000	950	60	15.8	60	6–15	80–140	1.2	IADC 115
1	2	17.½	115	2,100	1,100	110.0	10.0	110	10–18	120–160	1.4	IADC 115
2	2	17.½	115	3,300	1,200	120.0	10.0	230	10–15	80–160	1.5	IADC 115
3	3	14.¾	PDC	4,150	850	100.0	8.5	330	2–3	120–130	2.0	Bicenter Bit
4	3	14.¾	PDC	4,500	350	33.0	10.6	363	2–3	120–130	2.0	Pilot bit 8.5
5	4	12.¼	PDC	5,260	760	95.0	8.0	458	2–4	80–120	1.9	Pilot bit 8.5
6	4	12.¼	PDC	5,370	110	30.0	3.7	488	3–5	70–100	1.9	Pilot bit 8.5
7	4	12.¼	PDC	5,500	130	35.1	3.7	523	3–5	70–100	1.9	Pilot bit 8.5
8	5	8.½	537	5,610	110	70.0	1.6	593	8–10	80–100	1.0	Metal seals
9	5	8.½	537	5,720	110	70.0	1.6	663	8–10	80–100	1.0	Metal seals
10	5	8.½	517	5,830	110	70.0	1.6	733	6–8	80–100	1.0	Metal seals
11	5	8.½	517	5,940	110	70.0	1.6	803	6–8	80–100	1.02	Metal seals
12	6	6	517	6,090	150	90.0	1.7	893	4–6	80–100	0.67	Metal seals
13	6	6	517	6,237	147	90.0	1.6	983	4–6	80–100	0.67	Metal seals
14	6	6	517	6,384	147	90.0	1.6	1,073	2–5	80–100	0.67	Metal seals
15	6	6	517	6,531	147	90.0	1.6	1,163	2–5	80–100	0.67	Metal seals

(*Source:* R EM, Int. Eng. J. vol.72 no.4 Ouro Preto Oct./Dec. 2019 Epub Sep 16, 2019)

An example of actual day rate, Transocean signed a contract in December 2018 with Chevron to provide drilling services. The contract is for one rig, will span 5 years and is worth $830 million. The effective day rate for the rig is **$455,000.**

13.9.2 Economic Evaluation and Applications

The recovery of oil from underground, or offshore, reservoirs is a good application of the "principle of economic balance". The problem is one of determining the optimum number of wells to drill, and the accurate spacing of these wells, to get maximum profit.

The following considerations highlight the subject:

1. Actually, the greater the number of wells, the larger will be the ultimate recovery, provided that the recovery rate does not exceed the "most efficient engineering rate". But the most efficient engineering rate (economic balance) does not necessarily mean the optimum rate for maximum profits.
2. Economic balance, therefore, consists of a balance of:
   a. greater fixed costs for a larger number of wells drilled plus usually higher operating costs for higher production rates against
   b. greater ultimate recovery from the larger number of wells.

Thus, the principle of economic balance in the oil fields is:

To drill as many wells as possible and needed within fixed costs and operating cost limits relative to the greatest ultimate recovery in terms of the realizable value (sales value) for the recovery. There is an upper limit to the number of wells that can be drilled, however, because of technical considerations.

In other words, greater fixed costs plus higher operating costs must be considered when increasing the number of wells to be drilled in an attempt to obtain a greater ultimate recovery of oil.

3. Upon discovery of large enough reserves for commercial drilling, the concept of well spacing becomes important to the oil engineer. The characteristics of reservoirs largely control the well-spacing pattern. For example, reservoirs with thick or multiple zones of oil will usually require more wells, and possibly closer spacing between wells, to take advantage of natural drainage (gravity flow) at its maximum than those reservoirs with thin crude oil composition located in single zones. Furthermore, porous reservoirs will produce more barrels of oil than "tight" reservoirs.
4. Other factors of a technical nature, which should be considered in the spacing of wells, besides thickness vs thinness of the crude itself and the multiple zones vs single zones, include depth to the productive zones of the oil, viscosity of the oil, gravity of the oil, reservoir pressures, and reservoir properties. Therefore, in well spacing, economics of anticipated recoveries based on thickness of oil and saturation of the pay zone become important. Obviously, the greater the number of wells drilled in a

single reservoir, the greater will be the ultimate recovery per surface area of oil and/or gas.

5. There is a practical limit to the number of wells, and hence the spacing of wells, that can be drilled, however, which is controlled by the cost of drilling and operation. This limit to the number of wells to be drilled is based on estimated ultimate recovery, in barrels of oil, from each well. Since depth is the principal factor governing drilling costs, depth has a bearing on the problem of well spacing.
6. There is no hard and fast rule on spacing of wells; the technical and nontechnical factors relative to the oil reservoir must be considered separately.
7. Oil wells drilled in the United States are widely spaced and located at the centers of 40-acre tracts or at like ends of 80-acre tracts. For gas wells, on the other hand, spacing ranges between 160 and 640 acres per well.
8. The acreage assigned to each development well is known as a *drilling unit* prior to completion of the well and as a *production unit* upon successful completion.
9. Usually, the greater the depth to reach productive zones of oil, the wider the spacing of wells. Furthermore, since viscous oils do not possess the mobility of ready passage through reservoirs, as lighter, less viscous oils do, a closer spacing of wells is usually needed with oils of heavy viscosity properties in order to effect maximum efficient drainage. In the case of gravity, the lighter-gravity oils (with the higher API) contain more dissolved gases, have more mobility, and are less viscous than the lower-gravity oils, and so will require fewer wells and wider spacing to effect maximum efficient drainage. On reservoir pressures, reservoirs with high pressures, particularly if pressures are maintained by some recycling operations such as use of water, gas, or air, offer higher recovery per well. Thus, a wider spacing can be employed in reservoirs with high pressures.
10. Such reservoir properties as porosity, the ability to contain fluids and permeability have an influence on well spacing. Porous and permeable reservoirs, which allow fluids such as oil to flow through the reservoir to the well bore, means that reservoirs can be effectively drained, so fewer wells with wide spacing is suitable under such conditions. Closer spacing of wells is necessary when "tight" reservoirs, with low porosity and permeability, are involved.
11. Some nontechnical factors also affect well spacing. These include, for instance, the rate of production desired because of terms of the oil lease, market price of crude, market demand, etc. Also, proration laws of a government can dictate the amount of oil or gas an oil company can produce. When this is the case, the number of wells drilled, and the spacing, may be affected. Where the rate of payout desired is lengthened, and deferment of income over a wide period because of income tax problems is the objective, the number of wells drilled may be cut back. Thus, spacing will tend to be wider under such conditions. The opposite of this, where the rate of payout desired is for a short period dictates more wells drilled with closer spacing.

CASE STUDY - 13.1

The following simple example offers two alternatives relative to the number of wells to be drilled and spaced in a reservoir involving the following information:

	Alternative 1: Drill 2 Wells	Alternative 2: Drill 6 Wells
Total capital investment ($)	3,800,000	8,400,000
Annual operating costs	560,000	1,800,000
Total production (bbl/day)	20,000	100,000

REQUIRED

(a) Determine the spacing between wells. (b) Which alternative do you recommend: the wider spacing between two wells or the closer spacing between six wells?

SOLUTION

(a) Let us establish the following table using some common basis:

	Alternative 1	Alternative 2
1. Capital investment/well ($)	1,900,000	1,400,000
2. Annual operating cost/well ($)	280,000	300,000
3. Capitalized cost of item (2) using interest rate of 10%	2,800,000	3,000,000
4. Sum of items (1) + (3)	4,700,000	4,400,000
5. Production bbl/(day)(well)	10,000	16,667

Spacing is calculated on the assumption that a producing well is located on an area of *one acre*. Hence, daily oil production is reported on the basis of bbl/(well)(acre).

In addition, income is reported by assigning an arbitrary value for the drilled oil equal to 33% of the well-head value of produced oil.

Now, for *one day* of production, and taking *one well* as a basis for our calculation, we obtain:

$$\text{Spacing between wells in given by}: \frac{\text{Capital investment (\$)}}{\text{Revenue (\$ / acre)}}$$

For alternative 1, spacing = 17 acres

For alternative 2, spacing = 10 acres

Thus, a spacing of 17 acres between two wells is recommended for alternative 1, while 10 acres is to be used as a spacing for the six-well alternative.

(b) Although operating costs are greater in total and on a per-well basis with six wells, total production is greater, and hence total revenues earned, including profits, will be greater. Furthermore, the payout period favors the six-well alternative over the payout period of the alternative on two wells, since more overall production of six wells will increase total revenues received, sufficient to return investment more quickly.

Finally, capital investment per barrel produced per day favors alternative 2. Capital investment per barrel per day with six wells drilled is $84, whereas capital investment per barrel per day with two wells drilled is $190.

Obviously, Alternative 2, or six wells, is the selection, assuming everything else favors this alternative, including reservoir pressures, no limit on production, favorable permeability and porosity features, etc.

CASE STUDY - 13.2

Explorers for crude oil try to determine how often success will be gained from a given program of N well (wells drilled). "What are the odds of success?" a company might ask. A company drilling, say, 20 or 30 wells per year might want to know the odds of making one, two, three, or five discoveries, with discovery meaning simply a producing well and not profitability of the well. How much oil there is, is not part of discovery, but comes under field size distribution. To find these odds of success to total wells drilled, a mathematical technique called binomial (two numbers) expansion is used.

For simplicity, assume that each well in the program has the same chance of success with an assumed 10% success rate. Oil explorers know that some prospects have better "odds" or chances of success than others. For most exploration programs, we can assume an "average success" rate with reasonable safety.

F indicates probability of failure (a dry hole), and S indicates probability of success.

For one well (one outcome) $F + S = 1.00$, or we can write $F + S\ (F + S)^1$. For two wells, there are four possible outcomes, $FF + FS + SF + SS = 1.00$; and, of course, $FS + SF$ can be written 2FS. Then $F^2 + 2FS + S^2 = 1.00$.

Now, if you remember your algebra, $F^2 + 2FS + S^2$ is the product of $(F + S)$ $(F + S)$ and can be written as $(F + S)^2$. So $F^2 + 2FS + S^2 = (F + S)^2$. The left half of this equation is the expansion of the binomial $(F + S)$ to $(F + S)^2$.

Now, we can setup a cumulative binomial probability table as shown in Table 13.2, with an assumed 10% success rate, for any larger number of wells to be drilled and we will get some probabilities of success in number of discoveries to total number of wells drilled.

TABLE 13.2
Cumulative Binomial Probability (Using a 10% Success Rate)

No. of Wells Drilled	No. of Discoveries	Probability Success in No. of Discoveries (%)	Odds of Success
10	1	60	1 in 10
10	2	26	1 in 5
10	3	15	3 in 10
20	1	80	1 in 20
20	2	61	1 in 10
20	3	50	3 in 20
20	4	25	1 in 5
20	5	10	1 in 4
30	1	90	1 in 30
30	2	73	1 in 15
30	3	70	1 in 10

From Table 13.2, a graph can be drawn as shown in Figure 13.4, to illustrate tables of cumulative binomial probabilities. This graph provided the following information:

1. At least one discovery or more is 88% (or 88 chances of success in a total of 100 chances), or with 4.4 chances of S in five chances.
2. At least two discoveries is 60% (or 60 chances of success in 100 total chances), or 3 in 5 chances.

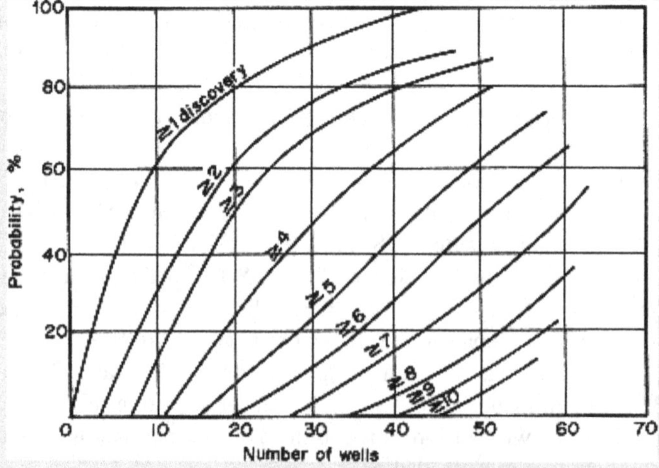

FIGURE 13.4 Commulative bionomial probablity, assuming 10% success.

3. At least three discoveries is 30% (30 chances of success in 100 chances), or about 1.5 in 5 chances.
4. At least four discoveries is 13% (13 chances of success in 100 chances), or about 1 in 8 chances.

The chance of drilling any number of dry holes in succession, like the chance of one dry hole "in succession", is 1.00 – 0.10, or 0.90 (90%). For additional wells, they are as follows:

2 dry holes in succession = 81%, or 4 in 5 chances
5 dry holes in succession = 69%, or 3 in 5 chances
10 dry holes in succession = 35%, or 1 in 3 chances
20 dry holes in succession = 12%, or 1 in 8 chances

Thus, even with a 10% success rate, even in drilling 20 holes, we still face a 12% chance that all holes will be dry.

The employment of such a table and graph is a possibility for explorers for crude oil in their efforts to predict success and failure, or discoveries to dry holes. It can also be useful to oil engineers in estimating probabilities, or odds of success.

2nd: Using the binomial distribution to find the probability of an exact number of successes (discovery wells) in several trials (number of wells to be drilled), the following relation could be applied:

$$p(x) = \begin{pmatrix} N \\ x \end{pmatrix} p^N q^{N-x}$$

$$= C_x^N p^N q^{N-x}$$

where:

$p(x)$ = probability of obtaining exactly x successes in N trials
N = size of the sample, or number of trials of an event
x = number of successes, or favorable outcomes within the N trials
p = probability of success
$q = 1 - p$ = probability of failure
$\binom{N}{x} = C_x^N$ = number of combination in which N objects can be displayed as groups of size x, where the order within the individual groups is unimportant

The mean, variance, and standard deviation of the binomial are given by:

$$m = NP$$

$$\sigma^2 = Npq$$

$$\sigma = (Npq)^{1/2}$$

Example 13.3

As an example, the probability of obtaining zero heads when a coin is tossed five times is calculated as follows; using Equations (12.2) and (12.3):

$$p(x) = \binom{N}{x} p^x q^{N-x},$$

$$p(0) = \binom{5}{0} (0.5)^0 \ (1-0.5)^{5-0} = (1)(1)(0.5)^5$$

Roughly, the probability is 3 of 100 times. That is, where successive tosses were gathered into groups of five tosses in each group, out of 100 such groups, about three would contain no heads.

Example 13.4

Ten wells are to be drilled. The probability of success is taken to be 0.15. What is the probability of there being more than two successful wells?

SOLUTION

The answer to this can be found in one of two ways: (1) the individual probabilities of 3, 4, 5, 6, 7, 8, 9, and 10 successes can be calculated and added together, or (2) the individual probabilities of 0, 1, and 2 successes can be added together and then subtracted from 1 to obtain the same answer. The second method is shorter, and is given as follows:

$$p(x) = \binom{N}{x} p^x q^{N-x}$$

$$p(0) = \binom{10}{0} (0.15)^0 \ (0.85)^{10} = 0.1969$$

$$p(1) = \binom{10}{1} (0.15)^1 \ (0.85)^9 = 0.3474$$

$$p(2) = \binom{10}{2} (0.15)^2 \ (0.85)^8 = \frac{0.2759}{0.8202}$$

$$p(\text{more than 2 producers}) = 1 - 0.8202 = 0.1798$$

Hence, probability is approximately 18%.

Most oil companies are not concerned with how far down drilling proceeds, but with how high the cost will be to get that deep and what the cost will be to go, say, another 100 ft or more. Marginal costs are some direct function of depth. If, then, we let Y be those costs which vary with depth, but no overhead costs, and let X be depth itself, a formula can then be written as:

$$\frac{dY}{dX} = C(X), \text{ the cost per foot}$$

Thus, depth affects marginal costs. For example, the rise of temperature with depth, among other things, increases the probability that a drilling bit will have to be replaced an additional time in a well drilled an additional 100 ft, because mechanical energy is lost as the drilling process continues. But also, some costs, such as the costs of additional "mud materials", needed to drill a deeper well may actually increase rather slowly in relation to increase in depth, thus giving a decreasing marginal cost in relation to depth.

Possibly the one factor that most affects the costs of drilling is the average footage drilled per hookup. As more information on drilling tendencies in any one oil field become available, the number of changes in drilling hookup is reduced and the speed of the drilling operation is increased. Also, feet per hour at the bottom of the well, combined with the amount of time spent at the bottom, is perhaps the best measure of the relative efficiency and speed of a drilling operation in a particular oil well and for a given amount of controlled footage.

In sum, costs of drilling increase because of the following, usually in some combination:

1. A poorly designed casing program
2. An inadequate rig or incompetent personnel on the test drill
3. Poor selection of proper drilling bits for the formations to be penetrated
4. Insufficient drilling bit weight for maximum penetration (economic balance here relative)

The following are some of the expressions and definitions used in "cost terminology" and reserves reporting; which are used in this chapter as well as in the following chapters.

13.10 SOME BASIC DEFINITIONS

Development costs: expense and capital costs incurred to bring on-stream a producing property (includes development well drilling and equipment, enhanced recovery, and extraction and treatment facilities).

Discoveries: newly found proven reserves, including production sharing type reserves, which may or may not be included (booked) in annual reserve estimates.

Exploration costs: expense and capital costs to identify areas that may warrant examination (includes geophysical, geological, property retention costs, dry hole expenses, exploration drilling).

Extensions: additions to existing fields, normally booked in the same year.

Finding oil: includes exploration (search) for oil, development of successful exploration discoveries, including the drilling of wells, and, finally, the drilling and preparing of oil for commercial production, including the laying of gathering pipelines and pump installation for the movement of oil to central points for gas separation.

Finding and development costs: used by securities analysts to measure and compare petroleum company performances in acquiring reserves.

Improved recovery: additions to reserves due to secondary and tertiary recovery, booked when production commences.

Property acquisition costs: those costs incurred to purchase or lease proven or unproven reserve properties, capitalized when incurred.

Punchback: deepening to new horizons or completing back to shallower horizons, the reserves of which may or may not be booked.

Purchase of reserves in place: proven reserves purchased from outside companies.

Revisions: additions or deletions to previous reserve estimates based on updated information on production and ultimate recovery.

Once the oil has been explored, developed, and produced, all costs involved in getting the oil to the surface, where it becomes a commodity as it is piped in gathering lines to central points for gas separation, are called the cost of oil field operation. The basic question "What does oil cost to find, to develop, and to ready for commercial production?" would be comparably simple to answer if, during a short period of time—say 1–3 years—an oil company could start in the oil-producing business, discover say 10 million bbl of oil, develop that 10 million bbl, and finally produce the 10 million bbl of crude. The cost of dinning, developing, and producing could then simply be found by dividing the total amount spent for exploratory, developing, and producing effort by 10 million bbl, which would give a cost per barrel of crude.

But this is just "grocery store accounting". Actual accounting for costs in the oil-producing industry is not that simple. When a company searches for oil, it may spend several years and millions of dollars on exploration and development before any substantial, and commercially feasible, amount of oil is located. In development alone, a company may work for several years and spend many dollars developing the oil reservoir which it is to produce over an even greater number of years; and also, all this time, the process is constantly repeating itself as more oil is being discovered, more oil is being developed, and more oil is being produced.

Finally, an oil company's success is measured by its ability to discover reserves. In its search for oil, it spends substantial amounts of money in many different ventures in widely scattered areas. The oil company does this knowing that many of these ventures will be nonproductive and will eventually be abandoned.

On the other hand, the oil company recognizes that successes in other areas must be large enough to recoup all money spent in order to break even or to provide a profit. Thus, the true assets are the oil reserves, and these costs are capitalized. But the costs of nonproductive exploration activities and of dry holes are also a necessary part of the full cost of finding and developing these oil reserves.

CASE STUDY: TO CHOOSE BETWEEN TWO ALTERNATIVES FOR DRILLING WELL IN A RESERVOIR

GIVEN

The following is a case for an offer for two alternatives relative to the number of wells to be drilled and spaced in a reservoir involving the following information. The case was presented to P E students as ca sort of a senior project at KFUPM, Dhahran, Saudi Arabia.

	Alternative 1: Drill 2 Wells	Alternative 2: Drill 6 Wells
Total capital investment ($)	3,800,000	8,400,000
Annual operating costs	560,000	1,800,000
Total production (bbl/day)	20,000	100,000

GIVEN

Determine the spacing between wells. (b) Which alternative do you recommend: the wider spacing between two wells or the closer spacing between six wells?

SOLUTION

(a) Let us establish the following table using some common basis:

	Alternative 1	Alternative 2
1. Capital investment/well ($)	1,900,000	1,400,000
2. Annual operating cost/well ($)	280,000	300,000
3. Capitalized cost of item (2) using interest rate of 10%	2,800,000	3,000,000
4. Sum of items (1) + (3)	4,700,000	4,400,000
5. Production bbl/(day)(well)	10,000	16,667

Spacing is calculated on the assumption that a producing well is located on an area of *one acre*. Hence, daily oil production is reported on the basis of bbl/(well)(acre).

In addition, income is reported by assigning an arbitrary value for the drilled oil equal to 33% of the well-head value of produced oil.

Now, for *one day* of production, and taking *one well* as a basis for our calculation, we obtain:

$$\text{Spacing between wells in given by}: \frac{\text{Capital investment (\$)}}{\text{Revenue (\$ / acre)}}$$

For alternative 1, spacing = 17 acres

For alternative 2, spacing = 10 acres

Thus, a spacing of 17 acres between two wells is recommended for alternative 1, while 10 acres is to be used as a spacing for the six-well alternative.

(b) Although operating costs are greater in total and on a per-well basis with six wells, total production is greater, and hence total revenues earned, including profits, will be greater. Furthermore, the payout period favors the six-well alternative over the payout period of the alternative on two wells, since more overall production of six wells will increase total revenues received, sufficient to return investment more quickly.

Finally, capital investment per barrel produced per day favors alternative 2. Capital investment per barrel per day with six wells drilled is $84, whereas capital investment per barrel per day with two wells drilled is $190.

Obviously, Alternative 2, or six wells, is the selection, assuming everything else favors this alternative, including reservoir pressures, no limit on production, favorable permeability and porosity features, etc.

14 (E&P) Exploration, Drilling, and Oil Production—Part 2

14.1 INTRODUCTION

The extraction of petroleum is the process by which usable petroleum is drawn out from beneath the earth's surface location.

Primary subsurface production methods include cold production (horizontal and multilateral wells, waterflood, and cold heavy oil production with sand) and thermal production (cyclic steam stimulation, steam flood, and steam-assisted gravity drainage).

Crude oil production is defined as the quantities of oil extracted from the ground after the removal of inert matter or impurities. It includes crude oil, natural gas liquids (NGLs), and additives. … NGLs are the liquid or liquefied hydrocarbons produced in the manufacture, purification, and stabilization of natural gas.

Petroleum production engineering covers the widest scope of engineering/operations in the petroleum industry. It starts with the selection, design, and installation of the well completion and ends with the delivery of the useful fluids (i.e., oil and natural gas) to the customer. Between the two ends lie a large number of engineering activities and operations. For example, the design and installation of the well tubing and surface flowline, the workover operations which keep the well at its best producing conditions, the selection and design of the oil/gas production method and the design, installation, and operation of the surface separation and treatment facilities are all the responsibility of the petroleum production engineer.

The economics of most of the above-mentioned operations have to be evaluated before they are executed. In some cases, several technically viable alternatives would exist for executing a particular operation. In such cases, the decision to select one alternative over the others would be based entirely on economic evaluation of the various alternatives.

Following the introduction of each major production operation, economic-based decisions are presented. Applications and case studies illustrating the economic analysis in this strategic phase of the oil operations are given, with examples of the economic evaluation of some operations.

Over millions of years, layer after layer of sediment and other plants and bacteria were formed. As they became buried ever deeper, heat and pressure began to rise. The amount of pressure and the degree of heat, along with the type of biomass, determined whether the material became **oil** or natural gas.

The seven steps of oil and natural gas extraction are summarized as follows:

- Preparing the Rig Site.
- Drilling.
- Cementing and **Testing**. …
- Well Completion. …
- Fracking. …
- Production and Fracking Fluid Recycling. …
- Well Abandonment and Land Restoration.

14.2 OIL EXTRACTION (PRODUCTION) AND RECOVERY: OVERVIEW

Oil production methods are classified into the following categories:

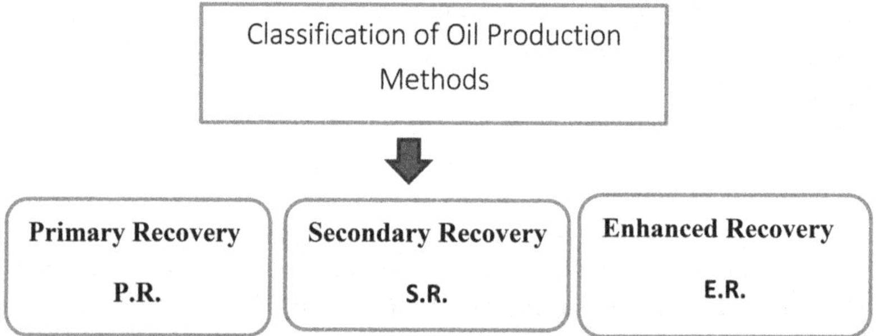

a. **PR:** Reservoir drive comes from a number of natural mechanisms. These include:
   - Natural water displacing oil downward into the well
   - Expansion of the associated petroleum gas at the top of the reservoir
   - Expansion of the associated gas initially dissolved in the crude oil
   - Gravity drainage resulting from the movement of oil within the reservoir from the upper to the lower parts where the wells are located. Recovery factor during the primary recovery stage is typically 5-1

b. **SR:** It is the stage where there is insufficient underground pressure to force the oil to the surface. After natural reservoir drive diminishes, *secondary recovery* methods are applied. These rely on supplying external energy to the reservoir by injecting fluids to increase reservoir pressure, hence increasing or replacing the natural reservoir drive with an artificial drive.

To summarize, one can say that Secondary recovery techniques increase the reservoir's pressure by water injection, gas reinjection, and gas.

c. **ER:** It is identified by increasing the mobility of the oil. Enhanced oil recovery methods (TEOR) are tertiary recovery techniques that heat the

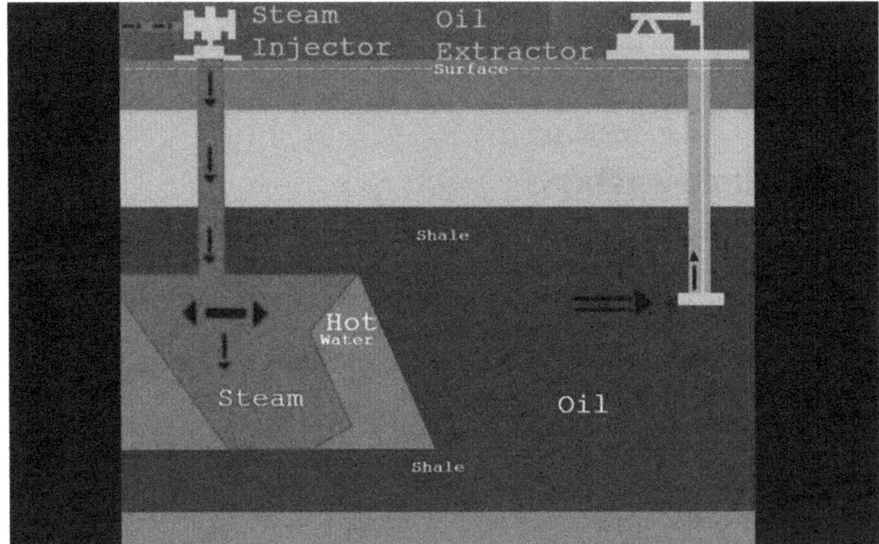

FIGURE 14.1 ER, steam is injected into many oil fields where the oil is thick and heavy.

Source: en.wikipedia.org › wiki › Extraction_of_petroleum

> oil, reducing its viscosity, and making it easier to extract. Steam injection is the most common form of TEOR, and it is often done with a cogeneration plant (Figure 14.1).

By production method we refer to the way in which the well fluids are delivered to the surface. Ideally, wells should be produced to deliver the fluids to the surface with a wellhead pressure sufficient to force the fluid flow through all surface facilities. There are two ways in which a well may be produced; these are described here.

14.3 NATURAL FLOW

A well is said to be produced naturally if it only utilizes the naturally stored energy, i.e., reservoir pressure, to lift the fluids to the surface. Most wells start their lives with natural flow. With time, the reservoir energy (pressure) is depleted, resulting in reduced production rates or reduced wellhead pressure or both. When this occurs, artificial lift may be implemented.

14.4 ARTIFICIAL LIFT

Steam is injected into many oil fields where the oil is thicker and heavier than normal crude oil.

Artificial lift refers to the use of external means to help lift the well fluids from the bottom of the well to the surface. Essentially, artificial lifting enables well production at lower bottom-hole pressures. It may be applied on a flowing well to increase its production in order either to meet market demands or to make the project economics more attractive. Artificial lifting is mostly applied, however, to wells which otherwise would not produce at all or would produce below the economic limit of operation.

14.5 WELL COMPLETIONS

After a well has been drilled, it must be completed before oil and gas production can begin. The first step in this process is installing casing pipe in the well.

Oil and gas wells usually require four concentric strings of pipe: conductor pipe, surface casing, intermediate casing, and production casing. The production casing or oil string is the final casing for most wells. The production casing completely seals off the producing formation from water aquifers.

The production casing runs to the bottom of the hole or stops just above the production zone. Usually, the casing runs to the bottom of the hole. In this situation the casing and cement seal off the reservoir and prevent fluids from leaving. In this case the casing must be perforated to allow liquids to flow into the well. This is a perforated completion. Most wells are completed by using a perforated completion. Perforating is the process of piercing the casing wall and the cement behind it to provide openings through which formation fluids may enter the wellbore.

14.6 FACTORS INFLUENCING WELL COMPLETION DESIGN

While safety and cost are of prime importance in selecting and designing a well completion, the engineer has to consider the following factors in finalizing his completion design:

- The type of reservoir and drive mechanisms
- The rock and fluid properties
- The need for artificial lift
- Future needs for stimulation and workover
- Future needs for enhanced recovery methods

Normally, the technical factors are first considered to determine possible completion designs; then, the economic aspects are considered to select the most economical design.

14.7 TUBING AND PACKERS

After cementing the production casing, the completion crew runs a final string of pipe called the tubing. The well fluids flow from the reservoir to the surface through the tubing. Tubing is smaller in diameter than casing—the outside diameter ranges from about 1 to 4-1/2 inches.

A packer is a ring made of metal and rubber that fits around the tubing. It provides a secure seal between everything above and below where it is set. It keeps well fluids and pressure away from the casing above it. Since the packer seals off the space between the tubing and the casing, it forces the formation fluids into and up the tubing.

14.8 SIZING PRODUCTION TUBING

The starting point in a completion design is determination of the production tubing (conduit) size. This is extremely important as it affects the entire drilling program and the cost of the project.

To determine the size of the tubing, the engineer has to conduct what is known as *well performance analysis*. This analysis requires the study of two relationships:

- The first one describes the flow of fluids from the formation into the wellbore; it is called the *inflow performance relation* (IPR). The IPR is represented, normally, as the relationship between the bottomhole flowing pressure (P_{wf}) and the flow (production) rate (q). Depending on the type of reservoir and the driving mechanism, the IPR may be linear or nonlinear, as illustrated in Figure 13.1. When the IPR is linear, it can be represented with what is called the *productivity index* (PI), which is the inverse of the slope of the IPR.
- The second relationship describes the relation between the flow rate of fluids and the pressure drop in the production tubing. It is called the *outflow performance* or the tubing *multiphase* flow performance. Several *multiphase* flow correlations exist for determining the relationship between flow rate and pressure drop in a well tubing. For a fixed wellhead pressure, the relationship between P_{wf} and q is as illustrated in Figure 14.2.

The interaction of the two relationships would provide several solutions, as shown in Figure 14.3. That is, several tubing sizes could be used, but each would yield a different production rate. Normally, higher production rates are obtained using larger tubing sizes; this means higher drilling and completion costs. The final selection of the tubing size should, therefore, be based on economic analysis of the various alternatives, as illustrated in Example 14.4.

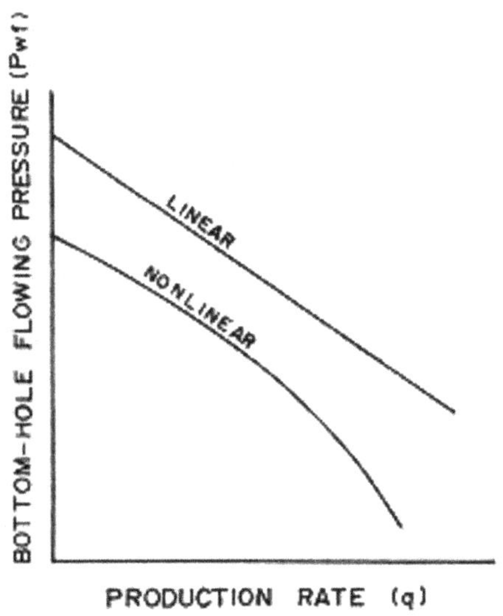

FIGURE 14.2 Inflow performance relations (IPR).

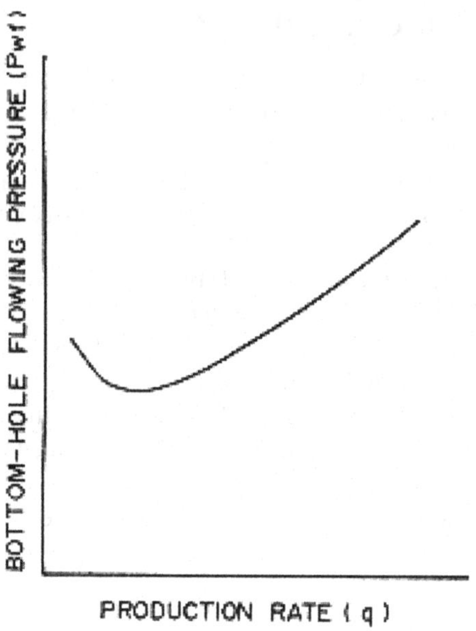

FIGURE 14.3 Outflow (vertical flow) performance.

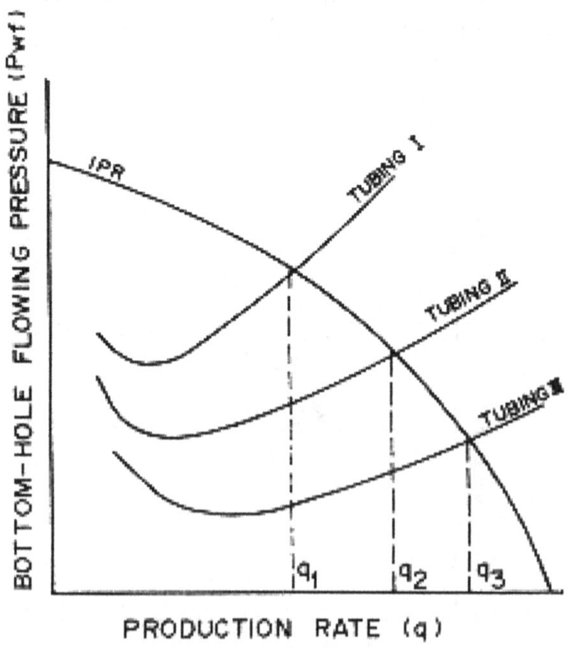

FIGURE 14.4 IPR and outflow performance for different production rate.

14.9 WORKOVER OPERATIONS

Example 14.1 (Another Case Study)

During field operations, the manager in charge is considering the purchase and the installation of a new pump that will deliver crude oil at a faster rate than the existing one.

The purchase and the installation of the new pump will require an immediate layout of $15,000. This pump, however, will recover the costs by the end of **1** year.

The relevant cash flows for the case as shown in Table 14.1.

If the oil company requires 10% minimum annual rate of return on money invested, which alternative should be chosen?

SOLUTION

The present worth method is applied in solving this problem (see Chapter 6).

Calculate the present worth for both alternatives, where:

$$\text{Present worth} = \text{Present values of cash flows, discounted at 10\%} - \text{Initial capital Investment}$$

a. For the new pump: P.V. = (190,000)/1.1 = $172,727

$$\text{Present W} = 172{,}727 - 15{,}000$$

$$= \underline{\mathbf{\$157{,}727}}$$

b. For the old pump: P.V. = (95,000)/1.1 + (95,000)/(1.1)^2

$$= 78{,}512 + 86{,}363$$

$$= \underline{\mathbf{\$164{,}875}}$$

Based on the above results, keep the old pump. It gives higher Present Value.

TABLE 14.1
Data for Example 14.1

	Year		
	0	**1**	**2**
Install new (larger pump)	–15,000	19,000	0
Operate existing (old pump)	0	95,000	95,000

TABLE 14.2
Data for Example 14.2

Year	Net Present

Example 14.2 (Case Study)

The XYZ oil production company was offered a lease deal for oil wells on which the primary reserves are close to exhaustion. The major condition of the deal is to carry out secondary recovery operation using water-flood at the end of the 5 years. No immediate payment by the XYZ Company is required. The relevant cash flows are estimated as given in Table 14.2

Example 14.3 (Case Study: Economic Evaluation of a Gas Lift)

Economic evaluation of a gas lift well: Perform an economic analysis of placing a well on gas lift given the following data:

Well depth = 8,000 ft
Reservoir pressure (PR) = 2,400 psi and decreases 100 psi for each 200,000 bbl of oil recovery
Productivity index = 4 BPD/psi (initially) and then changes as 0.00143 PR
Wellhead pressure = 120 psi (constant)
Injection gas pressure = 900 psi (from a central station)
Tubing size = 2.5 in
Oil price = $80.00/bbl
Injection cost = $0.5/MSCF
Production cost = $2.5/bbl
Maintenance cost = $1.0/bbl
Pulling the well = 600,000
New equipment = 415,000

SOLUTION

Based on the data given in Table 14.3, calculations are carried out as presented in Table 14.4

The payout period (P.O.P.) is calculated using the average annual cash flow over the 5 years period:

$$\text{P.O.P} = \text{Depreciable Capital Investment/Average Annual Cash Flow}$$

$$= 1.095 \times 10^6(\$)/5.54 \times 10^6(\$\text{per year})$$

$$= 0.1977 \text{ years}$$

$$= \mathbf{2.37\ month}$$

The return on investment, on the other hand, is = **<u>500</u>**%

TABLE 14.3
Comparison of Natural Flow and Gas Lift Wells

	Average Rate, BPD		Increased Production		Injection Gas, MMSCF/year
Year	Natural Flow	Gas Lift	Avg. Rate, BPD	Yearly bbl	
0–1	1,450	1,600	150	54,750	666
1–2	1,100	1,320	220	80,300	622
2–3	850	1,080	230	83,950	578
3–4	675	880	205	74,825	538
4–5	540	700	160	58,400	490

TABLE 14.4
Results of Calculations for Placing the Wells on Gas Lift

Year	Annual Gross Revenue × 10^6	Injection Costs ($)	(Product.+ Maint. Costs) × 10^6	Annual Net Revenue ×10^6	Net Cash ×10^6
0	—	—	—	—	−1.095
1	4.38	333	0.137	4.24	4.24
2	6.42	311	0.200	6.22	6.22
3	6.71	289	0.209	6.50	6.50
4	5.98	259	0.187	5.79	5.79
5	4.67	245	0.146	4.52	4.52

Example 14.4 (Case Study: Orit Mynde-Tullow Oil and Gas Industry)

Source T4 Case Study - May 2014 Oil and gas case - CIMA

www.cimaglobal.com › Documents › GBC › Case-Study

THIS CASE STUDY FITS CHAPTER 14.

May 2014. (*CIMA Global* Business Challenge)

This case study is concerned only with upstream operations within the oil and gas industry. Most large international oil and gas companies are known as being "integrated" because they combine upstream activities (oil and gas exploration and extraction), midstream (transportation and the refining process), and downstream operations (distribution and retailing of oil and gas products).

Industry Background

This case study is concerned only with upstream operations within the oil and gas industry. The oil and gas industry comprises a variety of types of company including the following:

- Operating companies—these hold the exploration and production licenses and operate production facilities. Most of these are the large multinational companies which are household names.

- Drilling companies—these are contracted to undertake specialist drilling work and which own and maintain their own mobile drilling rigs and usually operate globally.
- Major contractors—these are companies which provide outsourced operational and maintenance services to the large operating companies.
- Floating production, storage, and offloading vessels (FPSO's)—these companies operate and maintain floating production, storage, and offloading facilities and look like ships but are positioned at oil and gas production sites for years at a time.
- Service companies—these outsourcers provide a range of specialist support services including test drilling, divers, and even catering services for off-shore drilling facilities.
- Licenses—all companies operating in the exploration and production (E&P) sector need to have a license to operate each oil and gas field.
- Each country around the world owns the mineral rights to all gas and oil below ground or under the sea within its territorial waters. The country which owns the mineral rights will wish to take a share in the profits derived from any oil or gas produced. This generates enormous revenues for these mineral rich countries.
- When an E&P company has identified by survey work a potential site (but before any drilling has commenced) it needs to apply for a license.
- Fields into production its awareness and track record in respect of environmental issues he company's financial capacity in respect of the investment required to bring the oil and gas field into production. When an E&P company has identified by survey work a potential site (but before any drilling has commenced) it needs to apply for a license.

Licensing

- Licensing is conducted in differing ways in different areas of the world and there are a variety of alternative types of license that can be applied for.
- An E&P company could simply apply for a license to drill to identify whether an oil and gas field exists and to establish the size of it before selling the rights to another company to then apply for a production license.
- Alternatively an E&P company could apply for a production license, which allows it to drill and take the oil and gas fields into production. Licenses can be sold on to other companies but this is subject to approval by the government that had issued the license.
- The most commonly used form of licensing is through a "Production-Sharing Agreement" (PSA) license. A PSA license is where the government will take an agreed negotiated percentage share in the profits generated by the production of oil and gas, i.e., revenues from the sale of oil and gas less the amortized cost of drilling, any royalty taxes (see below) and all of the production costs.

Test Drilling

Once an oil and gas field has been test drilled to determine the proven size of oil and gas reserves, production drilling can commence. The time taken from identification of a potential oil and gas field to the start of oil being produced normally varies between 1 and 3 year.

Operation is on the Go

- Once a location has been identified and licenses obtained, then an off-shore installation is setup. Oil and gas off-shore installations are industrial "towns" at sea, carrying the people and equipment required to access the oil and gas reserves hundreds or even thousands of meters below the seabed. YJ uses outsourced drilling teams and outsourced service personnel for these off-shore installations. YJ hires mobile drilling platforms and FPSOs as the cost of owning drilling platforms is too prohibitive.
- Oil and gas fields can be classified according to the reasons for drilling and the type of well that is established.
- "Test" or "Exploration wells" are defined as wells which are drilled purely for information gathering purposes in a new area to establish whether survey information has accurately identified a potential new oil and gas reserve. Test wells are also used to assess the characteristics of a proven oil or gas reserve, in order to establish how best to bring the oil and gas into production.
- "Production wells" are defined as wells which are drilled primarily for the production of oil or gas, once the oil or gas reserve has been assessed and the size of the oil or gas reserve proved and the safest and most effective method for getting the oil or gas to the surface has been determined.

CONCLUSION

This case study is concerned only with upstream operations within the oil and gas industry. This case study gives an example of a real world project. It illustrates many steps that take place in the oil and gas industry, e.g., licensing. Oil and gas fields can be classified according to the reasons for drilling and the type of well that is established.

"Test" or "Exploration wells" are defined as wells which are drilled purely for information gathering purposes in a new area to establish whether survey information has accurately identified a potential new oil and gas reserve. Test wells are also used to assess the characteristics of a proven oil or gas reserve, in order to establish how best to bring the oil and gas into production.

The case study was presented to applicants as a sort of examination to answer.

Section III.III: Middle Stream Operations: "Surface Operations"

15 Principal Field Processing Operations and Field Facilities

15.1 FIELD PROCESSING OPERATIONS

Field processing of produced crude oil–gas mixture aims to separate the well stream into quality oil and gas saleable products in order to recover the maximum amount of each at minimum cost. Our objective in this chapter is to offer a basic understanding and to present a concise description for every processing surface unit from wellhead to finished quality stream products as shown in Figure 15.1.

15.1.1 SEPARATION OF GASES FROM OIL

The first step in processing the well stream is to separate the crude oil, natural gas, and water phases into separate streams. A gas–oil separator is a vessel that does this job. Gas–oil separators can be horizontal, vertical, or spherical.

Oil-field separators can be classified into two types based on the number of phases to separate:

- Two-phase separators, which are used to separate gas from oil in oil fields, or gas from water for gas fields
- Three-phase separators, which are used to separate the gas from the liquid phase, and water from oil

The liquid (oil, emulsion) leaves at the bottom through a level-control or dump valve. The gas leaves the vessel at the top, passing through a mist extractor to remove the small liquid droplets in the gas. Separators can be categorized according to their operating pressure. Low-pressure units handle pressures of 10–180 psi (69–1,241 kPa). Medium pressure separators operate from 230 to 700 psi (1,586–4,826 kPa). High-pressure units handle pressures of 975–1,500 psi (6,722–10,342 kPa).

Gravity segregation is the main force that accomplishes the separation, which means the heaviest fluid settles to the bottom and the lightest fluid rises to the top. The degree of separation between gas and liquid inside the separator depends on the following factors: separator operating pressure, the residence time of the fluid mixture, and the type of flow of the fluid (turbulent flow allows more gas bubbles to escape than laminar flow).

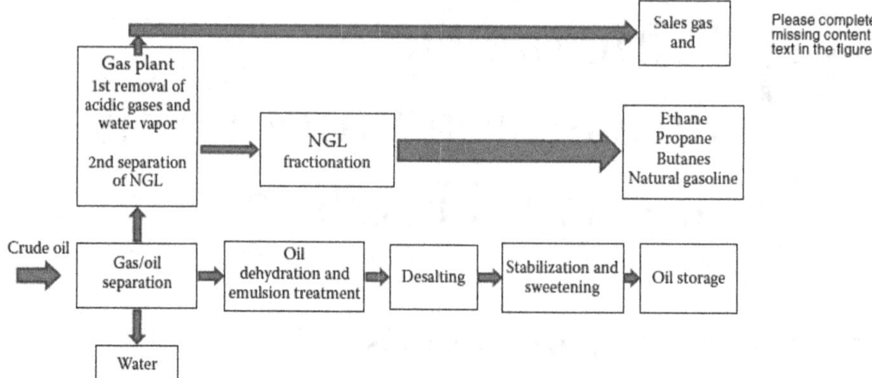

FIGURE 15.1 Overall flow diagram for crude oil processing.

15.1.2 Oil Dehydration and Emulsion Treatment

Once crude oil is separated, it undergoes further treatment steps. An important aspect during oil field development is the design and operation of wet crude handling facilities.

One has to be aware that not all the water is removed from crude oil by gravity during the first stage of gas–oil separation. Separated crude may contain up to 15% water, which may exist in an emulsified form. The objective of the dehydration step is a dual function: to ensure that the remaining free water is totally removed from the bulk of oil and to apply whatever tools necessary to break the oil emulsion. In general, free water removed in the separator is limited to water droplets of 500 μm and larger.

Produced crude oil contains sediment and produced water (BS&W), salt, and other impurities. These are readily removed from the crude oil through this stage. Produced water containing the solids and impurities is discharged to the effluent water treatment system. Clean, dehydrated oil flows from the top of the vessel. Depending on the salt specifications, a combination dehydrator followed by a desalter may be required.

A dehydration system, in general, comprises various types of equipment according to the type of treatment: water removal or emulsion breaking. Most common are the following:

- Free water knockout drum (FWKO)
- Wash tank
- Gunbarrel
- Flow treater
- Chemical injector
- Electrostatic dehydrator

It is very common to use more than one dehydrating aid, particularly for emulsion breaking. Examples are the heater-treater and chem-electric dehydrator.

The role played by adding chemicals to break emulsions should not be overlooked. These chemicals act as de-emulsifiers—once absorbed on the water–oil interface, they will rupture the stabilizing film causing emulsions.

15.1.3 Desalting

The removal of salt from crude oil is recommended for refinery feed stocks if the salt content exceeds 20 PTB (pounds of salt, expressed as equivalent sodium chloride, per thousand barrels of oil). Salt in crude oil, in most cases, is found dissolved in the remnant water (brine) within the oil. It presents serious corrosion and scaling problems and must be removed.

Electrostatic desalting, whether employed for oil field production dehydration and desalting or at oil refineries, is used to facilitate the removal of inorganic chlorides and water-soluble contaminants from crude oil. In refinery applications, the removal of these water-soluble compounds takes place to prevent corrosion damage to downstream distillation processes.

Salt content in crude oil (PTB) is a function of two parameters: the amount of remnant water in oil (R) and the salinity of remnant water (S). To put it in a mathematical form, we say:

$$\mathrm{PTB} = \mathrm{f}\ (\mathrm{R}, \mathrm{S}).$$

The electrostatic desalting process implies two important consecutive actions:

1. Wash water injection in order to increase the population density of small water droplets suspended in the crude oil (water of dilution)
2. Creating a uniform droplet size distribution by imparting mechanical shearing and dispersion of the dispersed aqueous phase (electrostatic coalescer)

15.1.4 Stabilization and Sweetening

Once degassed, dehydrated, and desalted, crude oil should be pumped to gathering facilities for storage. However, stabilization and sweetening are a must in the presence of hydrogen sulfide (H_2S). H_2S gas is frequently contained in the crude oil as it comes from the wells. It not only has a vile odor, it is also poisonous. It can kill a person if inhaled. It is also corrosive in humid atmosphere forming sulfuric acid. Pipeline specifications require removal of acid gases (carbon dioxide, CO_2) along with H_2S.

The stabilization process, basically a form of partial distillation, is a dual job process. It sweetens "sour" crude oil (removes the H_2S and CO_2 gases) and reduces vapor pressure, thereby making the crude safe for shipment in tankers. Vapor pressure is exerted by light hydrocarbons, such as methane, ethane, propane, and butane. As the pressure on the crude is light hydrocarbons vaporize and escape from the bulk of the oil. If a sufficient amount of these light hydrocarbons is removed, the vapor pressure becomes satisfactory for shipment at approximately atmospheric pressure.

15.1.5 Gas Sweetening

Having finished crude oil treatment, we turn now to the treatment and processing of natural gas. The actual practice of processing natural gas to pipeline dry gas quality levels can be quite complex but usually involves three main processes to remove the various impurities:

- Sulfur and CO_2 removal (gas sweetening)
- Water removal (gas dehydration)
- Separation of natural gas liquids (NGLs)

It should be pointed out that sweetening of natural gas almost always precedes dehydration and other gas plant processes before the separation of NGLs. Dehydration, on the other hand, is usually required for pipeline Sour natural gas composition can vary over a wide concentration of H_2S and CO_2. It varies from parts per million to about 50 volume percent. Most important are H_2S and CO_2 gases. Gas sweetening is a must for the following reasons: the corrosiveness of both gases in the presence of water; and the toxicity of H_2S gas and a heating value of no less than 980 Btu/SCF.

Some of the desirable characteristics of a sweetening solvent are:

- Required removal of H_2S and other sulfur compounds must be achieved.
- Reactions between solvent and acid gases must be reversible to prevent solvent degradation.
- Solvent must be thermally stable.
- The acid gas pickup per unit of solvent circulated must be high.
- The solvent should be noncorrosive.
- The solvent should not foam in the contactor or still.
- Selective removal of acid gases is desirable.
- The solvent should be cheap and readily available.

Amine gas treating, also known as gas sweetening and acid gas removal, refers to a group of processes that use aqueous solutions of various alkylamines (commonly referred to simply as amines) to remove H_2S and CO_2 from gases. They are known as *regenerative chemical solvents*. It is a common unit process used in refineries, and it is also used in petrochemical plants, natural gas processing plants, and other industries.

15.1.6 Gas Dehydration

Glycol dehydration is a liquid desiccant system for the removal of water from natural gas and NGLs. It is the most common and economical means of water removal from these streams. Triethylene glycol (TEG) is used to remove water from the natural gas stream in order to meet the pipeline quality standards. This process is required to prevent hydrates formation at low temperatures or corrosion problems due to the presence of CO_2 or H_2S (regularly found in natural gas). Dehydration, or water vapor removal, is accomplished by reducing the inlet water dew point (temperature at which vapor begins to condense into a liquid) to the outlet dew point temperature, which will contain a specified amount of water.

15.1.7 Recovery and Separation of Natural Gas Liquids

Although some of the needed processing of natural gas can be accomplished at or near the wellhead (field processing), the complete processing of natural gas takes place at a processing plant, usually located in a natural gas producing region. NGLs usually consist of the hydrocarbons ethane and heavier (C_2).

In order to recover and to separate NGLs from a bulk of a gas stream, a change in phase is to be induced. In other words, a new phase has to be developed for separation to take place.

Two distinctive operations are in practice for the separation of NGL constituents, dependent on the use of either *energy* or *mass* as a separating agent:

1. Energy separating agent (ESA)—Removing heat by refrigeration will allow heavier components to condense, hence a liquid phase is formed. Production of NGLs at low temperature is practiced in many gas processing plants. For example, to recover, say, C_2, C_3, and C_4 from a gas stream, demethanization by refrigeration is done.
2. Mass separating agent (MSA)—To separate NGLs a new phase is developed by using a liquid (solvent), MSA, to be introduced in contact with the gas stream, that is, absorption. This solvent is selective to absorb the NGL components.

15.1.8 Fractionation of Natural Gas Liquids

Once NGLs have been separated from a natural gas stream, they are further separated into their component parts, or fractions, using the distillation or fractionation process. This process can take place either in the field or at a terminal location hooked to a petrochemical complex. NGL components are defined as ethane, propane, butane, and pentanes plus natural gasoline.

Fractionation in gas plants has many common goals. As presented earlier in Figure 15.1, it is aimed at producing on-specification products and making sources available for different hydrocarbons. Fractionation is basically a distillation process leading to fractions or cuts of hydrocarbons. Examples of cuts or fractions are C_3/C_4, known as liquefied petroleum gas (LPG), and C_5, known as natural gasoline.

Liquid fractionation towers are used to separate and remove NGLs. They can be controlled to produce pure vapor-phase products from the overhead by optimizing the following factors:

- Inlet flow rate
- Reflux flow rate
- Rebolier temperature

- Reflux temperature
- Column pressure

15.2 FIELD FACILITIES

This include two important facilities: *Field Storage Tanks, Vapor Recovery System (VRS)*

15.2.1 Field Storage Tanks

Production, refining, and distribution of petroleum products require many different types and sizes of storage tanks. Small bolted or welded tanks might be ideal for production fields while larger, welded storage tanks are used in distribution terminals and refineries. There are many factors that should be considered in the selection of storage tanks in oil field operations.

Field operating conditions, storage capacities, and specific designs are most important. Storage tanks are often cylindrical in shape, perpendicular to the ground with flat bottoms, and with a fixed or floating roof. Atmospheric storage tanks, both fixed roof and floating roof tanks, are used to store liquid hydrocarbons in the field.

Storage tanks are needed in order to receive and collect oil produced by wells before pumping to the pipelines and to allow for measuring oil properties, sampling, and gauging.

The design of storage tanks for crude oil and other hydrocarbon products is a function of the following factors:

- The vapor pressure of the materials to be stored
- The storage temperature and pressure
- Toxicity of the petroleum material

15.2.2 Tank Classification and Types

According to the National Fire Protection Association (NFPA), *atmospheric storage tanks* are defined as those tanks that are designed to operate at pressures between atmospheric and 6.9 kPa gage. Such tanks are built in two basic designs: the cone-roof design where the roof remains fixed, and the floating-roof design where the roof floats on top of the liquid and rises and falls with the liquid level.

Pressure storage tanks, on the other hand, are used to store liquefied gases such as liquid hydrogen (LH) or a compressed gas such as compressed natural. They can be referred to as "high-pressure tanks". Storage tanks can also be classified as aboveground storage tanks (AST) and underground storage tanks (UST).

There are usually many environmental regulations and others that apply to the design and operation of each category depending on the nature of the fluid contained within.

As far as the types of storage tanks there are four basic types of tanks that are commonly used to store crude oil and its products:

- Floating Roof Tanks
- Fixed Roof Tanks
- Bullet Tanks
- Spherical Tanks (Storage Spheres)

15.2.3 Vapor Recovery System

In production operations, underground crude oil contains many lighter hydrocarbons in solution. When oil is brought to the surface it experiences drastic pressure drop by going through the gas–oil separator plant (GOSP). The evolution of hydrocarbon vapors is dependent on many factors:

- The product's physical characteristics
- The operating pressure of upstream equipment
- Tank storage condition

During storage, light hydrocarbons dissolved in the crude oil or in the condensate, including methane, other volatile organic compounds (VOCs), and hazardous air pollutants (HAPs), vaporize or flash out. These vapors collect in the space between the liquid and the fixed roof tank. As the liquid level in the tank fluctuates, these vapors are vented to the atmosphere, or flared. Alternatively, a vapor recovery compressor (or blower) may be installed to direct vapors vented from storage to downstream compressors for sales or injection. Significant economic savings are obtained by installing vapor recovery units (VRUs) on the storage tanks. These units are capable of capturing about 95% of the vapors. Losses of dissolved light are identified as:

- Flash losses due in the GOSP
- Working losses due change in the fluid level inside the tank during pumping, filling, or emptying
- Standing or breathing losses that occur with daily and seasonal temperature changes

Vacuum relief valves are needed to keep a vacuum from occurring because of tank breathing and pumping operations. If a vacuum develops, the tank roof will collapse.

15.2.4 Economic and Environmental Benefits

VOC and HAP emissions to the atmosphere cause pollution of the air we breathe. These emissions can be controlled by either destruction or by recovery using VRUs. VRUs are designed to comply with the U.S. Environmental Protection Agency (EPA)

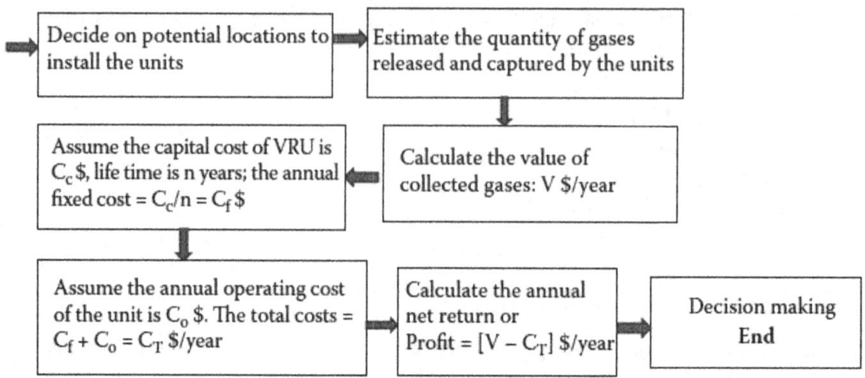

FIGURE 15.2 The economic return of installing VRU system.

standards, provide economic profits to the oil and gas producers, and eliminate of stock vapors to the atmosphere. Waste gas is the lost product, hence a lost revenue. Gases flashed from crude oil or condensate and captured by VRUs can be sold at profit or used at an oil field facility. Options for utilizing the recovered gases are to be used as a fuel for oil field operations, to be collected to natural gas gathering stations and sold, or to be used as a stripping agent.

In order to estimate the economic return when installing a VRU, one should follow the procedure as shown in Figure 15.2.

16 Fluids Separation

BACKGROUND

The main purpose of any oil and gas production facility is to separate the well effluent that consists of oil, water, and gas produced into their original phases. This is achieved by a stepwise reduction in pressure down to atmospheric pressure; using gas-oil-separation plant (GOSP), flashing off the gas and then dehydrating the crude oil to meet its export specification of less than 0.5% water in oil.

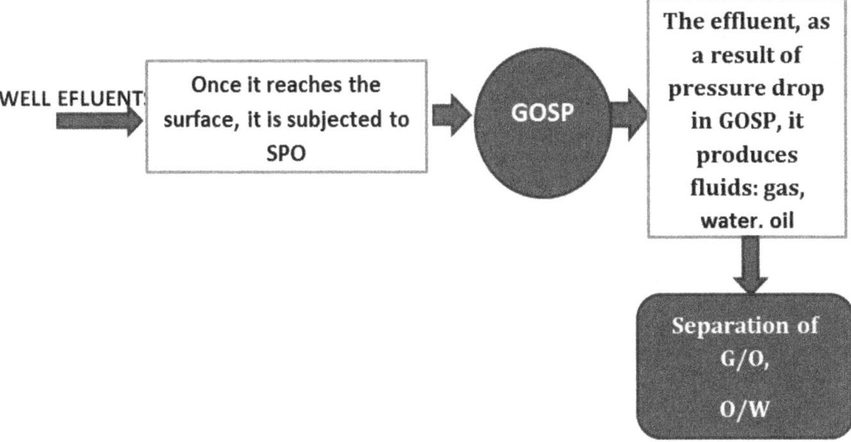

In the two-phase units, gas is separated from the liquid with the gas and liquid being discharged separately. Oil and gas separators are mechanically designed such that the liquid and gas components are separated from the hydrocarbon steam at specific temperature and pressure.

Separation of hydrocarbon liquids and gasses from water and sediments is a challenging operation.

Based on the configuration, the most common types of separator are horizontal, vertical, and spherical. Large horizontal gas–oil separators are used almost exclusively in processing well fluids in the Middle East, where the gas–oil ratio of the producing fields is high. Multistage GOSPs normally consist of three or more separators.

16.1 THE SEPARATION PROCESS: INTRODUCTION

Well effluents flowing from producing wells are usually identified as turbulent, high velocity mixtures of gases, oil, and salt water. As these streams flow reaching the surface, they undergo continuous reduction in temperature and pressure forming a two-phase fluid flow: gas and liquid. The gathered fluids emerge as a mixture of crude oil and gas that is partly free and partly in solution. They must be separated into their main physical components, namely: oil, water, and natural gas. The *separation system* performs this function which is usually made up of a free water knock-out (FWKO), flow line heater, and gas-oil (two-phase) separators, or gas-oil-water (three-phase separators). Gas-oil separators work on the principle that the three components have different densities, which allow them to stratify when moving slowly with gas on top, water on the bottom, and oil in the middle.

The physical separation of these three phases is carried out using what is called stage separation in which a series of separators operating at consecutively reduced pressures are used. The purpose of stage separation is to obtain maximum recovery of liquid hydrocarbons from the fluids coming from the wellheads and to provide maximum stabilization of both the liquid and gas effluents, as shown in Figure 16.1.

Case studies presented in the chapter include finding the "Optimum Separating Pressure for Three Stage Separators" and to investigate the causes of tight emulsions in GOSPs.

The process involved in a gas-oil separator encompasses two main stages in order to free oil from gas. These are recognized as: flash separation of the gas-oil mixture followed by oil recovery.

16.1.1 Flash Separation

In order to understand the theory underlying the separation of well-effluents of hydrocarbon mixtures, it is assumed that such mixtures contain essentially three main groups of hydrocarbons (Table 16.1).

- Light group, which consists of methane (CH_4) and ethane (C_2H_6)
- Intermediate group, which consists of two subgroups: propane (C_3H_8)/butane (C_4H_{10}) and pentane (C_5H_{10})/hexane (C_6H_{12})
- Heavy group, which is the bulk of crude oil and is identified as C_7H_{14+}

Basically, our objective in separating the gas-oil mixture is a dual function:

a. To get rid of all C_1 and C_2, i.e., light gases
b. Save the heavy-group components as our liquid product

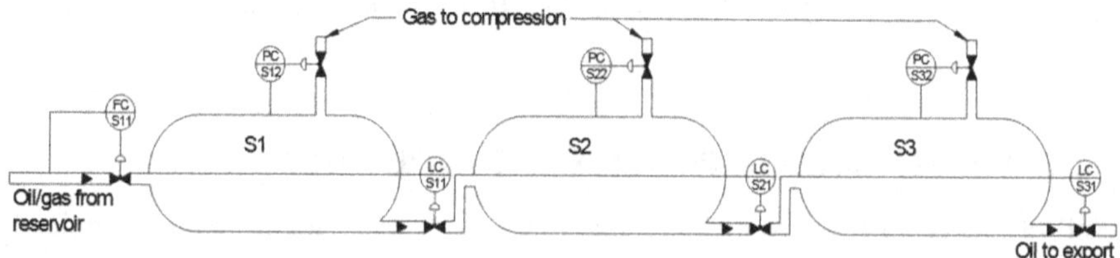

FIGURE 16.1 Three stage GOSP.

TABLE 16.1
Constituents of Crude Oil and Natural Gas

Hydrocarbons / Identification of the constituents			(i) in the field streams			(ii) as commercial products
Name	*Formula*	*Normal B.P. (°F)*	*Liquid Phase (at normal conditions)*	*Two Phases*	*Gaseous Phase (and liquefied gases)*	
Methane	CH_4	-259		Field separator gas (Methane–Octane); Gas condensate well effluent (Methane–Hexadecane); Crude oil well effluent (Methane–Asphaltene)	LNG (Methane–Ethane); Dry gas (Methane–Propane)	Natural Gas
Ethane	C_2H_6	-128			NGL (Ethane–Octane)	Natural Gas
Propane	C_3H_8	-44			LPG (Propane–n-Butane); (Propane–Octane)	Natural Gas, propane
Isobutane	i-C_4H_{10}	+11	Stock tank crude oil (Isobutane–Asphaltene); Stock tank condensate (Isobutane–Hexadecane); Natural gasoline (Isobutane–Octane)			Natural Gasoline, butane
n-Butane	n-C_4H_{10}	31				Natural Gasoline, motor fuel, butane
Pentanes	C_5H_{12}	90	Debutanized condensate (Pentanes–Hexadecane)			Natural Gasoline, motor fuel
Hexane	C_6H_{14}	145				Natural Gasoline, motor fuel
Heptane	C_7H_{16}	195				Natural Gasoline, motor fuel
Octane	C_8H_{18}	245				Natural Gasoline, motor fuel
Decanes	$C_{10}H_{22}$	345				Motor fuel
Tetradecane	$C_{14}H_{30}$	490				Kerosene, light furnace oil
Hexadecane	$C_{16}H_{34}$	549				Mineral seal oil, furnace oil
Triacontane	$C_{30}H_{62}$	855				Light lubricating oil, heavy fuel oil
Tetracontane	$C_{40}H_{82}$	1012				Lubricating oil, heavy fuel oil
Asphaltene	$C_{80}H_{162}$+	1200				Asphalt, road oil, bunker fuel oil

*LPG = Liquefied Petroleum Gases.
NGL= Natural Gas Liquids (normally C_3^+).
LNG= Liquefied Natural Gas.

In order to accomplish these objectives, we unavoidably loose part of the intermediate group in the gas stream, whose heavier components (C_5/C_6) would definitely belong to the oil product.

The problem of separating gases in general from crude oil in the well-fluid effluents breaks down to the well-known problem of flashing a feed mixture into two streams: vapor and liquid. This takes place using a flashing column (a vessel without trays). Gases liberated from the oil are kept in intimate contact. As a result, thermodynamic equilibrium is established between the two phases. This is the basis of flash calculations, which is carried out to make material balance calculations for the flashing streams.

16.1.2 Oil Recovery

Once flashing takes place, our concern centers next on recovering the crude oil. The effective method used implies two consecutive steps:

a. To remove oil from gas: Here, we are primarily concerned in recovering as much oil as we can from the gas stream. Density difference or gravity differential between oil and gas is the first means to accomplish separation at this stage. At the separator's operating condition of high pressure, this difference in density becomes large (gas law); and the oil is about eight times as dense as the gas. This could be a sufficient driving force for the oil particles to settle down and separate. This is true for large size having diameter of 100 microns or more. For smaller ones, mist extractors are needed.

Other means of separation would include change of velocity of incoming flow, impingement, and the action of centrifugal force. These methods would imply the addition of some specific designs for the separator to provide the desired method for achieving separation.

b. To remove gas from "locked" oil: The objective here is to recover and collect any non-solution gas that may be entrained or "locked" in the oil. The recommended methods are: settling, agitation, and applying heat chemicals.

16.2 GAS-OIL SEPARATOR AND CONTROL DEVICES

Regardless of their configurations, gas-oil separators usually consist of four functional sections:

Section A: Initial separation takes place in this section at the inlet of the separator. It is used to collect the entering fluid.

Section B: It is designated as the gravity settling section through which the gas velocity is substantially reduced allowing for the oil droplets to fall and separate.

Section C: Is known as the mist extraction section. It contains woven-wire mesh pad, which is capable of removing many fine droplets from the gas stream.

Section D: Is the final component in a gas-oil separator. Its main function is to collect the liquid recovered from the gas before it is discharged from the separator.

In addition to these main components, gas-oil separators normally include the following control devices:

- Oil level controlling system that consists of oil level controller (OLC) plus an automatic diaphragm motor-valve on the oil outlet. In case of a 3-phase separator, additional system is required for the oil-water interface. Thus, a liquid level controller plus a water discharge control valve is needed.
- An automatic back-pressure valve on the gas stream leaving the gas-oil vessel to maintain a fixed pressure inside it.
- Pressure relief devices.

These control devices are shown in Figure 16.2.

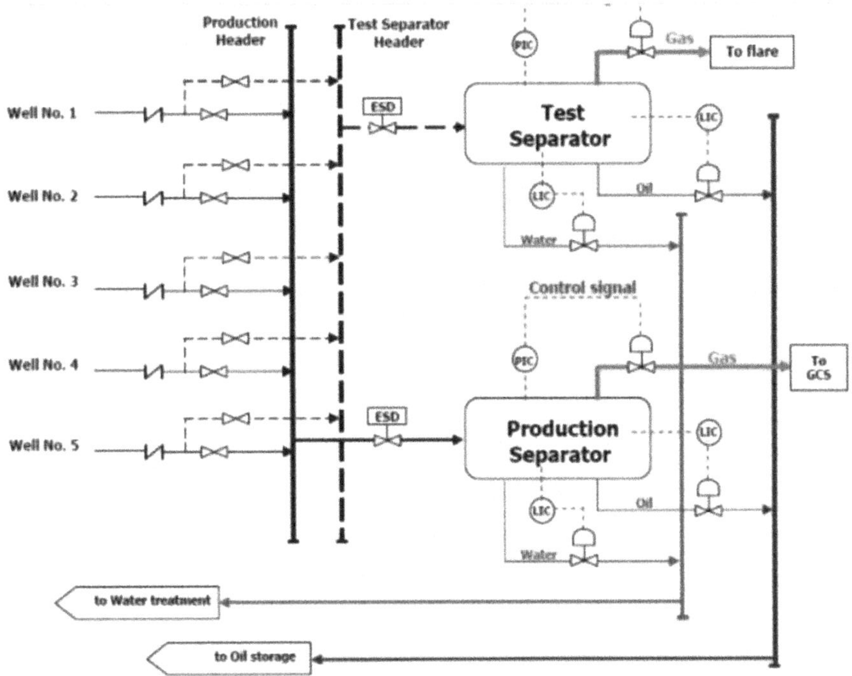

FIGURE 16.2 Gas oil separator fully automated.

16.3 METHODS AND EQUIPMENT USED IN SEPARATION

In the separator, crude oil separates out, settles, and collects in the lower part of the vessel. The gas lighter than oil fills the upper part of the separator. Crude oil with high gas-oil ratio (GOR) must be admitted to two or three stages as indicated in Figure 16.3. Movement of crude oil from one separator to the next takes place under the driving force of the flowing pressure. Pumps are needed for the final trip to transfer the oil to its storage tank.

The essential characteristics of a gas-oil separator are:

i. To cause a decrease in the flow velocity, permitting separation of gas and liquid by gravity.
ii. To operate at temperature above the hydrate point of the flowing gas.

The conventional method using multistage flash separators is recommended for relatively high pressure high GOR fluids. Separation takes place in a stage by what is known as Flash Distillation (unit operation). Generally speaking, the number of stages is a strong function of: the API gravity of oil, GOR, and flowing pressure. Based on configuration, three types of separators are known: horizontal, vertical, and spherical. It is most common to see large horizontal gas-oil separators used in processing well fluids in the Middle East, with three or more separators.

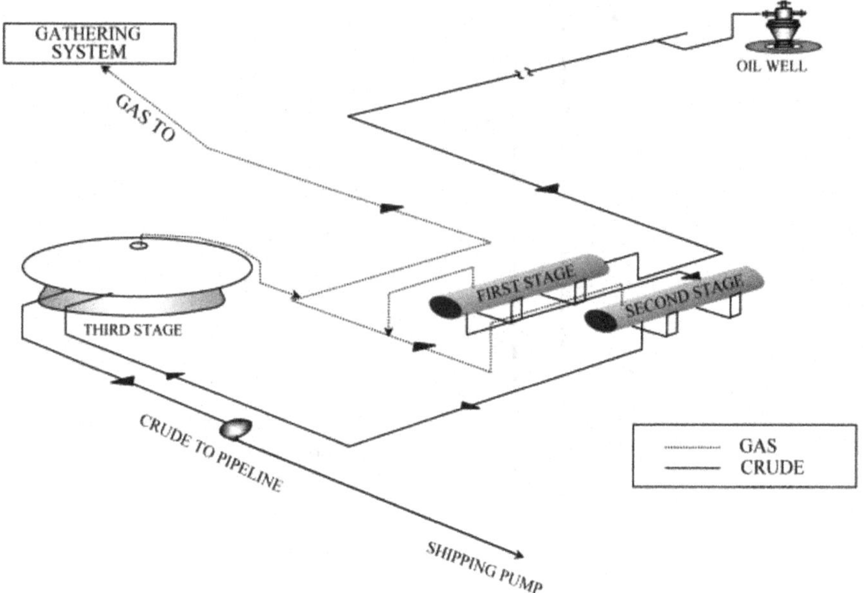

FIGURE 16.3 Flow of crude oil from well through GOSP.

The need for what is called "Modern GOSP" may arise as the water content of the produced crude increases. The function of such set-up is a multipurpose one. It will separate the hydrocarbon gases from oil. Remove water from crude oil. Finally, it will reduce salt content to the acceptable limits. Three phase separators are common in many fields in the Middle East.

If the effect of corrosion due to high salt content in the crude is recognized, then modern desalting equipment could be included as a third function in the GOSP design.

The functions of a modern GOSP could be summarized as follows:

- To separate the hydrocarbon gases from crude oil and remove water from crude oil.
- To reduce the salt content to the acceptable level (BS&W).

A GOSP can function according to one of the following process operations:

- Three-phase, gas–oil–water separation. Read also Three-Phase, oil–water–gas separators
- Two-phase, gas–oil separation. Read also Two-Phase, gas–oil separators
- Two-phase, oil–water separation
- De-emulsification
- Washing
- Electrostatic coalescence

16.4 DESIGN EQUATIONS FOR SIZING GAS-OIL SEPARATORS

Before presenting the design equations, it is necessary to state first some basic fundamentals and assumptions relevant to the sizing of gas-oil separators:

Fundamentals:

- The difference in densities between the liquid and gas is taken as a basis for calculating the gas capacity.
- In the gravity settling section, liquid drops will settle at a velocity determined by equating the gravity force acting on the drop with the drag force caused by its motion relative to the gas phase.
- A normal retention time to allow for the gases to separate from oil is considered to be between **30 seconds and 3 minutes.** Normally retention time is defined as the residence time or the time for a molecule of liquid is retained in the vessel.

Mathematically: Retention time = Volume of vessel/Liquid flow rate

- For vertical separators, liquid particles (oil) separate by settling downward against upflowing gas stream; while for horizontal ones liquid particles assume a trajectory-like path, while it flows through the vessel.
- For vertical separators, the gas capacity is proportional to the cross-sectional area of a separator; while for a horizontal one the gas capacity is proportional to the area available for disengagement. The volume of accumulation of either type will be the determining factor for the liquid capacity.

Assumptions:

- No oil foaming takes place during the gas-oil separation (otherwise retention time should be increased to 5–20 minutes).
- The cloud point of the oil and hydrate point of the gas are below the operating temperature of 60°F.
- The smallest separable liquid drops are spherical ones having diameter of 100 microns.
- Liquid carryover with separated gas does not exceed 0.10 gallon/MMSCF.

Sizing of gas-oil separators requires the calculation of two parameters:

- The oil capacity, a separator can handle
- The gas capacity to be processed by a separator

The equations needed to calculate the oil capacity and gas capacity are presented as follows:

$$\textbf{The rated oil capacity, } q = [50.54\ d^2 L]/t \textbf{ bbl/day} \tag{16.1}$$

where d is inside diameter of the vessel in ft, L is the shell height in ft, t is the retention time in minutes.

$$\textbf{The gas capacity, } Q = 86400[C_1C_2C_3 / z].A \text{ SCF / day} \tag{16.2}$$

where: $C_1 = [P_f/T_f].[520/14.7]$, C_2 = difference in densities of oil & gas/density of gas, C_3 = separation coefficient of the vessel with typical values of 0.167 and 0.5 for vertical and horizontal separators respectively, z is the gas compressibility factor, P_f and T_f designates the flowing pressure and flowing temperature respectively.

It is to be noted that Equation (16.1) is applicable for horizontal separators. Equation (16.2), on the other hand, applies for both horizontal and vertical separators depending on the value of A.

For horizontal, A = ½ the cross section area, while for vertical, A = the entire cross section = $\prod/4D^2$

Equation (16.2) relates the gas capacity of gas-oil separator, Q, to the corresponding cross section area, A. This enables finding the diameter of a separator needed to handle a given input of a gas flow rate.

ECONOMIC EVALUATION AND APPLICATION

16.5 PROCESS ECONOMICS AND DESIGN PARAMETERS

As we have seen earlier, GOSPs are needed for environmental reasons. It is not appropriate to burn off the gases associated with crude oil. The economic reasons for processing and treating the produced crude are obvious. Recovering associated gases prevents wasting a natural resource, which was originally flared off. There are also other economic reasons for using GOSP. Removing contaminants from the crude, such as salt and hydrogen sulfide, protects plants from corrosion damage caused by corrosion.

During crude-oil processing at the GOSP, one of the most important variables that determines the efficiency of oil/water/gas separation is the tightness of the incoming emulsion. The tighter the emulsion, the higher the dosage of demulsifier needed to break them. The performance of the GOSP is closely tied to the characteristics of the feed emulsions.

Another aspect of GOSP performance is related to the process facilities (hardware) and process variables. The hardware includes the number and type of separators, dehydrators and desalters, water/oil separators (WOSs), and other hardware at the GOSP. Process variables include oil and water-flow rates, temperatures, water cuts, and GOSP operating conditions. A higher residence time of fluids in the GOSP will generally lead to better separation and better performance, all other variables being constant. Besides the residence time, process retrofits in the vessels also tend to enhance performance.

Usually it is most economical to use three to four stages of separation for the hydrocarbon mixture. Five or six stages may payout under favorable conditions, when, for example, the incoming wellhead fluid is found at very high pressure. However, the increase in liquid yield with the addition of new stages *is not linear*. For instance, the increase in liquids gained by adding one stage to a single-stage system is likely to be substantial. However, adding one stage to a three- or four-stage system is not as likely to produce any major significant gain. In general, it has been found that a three-stage separating system is the most cost effective.

The following parameters are detrimental in evaluating the performance and the economics of GOSP:

- Optimum separation conditions: separator pressure and temperature
- Compositions of the separated gas and oil phases
- Oil formation volume factor
- Product GOR
- API gravity of the stock tank oil

16.6 CASE STUDY 1: OPTIMUM SEPARATING PRESSURE FOR THREE STAGE SEPARATORS

Objective: *Optimizing the gas-oil separation facility in order to find the optimal conditions of pressure and temperature under which we would get the most economical profit from the operation.*

16.6.1 Process Description

It is assumed that we have three separators: high, intermediate, and low pressure separators. It is the pressure of the second stage (intermediate) that could freely be changed and optimized. The pressure in the first separator (high pressure), on the other hand, is usually kept fixed either to match the requirement of a certain pressure gas injection facilities, or to meet a sale obligation through a pipe line, or it is the flow conditions of the incoming feed line. Similarly, the pressure in the third separator (low pressure) is fixed; usually it is the last stage functioning as the storage tank.

The optimum pressure is defined as the one that gives the desired separation of gases from crude oil, with the maximum recovery of oil in the stock tank. Under these conditions, we should have minimum gas/oil ratio.

If R designates the recovery of the oil and is defined:

R = O/G of oil per SCF gas, then the optimum operating pressure in the 2nd stage, $(P_2)_O$ should be the value that makes R maximum; or 1/R is minimum.

16.6.2 Approach

The method depends on using a pilot unit to do experimental runs, in which he pressure in the 2nd stage is to be changed from run to run. A sample of the gases leaving the three separators is to be analyzed for the content of some key component, say C_5+. It is established, therefore, to minimize the loss of C_5+ in the gas stream separated from the crude oil.

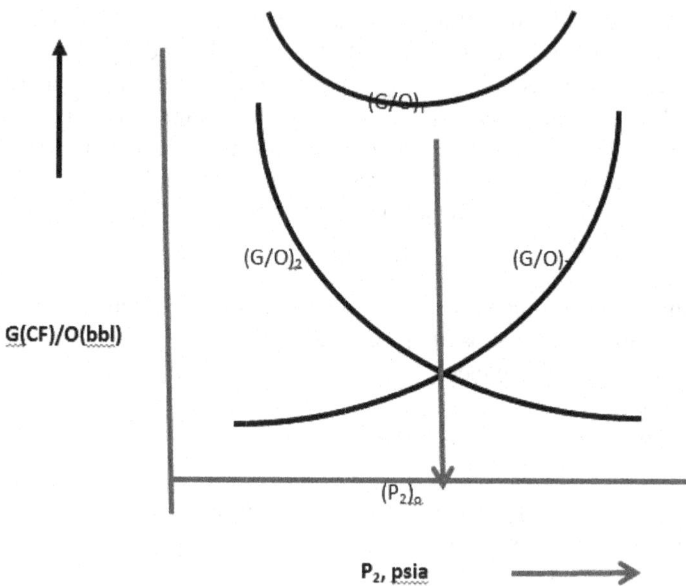

FIGURE 16.4 Variation of (G/O) with P_2.

The experimental runs will look as follows:

Run No	P_2 [psi]	$(G/O)_2$ [scf/bbl]	$(G/O)_3$ [scf/bbl]
1	—	—	—
2	—	—	—

The change in (G/O) for both separators with P_2 is plotted as shown in Figure 16.4. It is seen that with the increase in P_2, $(G/O)_2$ decreases indicating more condensation of heavier hydrocarbons. On the other hand, increasing P_2 will increase $(G/O)_3$, because the pressure difference between stages 2 and 3 will increase causing more hydrocarbons to vaporize from stage 3. The cumulative sum of $(G/O)_2$ plus $(G/O)_3$, named $(G/O)_T$ is plotted against P_2.

It is concluded right away, that the value of $(P_2)_O$ corresponds to the minimum $(G/O)_T$.

This minimum $(G/O)_T$ leads to 1/R or $(O/G)_T$, the maximum oil recovery, bbl per SCF of gas separated.

16.6.3 Conclusion

This optimization approach, would lead us to calculate the value of oil revenue for the system, by simply using the following formula:

Target Profit from Oil Sales \$/day = (\$/bbl) [price of oil] $(O/G)_T$[bbl oil/SCF] (Q) [SCF/day]

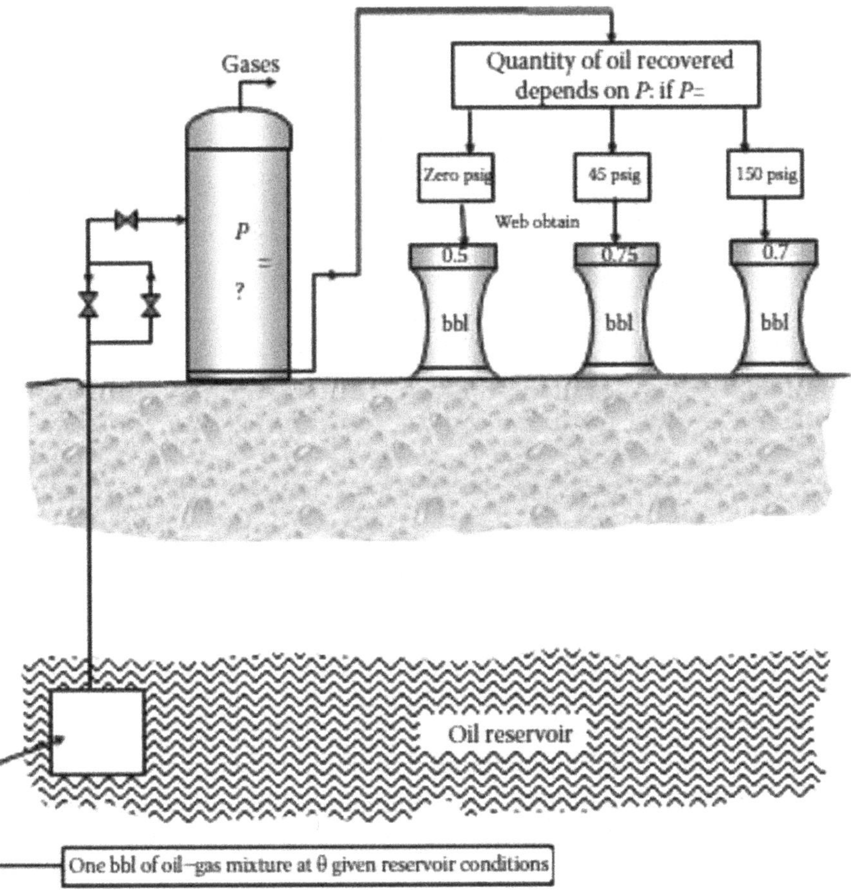

FIGURE 16.5 Effect of pressure in GOSP on crude oil yield.

The effect of operating pressure in gas/oil separation on crude yield has to be taken into consideration as indicated in Figure 16.5.

CASE STUDY: CAUSES OF TIGHT EMULSIONS IN GAS OIL SEPARATION PLANTS

INTRODUCTION

Problem Number 16.7 Case Study: Causes of Tight Emulsions in Gas Oil Separation Plants

The giant *Ghawar* field in Saudi Arabia has several wet crude handling facilities referred to as gas oil separating plants or GOSPs, located at Mubarraz area (These GOSPs process Arabian Light crude and their primary function is to separate oil, water, and gas.

Ghawar Field

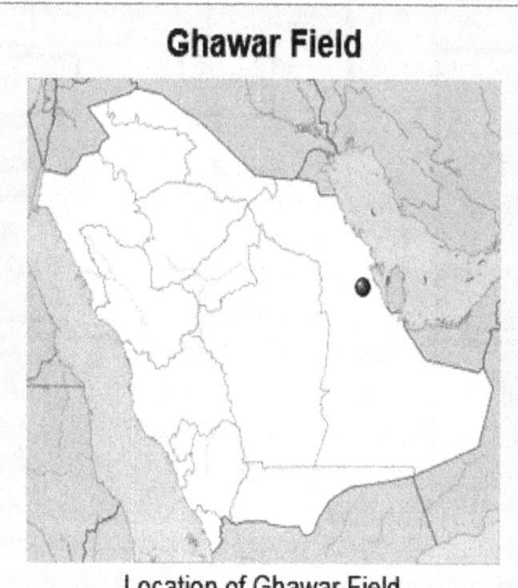

Location of Ghawar Field

Country	Saudi Arabia
Region	Eastern Province
Location	Al-Ahsa
Offshore/onshore	Onshore
Coordinates	25.43°N 49.62°E
Operator	Saudi Aramco

Saudi Arabia's Key Oil Fields

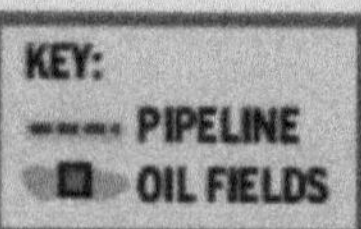

OIL FIELD	PRODUCTION CAPACITY	RESERVES
1. Ghawar	5 MILLION BBL./DAY	70 BILLION BBL
2. Safaniya	1.2 MILLION BBL./DAY	35 BILLION BBL
3. Shaybah	0.55 MILLION BBL./DAY	15.7 BILLION BBL

Data: Saudi Aramco, Wood MacKenzie

OBJECTIVE

To evaluate the relative performance of de-emulsifiers and to optimize their usage in GOSPs while meeting crude and water specifications. The case study was brought to the attention of the Ch.E. department at KFUPM, Saudi Arabia.

PROCESS DESCRIPTION

Formation of emulsions during oil production is a costly problem, both in terms of production losses and chemical costs. In these days of high oil prices and the need to reduce production costs, there is an economic necessity to control, optimize, or eliminate the problem by maximizing oil-water separation.

Analysis of crude oils from wells in Ghawar indicates that these oils are produced in the form of tight water-in-oil emulsions. Tight or strong emulsions are difficult to separate and cause production and operational problems. These problems have led, at times, to an increase in de-emulsifier usage, production of off-spec crude, and occasionally caused equipment upsets in the GOSP.

The main causes of emulsion problems are (a) The presence asphaltenes and fine solids in the crude, (b) lower temperatures in the winter time, and (c) an increase in water production.

APPROACH

In this case study of tight emulsions, once you have collected all the positive and negative factors and have quantified them you can put them together into an accurate Cost-Benefit analysis.

On the cost side, one can envisage the following:

- Cost of de-emulsifier.
- The addition of asphaltenes dispersants and surfactants to the crude oil.
- Using elaborate techniques to quantify the oil-water separation process, such as Emulsion Separation Index (ESI) (method developed by Suadi Aramco).

On the benefit side, we get:

- A reduction in the quantity of de-emulsifiers used
- Less production losses
- Less operation problems
- An increase in oil revenue
- Fast rate of separation in the GOSP, which gives less residence time. This reduces the diameter of the separator

17 Operations Handling Crude Oil: Treatment, Dehydration, and Desalting

OVERVIEW

This chapter deals first with the *dehydration stage* of crude oil to free it from the emulsified water. Depending on the original water content of the oil as well as its salinity, oil field treatment could produce oil with a remnant water content of 0.2–0.5 of 1%.

The next stage in the treatment process of crude oil is *desalting*. The removal of salts found in the form of what we may name it "*remnant brine*" is carried out in the desalting process. This will reduce the salt content in the crude oil to the acceptable limits of 15–20 PTB (Pounds per Thousand Barrel). After treating the oil by the dehydration and the desalting process the possibility of stabilizing the crude oil and sweetening exists in the case of sour oil. This represents the final stage in this chapter. Figure 17.1 represents this process.

17.1 INTRODUCTION

Oil leaving the gas-oil separators may or may not meet the purchaser's specifications. As presented in Chapter 15, associated gas and most of the free water in the well stream are removed in the separators. The free water separated is normally limited to water droplets of 500 μm and larger. Oil stream leaving the separators would normally contain water droplets of smaller size along with water emulsified in the crude oil.

Dewatering, or dehydration, followed by desalting of crude oil upstream of crude distillation unit is considered a key process operation for the removal of saline water, salts, and other contaminants from crude oil before it reaches any major unit operation. As stated above, dehydration of crude oil is simply to free it from the emulsified water.

The following steps are part of the dehydration process of crude oil.

17.1.1 How Emulsion Is Formed?

Crude oil emulsions form when oil and saline water (brine) come into contact with each other, when there is sufficient mixing, and when an emulsifying agent or emulsifier is present. The amount of mixing and the presence of emulsifier are

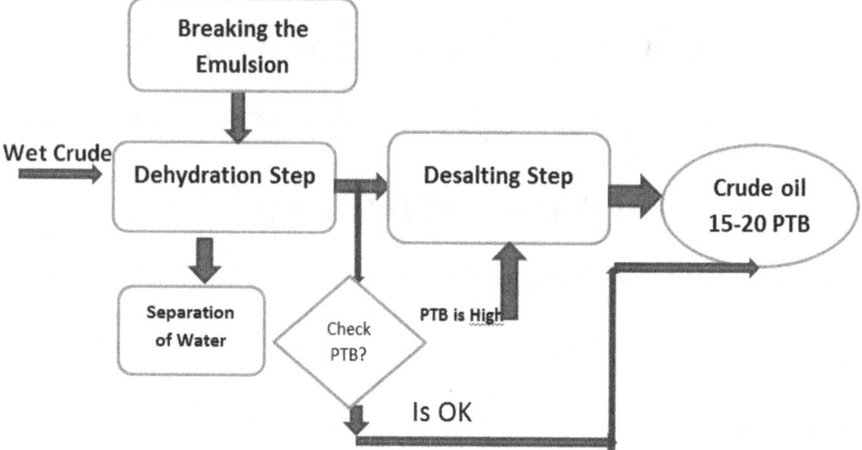

FIGURE 17.1 Flow diagram for the treatment of wet crude oil.

critical for the formation of an emulsion. During crude oil production, there are several sources of mixing, often referred to as the amount of shear, including flow through reservoir rock; flow through tubing, flow lines all the way to reach the surface equipment.

When it comes to emulsifiers, the presence, amount, and nature of the emulsifier agents determines, to a large extent, the type and tightness of an emulsion.

Produced oilfield water-in-oil emulsions contain oil, water, and an emulsifying agent. Emulsifiers stabilize emulsions and include surface-active agents and finely divided solids. Figure 17.2 depicts water-in-oil emulsion.

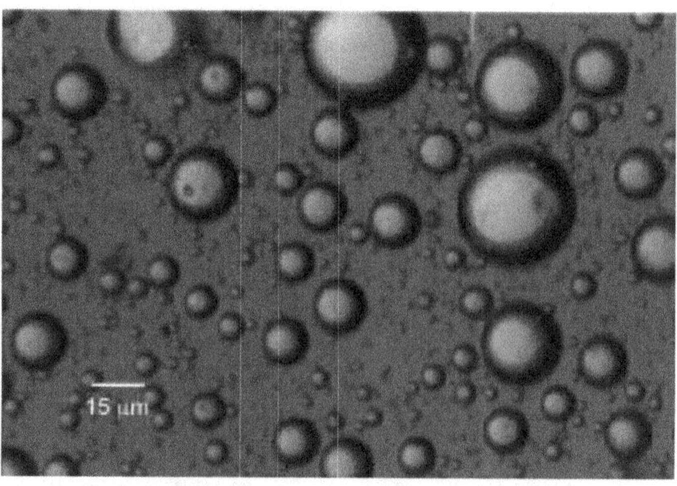

FIGURE 17.2 Water-in-oil emulsions.

17.1.2 Emulsion Treatment

The resolution of emulsified oil follows a three-step procedure:

a. To reduce or to rupture the stabilizing films surrounding the water droplets. This step may be named destabilization process, which could be effectively carried out by adding chemicals and heating the emulsified oil.
b. Coalescence of the liberated water droplets occur forming larger drops of water. This process is enhanced by electric field and heating as well. It is also a function of residence time in the vessel.
c. Gravitational settling with subsequent separation of water drops from oil (time element).

Basically, water-oil emulsions are resolved by treatment equipment utilizing a combination of two or more of the dehydration aids, known as:

- Heating indispensible variable or element. It is the element that determines the size of the equipment, which in turn determines its cost.

17.1.3 Heating

The most pronounced effect is the reduction of oil viscosity. In addition, other advantages are contributed to heat. These are:

a. An increase in the difference in specific gravity between oil and water.
b. An increase in the droplet size as demonstrated by its molecular movement which enhances coalescence.
c. Heat will help destabilization of the emulsifying film.

On the other hand, heat has some disadvantages such as:

a. Loss of valuable hydrocarbons
b. Consumption of fuels for heaters
c. Heating equipment is costly
d. Gases liberated during heating will add additional problems in handling and represents a safety hazards

Field heaters are of two types:

- Direct, in which the crude oil is passed through a coil exposed to the hot gases used as a fuel.
- Indirect, in which water is used as a transfer medium for heat from hot flue gases to the oil to be heated and immersed in the water. Both methods are illustrated as shown in Figure 17.3. Examples of some industrial field heaters are: line heaters, wash tanks, gun-barrel treaters.

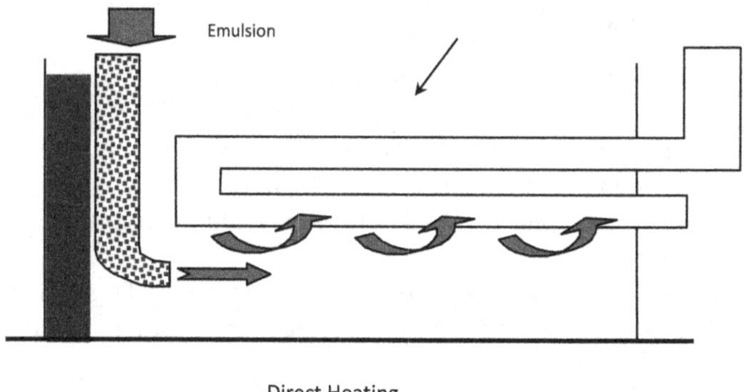

Emulsion

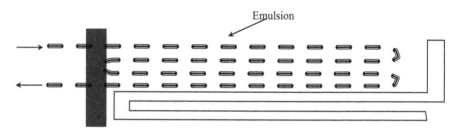

Indirect Heating

FIGURE 17.3 Methods of heating oil emulsions.

17.1.4 Chemical Treating

Chemical additives function to break crude oil emulsions by adding agents comprising high molecular weight polymers adsorbed at the water-oil interface. These chemicals (called de-emulsifiers) can either rupture the film and/or displace the stabilizers due to reduction in surface tension on the inside of the film. They are complex organic compounds with surface active characteristics such as: sulfonates, polyglyocol esters, polyamine compounds, and many others. They are usually added using a small chemical pump up-stream of the choke. Dosage is estimate to be about 1.0 quart of the chemical for each 100 barrels of oil.

The principle of breaking oil-water emulsions using electric current, which is known as electrostatic separation will be discussed next in the desalting of crude oil.

17.2 DESALTING OF CRUDE OIL

17.2.1 Introduction

In the desalter, the crude oil is heated and then mixed with 5–15% volume of fresh water so that the water can dilute the dissolved salts. The oil-water mix is fed into a settling tank to allow the salt-containing water to separate and be drawn off. Frequently, an electric field is used to encourage water separation.

The removal of salt from crude oil for refinery feed stocks is implied by most of the refiners, particularly if the salt content exceeds the range 15–20 PTB. Values for the salt content of some typical crude oils could be as low as 8–10 PTB for Middle East, while it could reach a high value above 70 PTB for Oklahoma, USA. Crude oil arriving from oilfield generally contains 1% or more of saline water and organic salts. The salinity of the water could be in the range of 15,000–30,000 ppm (parts per million) or even much higher. Part of the salts contained in the crude oil, particularly magnesium chloride, is hydrolyzed at temperatures above 120°C. Upon hydrolysis, the chlorides get converted into hydrochloric acid and corrode the distillation column's overhead and the condensers.

The most economical place for desalting is the refinery. But, in many situations, when marketing and/or pipeline requirements are imposed, field treatment is applied. Principles stay the same, using unit-operations fundamentals.

Salt in crude oil is in most cases found dissolved in the remnant water within the oil. It is evident that the amount of salt found in crude oil depends on two factors:

- The quantity of remnant water that is left in oil after normal dehydration.
- The salinity or the initial concentration of salt in the source of this water.

Economically, there is limit on reducing the salinity by lowering the quantity of remnant water, by dehydration only. The other alternative is to substantially decrease salt content of the remnant water by mixing it with water with much less concentration of salts in it. This is what we accomplish in desalting of crude oil.

17.2.2 Description of Desalting Process

The desalting process involves basically two steps as given in Figure 17.4. The first is adding fresh water to the crude oil, to be thoroughly mixed. This is followed next by separation or a dehydration step. In other words, you think of the process as "washing" the salty crude oil with water followed by separating the water phase from crude oil.

17.3 BASIC CONCEPTS IN CRUDE OIL DESALTING

The mixing step in the desalting step is normally accomplished by pumping the crude oil and wash water, each separately through a mixing device (could be throttling valve or orifice plate mixers). In the electrical desalting process, a high potential field (16,500–33,000 volt) is applied across the settling vessel to help the coalescence of water drops. In the process, 2–5% by volume of water is emulsified in the untreated crude oil, heated to a temperature of 180–300°F. In the desalting process, it is a common practice to apply pressure to suppress losses of hydrocarbons from the oil. Pressure used is normally in the range of 50–250 psi.

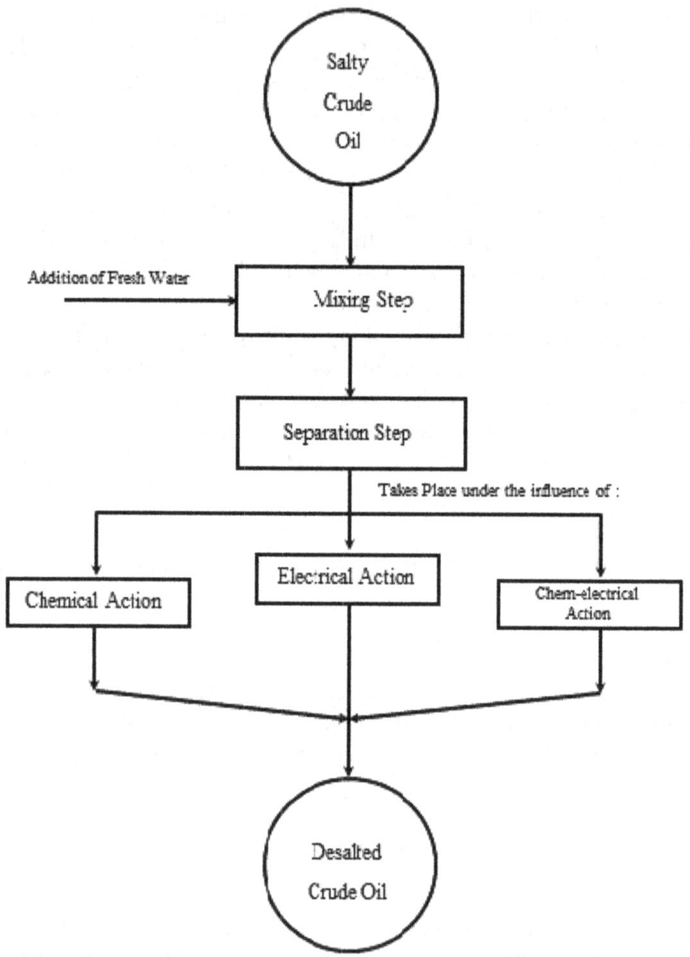

FIGURE 17.4 Illustration of the basic concept in the desalting operation of crude oil.

The steps involved in the desalting of crude oil with preliminary dehydration are summarized as follows:

1. Adding a de-emulsifier to the feed oil to enhance breaking emulsions.
2. Pumping the feed oil through heat exchangers to heat it to 200–300°F to enhance separation of the water from oil.
3. Adding wash water to the feed ensuring a thorough and effective mixing.
4. Allowing the emulsion that is formed between wash water and remnant water in oil to settle in the desalter, subjected to high voltage electric field. This will help separation of the two phases.
5. Removing effluent water and contaminants from the desalter.
6. Obtaining "dry" oil from the top to be shipped to destination.

17.4 STABILIZATION AND SWEETENING OF SOUR CRUDE OIL

17.4.1 Introduction

Previous discussions have dealt with the separation of water and the removal of salts from the liquid phase comprising crude oil. Our next objective is to present methods for stabilizing the crude oil relative to specified vapor pressure and allowable concentration of hydrogen sulfide (H_2S). Some produced crude oils contain H_2S and other sulfur products. When it contains more than 400 ppm of H_2S gas, the oil is classified as sour crude. Sour crude oils present serious safety and corrosion problems. In such cases, another treatment known as the sweetening process is needed to remove H_2S or reduce its content to acceptable limits.

While this is true, maximization of yield of production by minimizing the loss of valuable lighter hydrocarbons should be also a target. The series of hydrocarbons that is distributed between the gas phase and liquid phase has a wide spectrum. Cuts can be identified as finished products, depending on the individual hydrocarbons that are included.

Dual operation of stabilization and sweetening of crude oil targets the above objectives. Retention of too many light ends in the presence of H_2S can cause many problems. Refiners and shipping tankers impose restriction on crude oil to have a vapor pressure of 5–20 RVP (Reid vapor pressure) and a maximum of 10–100 ppmw (parts per million by weight) of H_2S. It should also be mentioned that this dual operation will lead to an increase in the API gravity of the oil; an advantage in its sales value.

The environmental effect from the exposure to H_2S as well as some exposure standards as reported in the oil industry are presented as follows:

H_2S Concentration	Standard	Health Effect
15 ppm	TLV-STEL[1]	A small % of workers may experience eye irritation
300 ppm	IDLH[2]	Maximum concentration from which one could escape within 30 minutes without a respirator
700 ppm		Quick loss of consciousness, breathing will stop and death will result if not rescued promptly

[1] Threshold-limit value for 15 minutes. Short Term Exposure Limit

[2] Generally recognized Immediately Dangerous to Life and Health concentration

17.4.2 Process Description

The total pressure exerted by the crude oil is contributed by the partial pressure of low boiling compounds which may be present in small quantities. Examples are methane and H_2S. The maximum volume of hydrocarbon liquid that is stable under stock tank conditions can be obtained by using what is known as trayed stabilizer. It is a fractionating column, but with no reflux pumps, and no condensers. Cold feed is introduced to the top plate of the column. This provides internal reflux, where the falling liquid contacts the warm vapors rising from the bottom of the column. The rising vapors strip the lighter ends from the crude; while the crude absorbs and

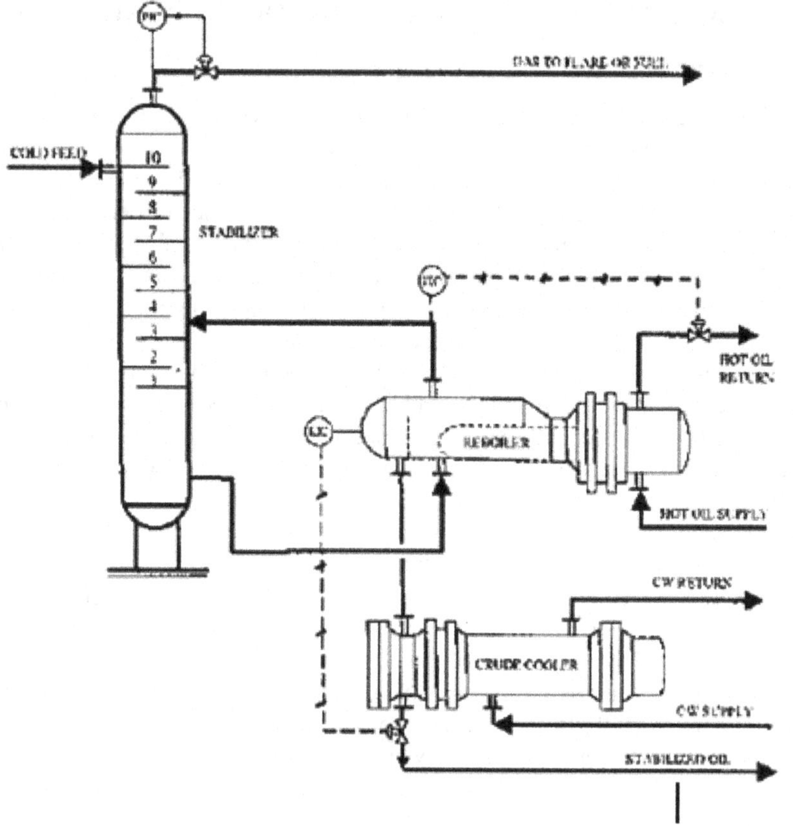

FIGURE 17.5 Typical trayed stabilizer.

dissolves some of the heavy ends from the vapors. A flow diagram for the process is given in Figure 17.5. Stabilization generally increases the recovery of stock tank of crude by 3–7% over simple stage stabilization or separation.

CASE 1: STATIC MIXER IMPROVES DESALTING EFFICIENCY: (SOURCE: CHEMICAL ONLINE NEWSLETTER, OCTOBER 13, 2000 AND SPE 124823 PAPER)

Objective: To study the economic feasibility of replacing a typical globe-type mix valve by a static mixer in a crude oil desalter.

Approach: The case could be handled using the method presented in Chapter 8: "The analysis of alternative selections and replacements". In this method, all costs incurred in buying, installing operating and maintaining an asset are put on annual basis. Selection is then based on what we call "differential approach", or the return on extra investment.

Process description: The mixer was installed at a 150,000 bbl/d crude distillation unit's desalter. Crude at this refinery is a mixture of local production and imports from Indonesia and Alaska. The crude oil and water are then simultaneously mixed though two-by-two division, cross-current mixing, and back-mixing, which improves turbulence and increases mixing efficiency without requiring high fluid shear velocities.

Merits of the static mixer: Table 17.1 summarizes the main performance of the static mixer as compared to the globe valve.

Conclusion: The modified desalter system has operated well on 14° and 22° API naphthenic crudes, with less that 5% oil in the effluent water. At the same time, the mixer has helped reduce emulsions formed by too much pressure drop created by the mix valve. With less oil carry under, less fuel is consumed from having to reheat recycled oil up to 300°F before it re-enters the crude unit. Salt removal also increased as a result of using the static mixer (Table 17.1). Depending on the type of crude oil, the refiner has been able to remove between 5% and 10% more salt than by the mix valve method. With less salt carried over out of desalter, less corrosive HCI will be generated in the crude unit furnaces. This will require less ammonia to neutralize the atmospheric column overhead stream. Also, pressure drop due to the mixing device was decreased from 10 psi to 1.5 psi.

The payout period of the new mixer was calculated and found to be 1 year. In other words, the mixer will pay for itself in its first year of operation with combined savings of $4,000/year in power consumption and chemical costs and $1,000/year in fuel costs.

TABLE 17.1
Desalter Performance*

	Salt in⁺	Salt out⁺	% Removal
Mix Valve (Globe Valve)			
90,000 b/d 22° API Crude	42	4.4	89
14° API Crude§	-----	-----	-----
Static Mixer (New Mixer)			
90,000 b/d 22° API Crude	41	1.6	96
45,000 b/d 14° Crude	43	¼	97

* Desalter mix valve and static mixer all designed for full design crude unit feed rate of 150 MBPD

⁺ PTB

CASE 2: UPGRADING OF THE QUALITY OF CRUDE OIL BY A DESALTING UNIT

Objective: Evaluation of the Economic Feasibility of a Desalting Unit.

Approach: Calculation of the Return on Investment (ROI) and Pay-Out Period (POP).

Process description: The following results were obtained from field desalting of a crude oil in the Middle East using one stage (Source: Abdel-Aal 1998):

- Crude oil flow rate (feed)........... = 120,000 BPD
- BS&W, vol % of feed.............. = 1.6
- Salt content of feed (PTB)........... = 900
- Water of dilution, vol. % .... = 2
- Salt content of desalted oil (PTB)... = 46

Eventually, for this type of crude oil, a two-stage desalting unit, shown above, should be applied to bring the salt content in the final product to 15–20 PTB. This upgrading process is to be investigated along the following guidelines.

Given:

- The upgrading of crude oil to an acceptable PTB could realize a saving in the shipping costs of 0.1 $/bbl in the shipping costs of the oil.
- The crude oil desalting unit has a design capacity of 120,000 bbl/day.
- The capital investment is estimated to be $5.0 million, service life is 10 years and operating factor is 0.95.
- The total annual operating expenses are $10/1000 bbl and the annual maintenance expenses are 10% of the capital investment.

Find:

a. The return on investment (ROI)
b. The payout period (POP)

SOLUTION:

Annual savings in shipping costs of upgraded crude oil = $\$4.1610 \times 10^6$

Total annual expenses incurred by installing the desalting unit = $\$1.4161 \times 10^6$

Net savings = $\$2.7449 \times 10^6$

ROI = Net savings/Capital investment = 55%

POP (number of years to recover the capital investment) = 1.8 years

18 Operations for Gas Handling (Conditioning), Treatment, and Separation of NGL

OVERVIEW

Natural gas associated with oil production or produced from gas fields (free) generally contains undesirable components such as H_2S, CO_2, N_2, and water vapor. In this chapter, natural gas conditioning is detailed. This implies the removal of such undesirable components before the gas can be sold in the market. Specifically, the gas contents of H_2S, CO_2, and water vapor must be removed or reduced to acceptable concentrations. N_2, on the other hand, may be removed if it is justifiable. Gas compression is usually needed after these treatment processes. Once the gas is treated, then the recovery of natural gas liquid (NGL) becomes a feasible and a profitable avenue.

In this chapter, for convenience, a system involving field treatment of a gas plant could be divided into two main stages, as shown in Figure 18.1.

The gas treatment operations carried out in stage I involve the removal of gas contaminants (acidic gases), followed by the separation of water vapor (dehydration). Gas processing, stage II, on the other hand, comprises two operations: NGL recovery and separation from the bulk of gas and its subsequent fractionation into desired products. The purpose of a fractionator's facility is simply to produce individual finished streams needed for market sales.

Gas field processing in general is carried out for two main objectives:

- The necessity to remove impurities from the gas.
- The desirability to increase liquid recovery above that obtained by conventional gas processing.

18.1 GAS TREATMENT AND CONDITIONING

18.1.1 Background

In its broad scope, gas field processing (G.F.P.) includes dehydration, acidic gas removal (H_2S and CO_2), and the separation and fractionation of liquid hydrocarbons (known as NGL). Sweetening of natural gas almost always precedes dehydration

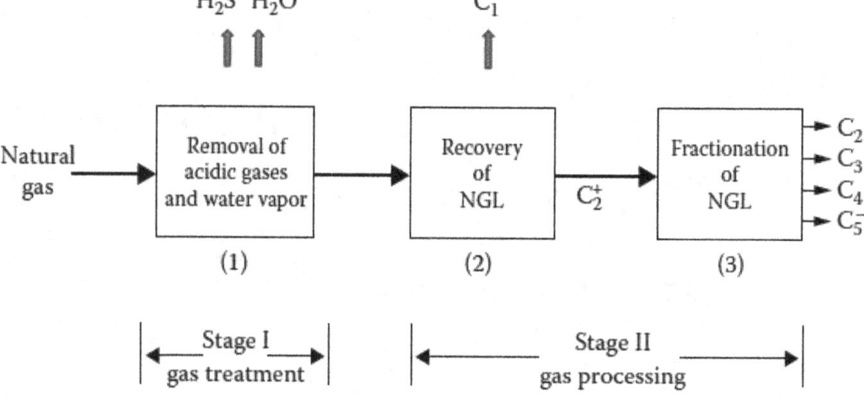

FIGURE 18.1 Division of gas system into two stages: gas treatment (conditioning) and gas processing.

and other gas plant processes carried out for the separation of NGL. Dehydration, on the other hand, is usually required before the gas can be sold for pipeline marketing and it is a necessary step in the recovery of NGL from natural gas.

18.1.2 Effect of Impurities (Water Vapor, H_2S/CO_2), and Liquid Hydrocarbons Found in Natural Gas

The effect each of these components has on the gas industry, as end user, is briefly outlined:

i. Water Vapor

It is a common impurity. It is not objectionable as such. If it condenses to liquid, it accelerates corrosion in the presence of H_2S gas. In case it leads to the formation of solid hydrates (made up of water and hydrocarbons), it will plug valves and fittings in the pipe.

ii. H_2S/CO_2

Both gases are harmful, especially H_2S, which is toxic if burned to give SO_2 and SO_3, which are nuisance to consumers. Both gases are corrosive in the presence of water. In addition, CO_2 contributes a lower heating value to the gas.

iii. Liquid Hydrocarbons

Their presence is undesirable in the gas used as a fuel. The liquid form is objectionable for burners designed for gas fuels. In case of pipelines, it is a serious problem to handle two-phase flow: gas and liquid.

18.2 SOUR GAS TREATING

18.2.1 Selection of Gas Sweetening Process

There are many key parameters to be considered in the selection of a given sweetening process. These include the following:

- Type of impurities to be removed (H_2S and mercaptans)
- Inlet and outlet acid gas concentrations
- Gas flow rate, temperature, and pressure
- Feasibility of sulfur recovery
- Acid gas selectivity required
- Presence of heavy aromatic in the gas
- Well location
- Environmental consideration
- Relative economics

Generic and specialty solvents are being divided into three different categories to achieve sales gas specifications:

- Chemical solvents
- Physical solvents
- Physical-Chemical (hyprid) solvents

The selection of the proper gas sweetening process depends on the sulfur content in the feed and the desired product as illustrated in Figure 18.2. Several commercial processes are available and shown in the schematic flow sheet of Figure 18.3.

18.2.1.1 Acid Gas Concentration in Outlet Gas

18.2.2 Amine Processes

Amine gas sweetening is a proven technology that removes H_2S and CO_2 from natural gas and liquid hydrocarbon streams through absorption and chemical reaction. Aqueous solutions of alkanolamines are the most widely used for sweetening natural gas. Each of the amines offers distinct advantages to specific treating problems:

a. Monoethanolamine (MEA): it is used in low-pressure natural gas treatment applications requiring stringent outlet gas specifications.

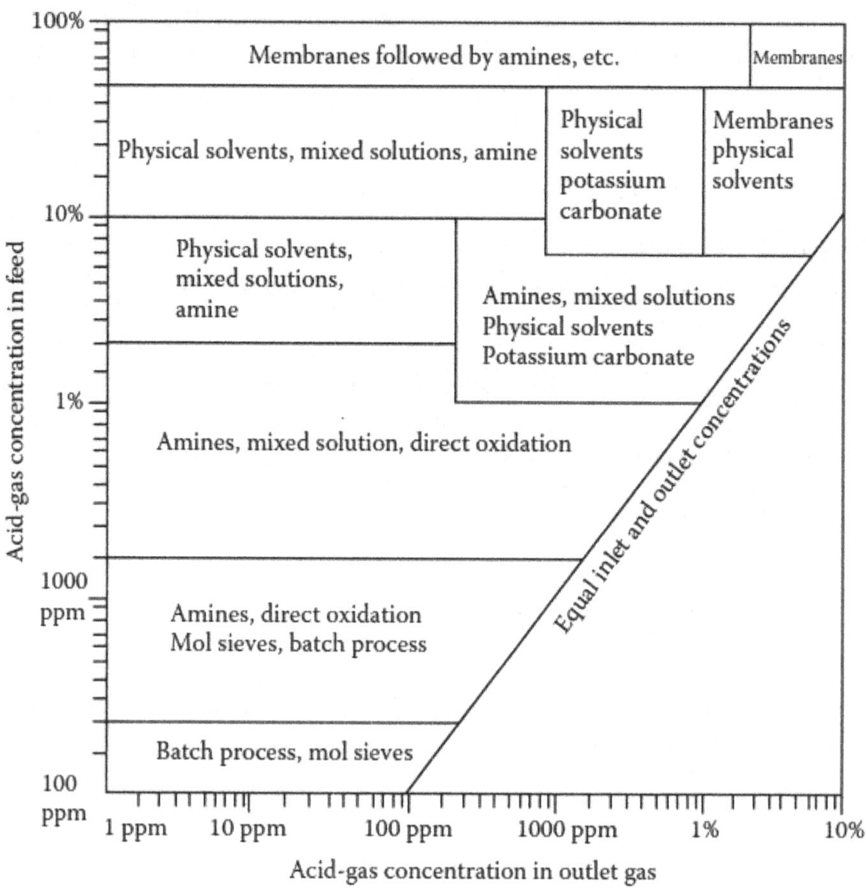

FIGURE 18.2 Selection of gas sweetening processes.

b. Methyl diethanolamine (MDEA): it has a higher affinity for H_2S than CO_2, which allows some CO_2 "slip" while retaining H_2S removal capabilities.
c. Diethanolamine (DEA): it is used in medium to high pressure treating and does not require reclaiming, as MEA and diglycolamine (DGA) systems do.
d. Formulated (specialty) solvent: a variety of blended or specialty solvents are available on the market.

A typical amine process is shown in Figure 18.4. The acid gas is fed into a scrubber to remove entrained water and liquid hydrocarbons. The gas then enters the bottom of absorption tower which is either a tray (for high flow rates) or packed

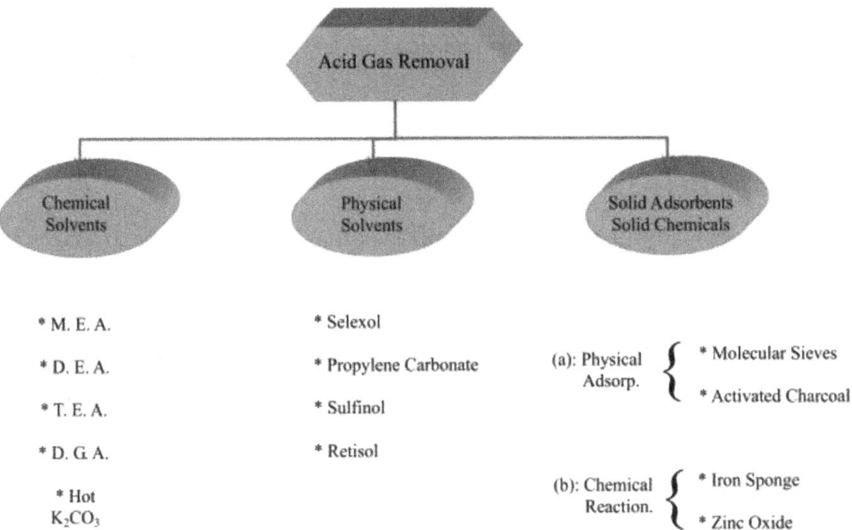

FIGURE 18.3 Classification of gas sweetening processes.

(for lower flow rate). The sweet gas exits at the top of tower. The regenerated amine (lean amine) enters at the top of this tower and the two streams are contacted counter-currently. In this tower, CO_2 and H_2S are absorbed with the chemical reaction into the amine phase. The exit amine solution, loaded with CO_2 and H_2S, is called rich amine. This stream is flashed, filtered, and then fed to the top of a stripper to recover the amine, and acid gases (CO_2 and H_2S) are stripped and

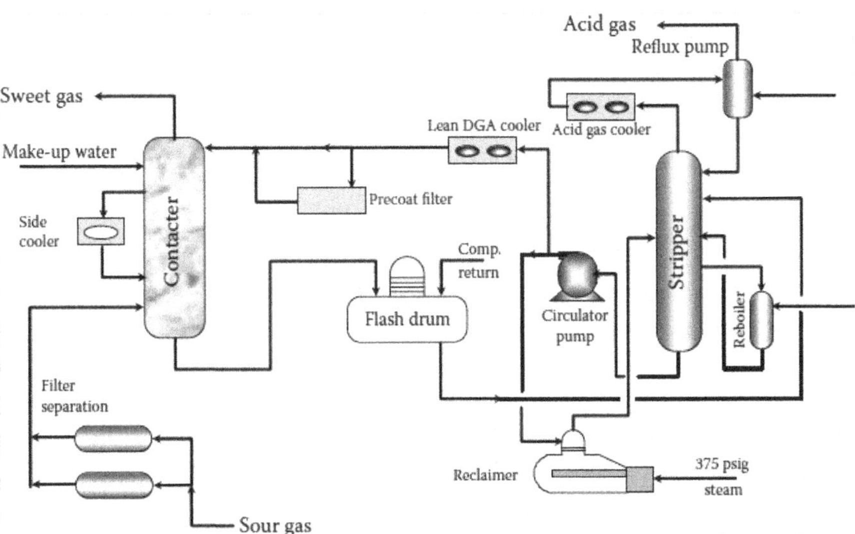

FIGURE 18.4 Flow sheet for the amine process.

exit at the top of the tower. The refluxed water helps in steam stripping the rich amine solution. The regenerated amine (lean amine) is recycled back to the top of the absorption tower.

18.3 GAS DEHYDRATION

Natural gas usually contains significant quantities of water vapor. Changes in temperature and pressure condense this vapor altering the physical state from gas to liquid to solid. This water must be removed in order to protect the system from corrosion and hydrate formation. The wet inlet gas temperature and supply pressures are the most important factors in the accurate design of a gas dehydration system. Without this basic information the sizing of an adequate dehydrator is impossible.

Natural gas dehydration is defined as the process of removing water vapor from the gas stream to lower the dew point of the gas. There are three basic reasons for the dehydration process:

1. To prevent hydrate formation: Hydrates are solids formed by the physical combination of water and other small molecules of hydrocarbons. They are icy hydrocarbon compounds of about 10% hydrocarbons and 90% water.
2. To avoid corrosion problems: Corrosion often occurs when liquid water is present along with acidic gases, which tend to dissolve and disassociate in the water phase, forming acidic solutions.
3. To avoid side reactions, foaming, or catalyst deactivation during downstream processing in many commercial hydrocarbon processes.

18.3.1 Prediction of Hydrate Formation

Methods for determining the operating conditions leading to hydrate formation are very essentials in handling natural gas. In particular, we should be able to find:

1. Hydrate formation temperature for a given pressure
2. Hydrate formation pressure for a given temperature
3. Amount of water vapor that saturates the gas at a given pressure and temperature (i.e., at the dew point)

At any specified pressure, the temperature at which the gas is saturated with water vapor is being defined as the "dew point". Cooling of the gas in a flow line due to heat loss can cause the gas temperature to drop below the hydrate formation-temperature. Elaborate discussion of both approximate methods and analytical methods are presented by Abdel-Aal et al.

18.3.2 Methods Used to Inhibit Hydrate Formation

Hydrate formation in natural gas is promoted by high-pressure, low temperature conditions, and the presence of liquid water. Therefore, hydrates can be prevented by adopting one (or more than one) of the following procedures:

1. Raising the system temperature and/or lowering the system pressure (temperature/pressure control).
2. Injecting a chemical such as methanol or glycol to depress the freezing point of liquid water (chemical injection).
3. Removing water vapor from the gas (liquid-water drop out); in other words depressing the dew point by dehydration.

18.3.3 Dehydration Methods

The most common dehydration methods used for natural gas processing are the following:

1. Absorption, using the liquid desiccants (e.g., glycols and methanol)
2. Adsorption, using solid desiccants (e.g., alumina and silica gel)
3. Cooling/condensation below the dew point, by expansion and/or refrigeration

This is in addition to the hydrate inhibition procedures described earlier.

Different dehydration methods are classified as shown in Figure 18.5

18.3.4 Dehydration Using Absorption System

The absorption process is shown schematically in Figure 18.6. The wet natural gas enters the absorption column (glycol contactor) near its bottom and flows

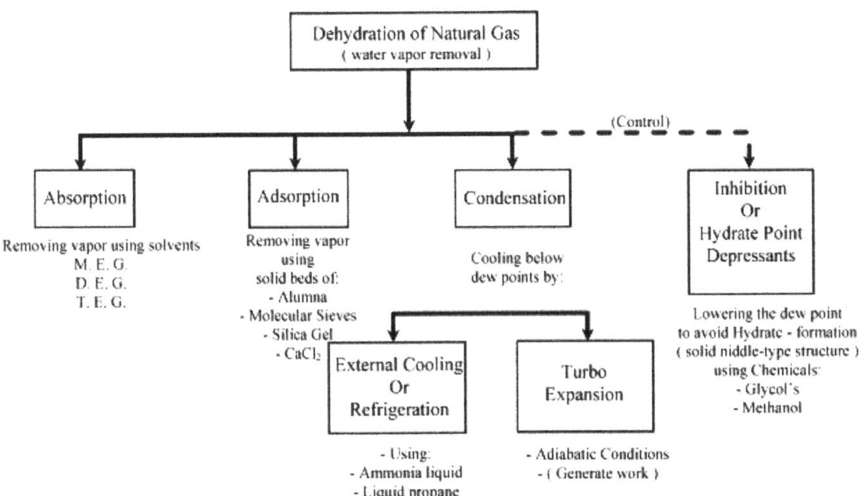

FIGURE 18.5 Gas dehydration methods.

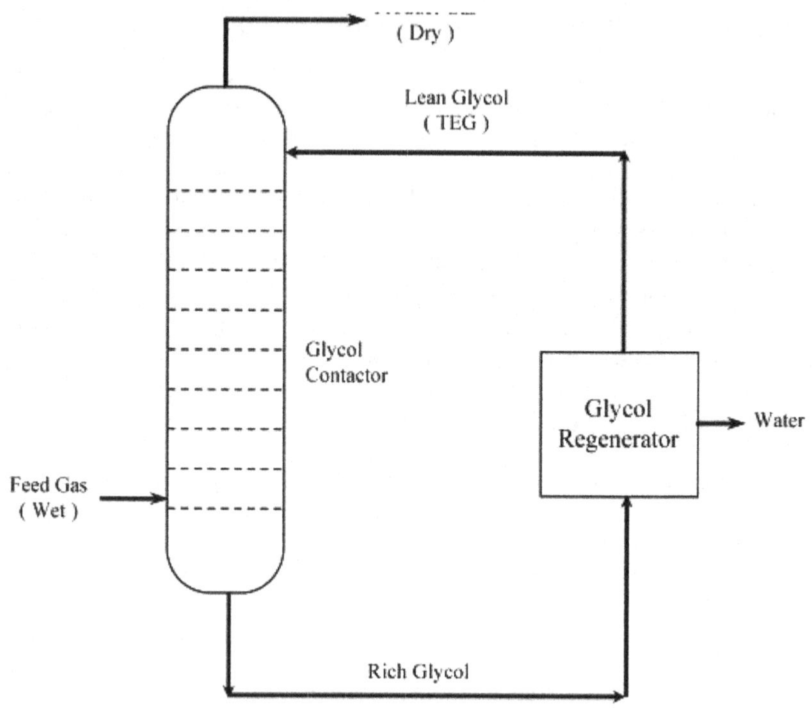

FIGURE 18.6 Glycol dehydration unit.

upward through the bottom tray to the top tray and out at the top of the column. Usually six to eight trays are used. Lean (dry) glycol is fed at the top of the column and it flows down from tray to tray, absorbing water vapor from the natural gas. The rich (wet) glycol leaves from the bottom of the column to the glycol regeneration unit. The dry natural gas passes through mist mesh to the sales line. The rich glycol is preheated in heat exchangers, using the hot lean glycol, before it enters the still column of the glycol reboiler. This cools down the lean glycol to the desired temperature and saves the energy required for heating the rich glycol in the reboiler.

18.3.5 Dehydartion Using Adsorption (Solid-Bed Dehydration)

When very low dew points are required, solid-bed dehydration becomes the logical choice. It is based on fixed-bed adsorption of water vapor by a selected desiccant. A number of solid desiccants could be used such as silica gel, activated alumina, or molecular sieves. The selection of these solids depends on economics. The most important *property is the capacity of the desiccant, which determines the loading design expressed as the percentage of* water to be adsorbed by the bed. The capacity decreases as temperature increases. Figure 18.7 represents solid bed dehydration.

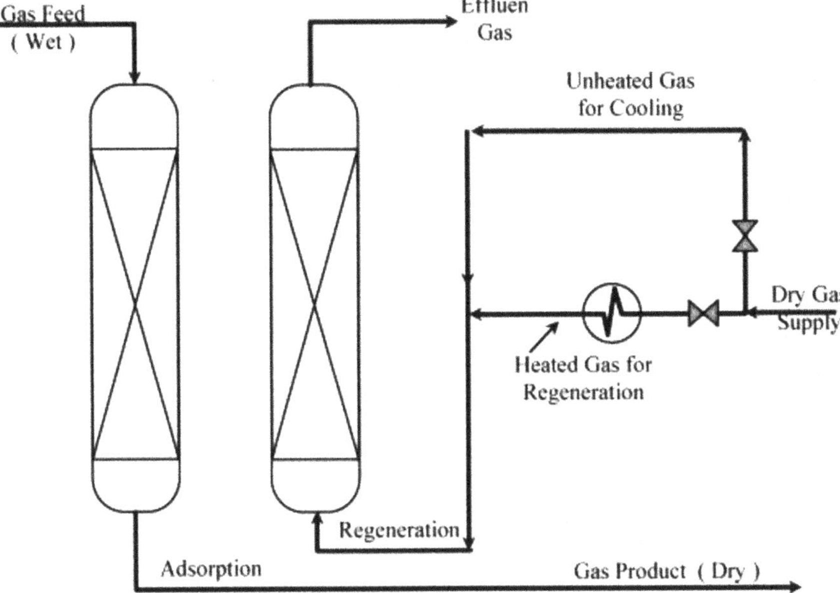

FIGURE 18.7 A solid desiccant unit for natural gas dehydration.

ECONOMIC EVALUATION AND APPLICATION

CASE 1: UTILIZATION OF NATURAL GAS RECOVERED FROM GAS PLANT

Objective: To investigate the economics of utilizing natural gas as a fuel for heating crude oil.

Process: Natural gas is recovered from gas–oil separator plant (GOSP) using an absorber de-ethanizer system, along with an amine treating unit and a gas dryer to have available desulfurized gas that can be used or sold as a fuel gas.

Given: The total cost for the recovery of this gas is estimated to be $0.75/MCF. It has been suggested to use this gas as a fuel for heating 5,000 bbl/day of 40° API crude oil from 80°F to 250°F.

Find:

1. The cost of heating the crude oil using this gas.
2. Compare it with the cost of heating using fuel oil at $2.2/MM Btu.
3. Do you recommend change in operation to use the fuel gas as a heating fuel instead of using the fuel oil?

SOLUTION

The heat duty required is calculated using the well-known equation: $Q = m\ Cp\ \Delta T$

$$= 127.7 \text{ MM Btu/day}$$

Assuming the heating value of the gas is 960 Btu/ft^3 and the heat efficiency is 60%; then the fuel gas consumption will be 221,700 ft^3/day.

$$\text{The cost of using this fuel gas for heating} = 2217000 \text{ ft3/day} \times \$0.75/\text{MCF}$$

$$= \$166.28/\text{day}$$

$$\text{The cost of using the fuel oil for heating} = [127.7 \text{ MM Btu} \times \$2.2/\text{MM Btu}] / 0.6$$

$$= \$468.23/\text{day}$$

A daily savings in the cost of fuel of about $300 is realized if the change to fuel gas takes place. One has to consider other economic factors in making this analysis. The capital cost involved in changing the burner system has to be considered.

CASE 2: HOW TO CONTROL THE CO_2 SPECS IN THE SWEET GAS? (DISCUSSION TYPE)

MDEA has become the amine molecule chosen to remove H_2S, CO_2, and other contaminants from hydrocarbon streams. Amine formulations based on MDEA can significantly reduce the costs of acid gas treating. Under the right circumstances, MDEA based solutions can boost plant capacity, lower energy requirements, or reduce the capital required.

The ultimate goal of amine sweetening is to produce specification quality product as economically as possible. Amine technology has produced selective absorbents which remove H_2S in the presence of CO_2. The use of selective amines results in:

- Lower circulation rates.
- Reduction in reboiler sizes and duties, while meeting the H_2S specification. Unfortunately, many operators now are exceeding the CO_2 specification in their sweet gas streams due to changes in inlet composition or increased throughput. Achieving specifications within the constraints of the process equipment is most cost effective and desirable.

In general, if the objective is to slip as much CO_2 as possible, the engineer should consider using the most selective amine at the lowest concentration and circulation rate with the fewest number of equilibrium stages in the absorber to achieve the H_2S specification. Cold absorber temperatures tend to increase the CO_2 slip and enhance H_2S pickup. If the objective is to achieve a certain CO_2 concentration, then the problem is more complicated. Variables to consider include increasing the amine concentration and using mixtures of amines. However, equipment size may have to be reevaluated. Increasing the lean amine temperature increases CO_2 pickup for the selective amines to a point. The maximum temperature depends on amine concentration, inlet gas composition, and loading. Higher lean amine temperature also increases water and amine losses and decreases H_2S pickup.

Alternatively, solvents that are designed for CO_2 removal are also available. For example, DOW's Specialty Amines cover the full range from the maximum CO_2 slip, to nearly complete CO_2 removal.

CASE 3: NON-CATALYTIC PARTIAL OXIDATION (NCPO) OF SOUR NATURAL GAS (DISCUSSION TYPE)

In order to exclude the costly DGA treatment of sour natural gas, the NCPO approach proposed by Abdel-Aal and Shalabi is recommended to produce synthesis gas from sour gas by direct partial oxidation, as illustrated in Figure 18.8.

Currently, synthesis gas is produced by steam reforming of sweet natural gas. This is a catalytic process in which the feed gas has to be sulfur free to avoid catalyst poisoning. As a result, acidic gas removal is a prerequisite for the steam-reforming process as shown in Figure 18.9. H_2S is separated from the natural gas by one of the physiochemical separation methods. The separation process is expensive and involves the use of amine solvents. The chemisorption of acidic gas into the solvents is followed by regeneration of these solvents. Although the

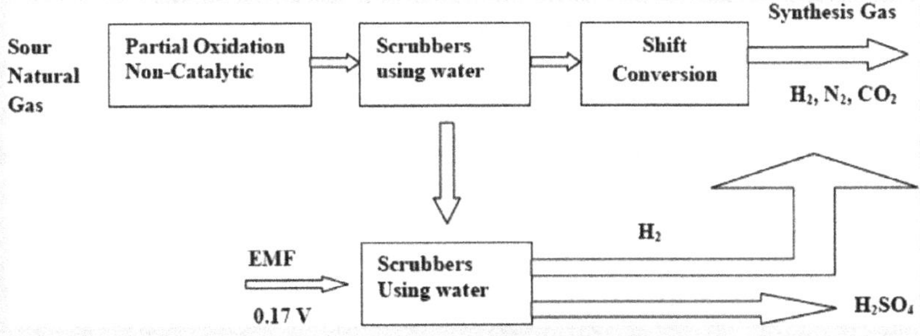

FIGURE 18.8 Non-catalytic partial oxidation of sour natural gas.

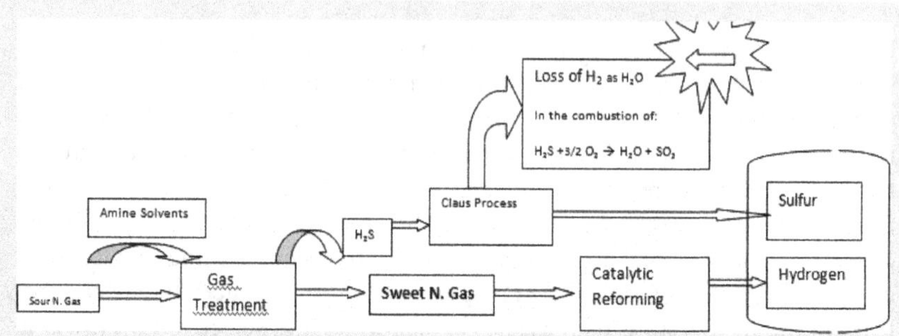

FIGURE 18.9 Current technology to produce synthesis gas from sour natural gas.

bulk production of synthesis gas is done via catalyzed steam reforming of sweet natural gas, non-catalyzed partial oxidation of sour natural gas with appropriate conditions may prove to be more attractive.

Compare between the two systems: the current technology and the NCPO from the technical and economic point of views.

18.4 GAS PROCESSING AND SEPARATION OF NGL

18.4.1 Background

Natural gas procrssing comprises two consecutive operations: NGL recovery (extraction) and separation from the bulk of gas followed by subsequent fractionation into desired products. The purpose of a fractionator's facility is simply to produce individual finished streams needed for market sales. Fractionation facilities play a significant role in gas plants.

Case study involving the optimum recovery of butane using lean oil extraction is presented.

Natural gas leaving the field can have several components which will require removal before the gas can be sold to a pipeline gas transmission company. All of the H_2S and most of the water vapor, CO_2, and N_2 must be removed from the gas. Gas compression is often required during these various processing steps.

The condensable hydrocarbons heavier than methane, which are recovered from natural gas, are called NGLs. Usually associated gas produces higher percentage of NGLs. It is generally desirable to recover NGL present in the gas in appreciable quantities. This normally includes the hydrocarbons known as C_3^+. In some cases, ethane C_2 could be separated and sold as a petrochemical feed stock. NGL recovery is the first operation in gas processing, as explained earlier. To recover and separate NGL from a bulk of a gas stream would require a change in phase; that is, a new phase has to be developed for separation to take place by using one of the following:

- An energy-separating agent; examples are refrigeration (cryogenic cooling) for partial or total liquefaction and fractionation.

- A mass-separating agent; examples are adsorption and absorption (using selective hydrocarbons, 100–180 molecular weight).

The second operation is concerned with the fractionation of NGL product into specific cuts such as liquefied petroleum gas (LPG) (C_3/C_4) and natural gasoline. It should be pointed out that the fact that all of the field processes do not occur at or in the vicinity of the production operation does not change the plan of the system of gas processing and separation.

The principal market for natural gas is achieved via transmission lines, which distribute it to different consuming centers, such as industrial, commercial, and domestic. Field processing operations are thus enforced to treat the natural gas in order to meet the requirements and specifications set by the gas transmission companies. The main objective is to simply obtain the natural gas as a main product free from impurities. In addition, it should be recognized that field processing units are economically justified by the increased liquid product (NGL) recovery above that obtained by conventional separation.

Description of a typical natural gas processing plant is shown in Figure 18.10.

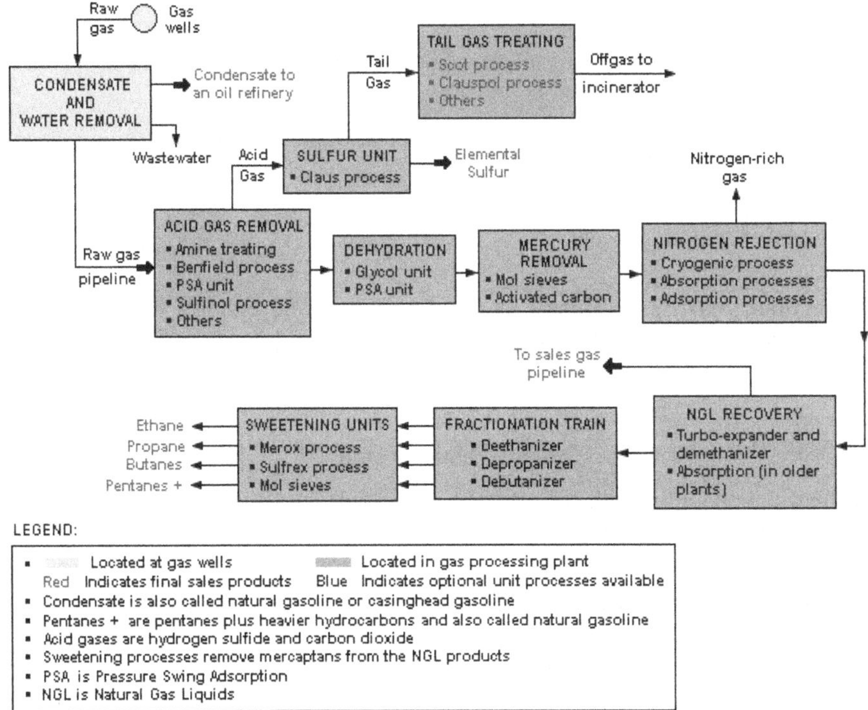

FIGURE 18.10 Description of a typical natural gas processing plant.

(Source: Wikipedia the free encyclopedia)

18.4.2 Recovery and Separation of NGL

18.4.2.1 Options of Phase Change

To recover and separate NGL from a bulk of gas stream, a change in phase has to take place. In other words, a new phase has to be developed for separation to occur. Two distinctive options are in practice depending on using Energy Separating Agent (ESA) or Mass Separating Agent (MSA).

i. Energy Separating Agent

The distillation process best illustrates a change in phase using ESA. To separate, for example, a mixture of alcohol and water heat is applied. A vapor phase is formed in which alcohol is more concentrated, and then separated by condensation. This case of separation is expressed as follows:

A mixture of liquids + Heat - → Liquid + Vapor

For the case of NGL separation and recovery in a gas plant, removing heat (by refrigeration) on the other hand, will allow heavier components to condense; hence, a liquid phase is formed. This case is represented as follows:

A mixture of hydrocarbon vapor - Heat - → Liquid + Vapor

Partial liquefaction is carried out for a specific cut, whereas total liquefaction is done for the whole gas stream.

ii. Mass Separating Agent

To separate NGL, a new phase is developed by using either a solid material in contact with the gas stream (adsorption) or a liquid in contact with the gas (absorption).

18.4.3 Parameters Controlling NGL Separation

A change in phase for NGL recovery and separation always involves control of one or more of the following three parameters:

1. Operating pressure, P
2. Operating temperature, T
3. System composition or concentration, x and y

To obtain the right quantities of specific NGL constituents, a control of the relevant parameters has to be carried out.

First: For separation using ESA, pressure is maintained by direct control. Temperature, on the other hand, is reduced by refrigeration using one of the following techniques:

a. Compression refrigeration
b. Cryogenic separation; expansion across a turbine
c. Cryogenic separation; expansion across a valve

In cryogenic cooling process to recover NGL, gas is cooled to very low temperature (–100 to –120°F) by adiabatic expansion of the gas mixture by turbo expanders. The water and acid gases are removed before chilling the gas to avoid ice formation. After chilling, the gas is sent to demethanizer to separate methane from NGL.

Second: For separation using MSA, a control in the composition or the concentration of the hydrocarbons to be recovered (NGL); x and y are obtained by using adsorption or absorption methods.

Adsorption provides a new surface area, through the solid material, which entrains or "adsorbs" the components to be recovered and separated as NGL. Thus, the components desired as liquid are deposited on the surface of the selected solid; then regenerated off in a high concentration; hence, their condensation efficiency is enhanced. About 10–15% of the feed is recovered as liquid. Adsorption is defined as a concentration (or composition) control process that precedes condensation. Therefore, refrigeration methods may be coupled with adsorption to bring in condensation and liquid recovery.

Absorption, on the other hand, presents a similar function of providing a surface or "contact" area of liquid–gas interface. The efficiency of condensation, hence NGL recovery, is a function of P, T, gas, and oil flow rates, and contact time. Again, absorption could be coupled with refrigeration to enhance condensation.

In lean oil extraction method, the treated gas is cooled by heat exchange with liquid propane and then washed with a cold hydrocarbon liquid, which dissolves most of the condensable hydrocarbons. The uncondensed gas is dry natural gas and contains mainly methane with small amounts of ethane and other heavier hydrocarbons. The condensed hydrocarbons or NGLs are stripped from the rich solvent, which is recycled back to the process.

To summarize the above, a proper design of a system implies the use of the optimum levels of all operating factors plus the availability of sufficient area of contact for mass and heat transfer between phases.

18.4.4 Fractionation of NGL

Due to their added value, heavier hydrocarbons are often extracted from natural gas and fractionated by using several tailor made processing steps. In general, and in gas plants in particular, fractionating plants have common operating goals:

1. The production of on-specification products
2. The control of impurities in valuable products (either top or bottom)
3. The control in fuel consumption

As far as the tasks for system design of a fractionating facility, these goals are as follows:

- Fundamental knowledge on the process or processes selected to carry out the separation; in particular, distillation.
- Guidelines on the order of sequence of separation (i.e., synthesis of separation sequences).

NGL are normally fractionated into the following three streams:

1. An ethane rich stream used for producing ethylene
2. LPG. It is propane-butane mixture and is important feedstock for olefin plants
3. Natural gasoline

NGLs may contain significant amounts of cyclohexane.

18.4.5 Shale Gas

Conventional gas reservoirs are areas where gas has been "trapped". After natural gas is formed, the earth's pressure often pushes the gas upward through tiny holes and fractures in rock until it reaches a layer of impermeable rock where the gas becomes trapped. This gas is relatively easy to extract, as it will naturally flow out of the reservoir when a well is drilled. Unconventional gas occurs in formations where the permeability is so low that gas cannot easily flow (e.g., tight sands), or where the gas is tightly adsorbed (attached) to the rock (e.g., coalbed methane). Gas shales often include both scenarios—the fine-grained rock has low permeability; and, gas is adsorbed to clay particles. The pore spaces in shales are typically not large enough for even tiny methane molecules to flow through easily. Consequently, gas production in commercial quantities requires fractures to provide permeability.

Shale gas is defined as natural gas from shale formations, i.e., natural gas trapped within shale (fine grained sedimentary rocks) formations. Shale has low matrix permeability to allow significant fluid flow to well bore; therefore, commercial production requires mechanically increasing permeability. Shale gas reserves are known for long but natural fracture technology used earlier was uneconomical to produce shale gas. The recent developments in horizontal drilling and hydraulic fracturing (called fracking) made it viable. Mitchell energy, a Texas gas company, first achieved economical shale gas fracture in 1998. Shale gas is currently under evolutionary stage and so far is largely confined to North America. The complete technology and economic factors are yet to get matured. Several high profile shale gas drilling efforts in Europe have already failed.

Shale gas costs more to produce than natural gas (NG) from conventional wells. The high cost is mainly due to expense of massive hydraulic fracturing treatments required to produce shale gas and horizontal drilling. Drilling a vertical and horizontal well cost about \$1 million and \$4 million respectively. The huge requirement of water for hydraulic fracking and then the waste water treatment are major cost inhibitors. Overall, addressing environmental concerns associated with shale gas hugely adds up to its cost. The shale gas production may be feasible only in those regions where energy/NG prices are high. The shale gas production cost in the USA is estimated to be between \$4 and \$7 per MMBtu but it's termed as "foggy economics" since all factors not considered. Earlier it was thought that shale gas will produce less green-house gases but scientists recently concluded otherwise and opine that it will accelerate global warming. Shale gas production requires large amounts of water and chemicals added to it to facilitate underground fracturing process that releases gas. A maximum of 70% of used water is recovered and rest remains underground which can lead to contamination. Significant use of water for shale gas production may affect the availability of water for other uses and can affect aquatic habitat. The treatment of large amount of recovered waste water before reuse or disposal is an important and challenging issue. There are some evidences of groundwater contamination in areas of fracking. The environmental impacts of shale gas production are therefore challenging but still considered to be manageable.

So far shale gas is mostly confined to North America. There is little drilling progress in China, Australia, and Poland. In other countries, it's still in pilot stages. Canada has huge shale gas reserves but exploration is restricted due to strict environmental regulations and related issues. In the USA, BP predicted NG self-sufficiency and NG share of total energy consumption to double to 40% with 4% anticipated annual growth in shale gas production by 2030. The Energy Information Administration (EIA) however slashed BP shale gas forecast reserves by 41% in Jan 2012. The energy demand (dominated by oil) will still grow in next two decades by 39%, but most of the growth in demand will be from Asian countries, especially China and India. In Saudi Arabia, evaluation of shale gas reserves is in progress and production may start in 2020, but low NG price remains a major issue in developing the prospects.

ECONOMIC EVALUATION OF SELECTED PROBLEMS

CASE 1: RECOVERY OF BUTANE USING LEAN OIL EXTRACTION

Associated natural gas is passed through an absorption unit to recover heavier hydrocarbons (butane plus), which can be sold for a value of \$7.5/gal. Calculations show that the minimum total cost for the recovery and the extraction of the butanes in the plant is estimated to be \$1.2/gal of *butane* recovered. Other additional costs for processing the absorbing oil used in the recovery are estimated to be \$27/million gal of the lean oil circulated.

The engineering group in the plant developed the following empirical relationship for the rate (R) of the absorber oil used as a function of the rate of butane produced (P):

$$\text{R, millions of gal/hr} = 0.004\ P^{1.3}, \text{ where P is in gal/hr}$$

1. Compute the optimum butane recovery, P_o, and the optimum circulating oil rate, R_o, applicable to this plant.
2. What is the value of P at which the process of recovery breaks even?

SOLUTION

$$\text{Profit} = \text{Income} - \text{Expenses}$$
$$\text{Income} = 7.5\ \$/\text{gal} * P\ \text{gal/hr} = 7.5\ P\ \$/\text{hr}$$
$$\text{Expenses} = 1.2\ \$/\text{m gal} * P\ \text{gal/hr} + 27\ \$/\text{m gal} * R$$
$$= 1.2\ P + 27\ (0.004\ P^{1.3}) = 1.2\ P + 0.108\ P^{1.3}\ \$/\text{hr}$$
$$\text{Profit} = 7.5\ P - 1.2\ P - 0.108\ P^{1.3} = 6.3\ P - 0.108\ P^{1.3}$$

d/dp (profit) = $6.3 - 1.3\ (0.108)\ P^{0.3}$; setting this derivative equal to zero:

$$P_{opt.} = [6.3/1.3(0.108)]^{1/0.3} = \underline{320{,}647}\ \text{gal butane/hr}$$
$$R_{opt.} = 0.004(320{,}647)^{1.3} = \underline{57{,}530}\ \text{million gal oil/hr}$$

At the break-even point, profit = 0

$$6.3\ P - 0.108\ P^{1.3} = 0;\ \text{hence,}\ P^{0.3} = 6.3/0.108$$
$$P_B = (6.3/0.108))^{1/03} = \underline{768{,}777}\ \text{gal butane/hr}$$

CASE 2: THE PROBLEM OF FINDING THE OPTIMUM DIAMETER OF AN ABSORPTION TOWER (DISCUSSION TYPE)

The tower must process a gas feed stream at affixed rate to remove a soluble gas component by absorption in a liquid phase. Here we have the two scenarios:

- Increasing the diameter of the tower, lowers the gas velocity in the bed, reducing the pressure drop, hence lowering the **pumping** costs of the feed gas. But a large diameter tower is more costly to construct.
- Choosing a smaller diameter will cause **flooding** inside the column to occur, and liquid is carried up the gas stream, making the tower **inoperative**.

CONCLUSION

Some balance must be reached between the pumping costs and the construction costs, in order to lower the total costs of operation. Also, it is not practical to construct a tower of extremely large diameter because of liquid distribution problems.

Apparently, there are constraints on the tower diameter. Solution is reached by optimization technique in order to minimize the total annual costs of operating the tower as a function of the tower diameter.

The total annual costs of operation = Capital cost of the tower, depreciated over the life time ($/year) + annual operating (pumping) costs ($/year).

Section III.IV: Downstream Operations: Petroleum Refining of Crude Oil Into Useful Products (Three Parts) and Oil and Gas Transportation (One Chapter)

19 Crude Oil Refining—Part 1

19.1 OIL FRACTIONATION (DISTILLATION)

19.1.1 Overview

Petroleum is of little use when it first comes from the ground. It is a raw material, much as newly fallen trees are raw materials for furniture, construction, etc. Thus, crude oil must be put through a series of processes to be converted into the hundreds of finished oil products derived from it. These processes, collectively, are known as refining. However, by today's technological standards, the term *refining* is a misnomer. In the early petroleum industry, the refining process involved nothing more than the use of a crude still (pipe still) that produced useful oil products by physical separation only.

Currently the expression *crude oil processing* is more appropriate, since more than 85% of petroleum products are produced by processes involving chemical changes along with the basic physical separation.

The first step in refining is distillation. This step roughly separates the molecules in crude according to their size and weight. The process is analogous to taking a barrel of gravel containing stones of many different sizes and running the gravel through a series of screens to sift out first the small stones, next those slightly larger, and so on up to the very largest stones. As applied to crude oil, the distillation process "sifts out" progressively such components as gas, gasoline, kerosene, home heating oil, lubricating oils, heavy fuel oils, and asphalt.

The oil refining process is the central activity of downstream oil and gas operations. The Energy Information Administration (EIA) explains the breakdown this way: "Inside the distillation units, the liquids and vapors separate into petroleum components called fractions according to their weight and boiling point. Heavy fractions are on the bottom, and light fractions are on the top".

The objective of petroleum refining processes to transform crude oil into useful products such as liquefied petroleum gas (LPG), gasoline or petrol, kerosene, jet fuel, diesel oil, and fuel oils.

The basic aspects of current refining operations, involving physical separation are presented in this chapter along with the application of economic techniques and analysis to many problems encountered in the petroleum industry.

The physical separation of crude oil into valuable products (cuts) is highlighted. Crude oil separation is accomplished in two consecutive steps: first by fractionating the total crude oil at essentially atmospheric pressure; then feeding the bottom residue from the atmospheric tower to a second fractionator, operating at high vacuum. Types of oil refineries and their classification are given.

Economic analysis is presented for the refining operations in various ways to determine the most economical refining scheme to find out, for example, whether to use new or existing equipment.

19.1.2 Illustrative Example

As we have seen the objective of petroleum refining processes to transform crude oil into useful products such as LPG, gasoline or petrol, kerosene, jet fuel, diesel oil, and fuel oils.

The oil refining process is the central activity of downstream oil and gas operations. The EIA explains the breakdown this way.

Crude oil is to be fractionated into straight-run products such as gasoline, gas oil, and others. It is heated first by heat exchangers and then desalted. Its temperature is raised next using fire heaters, before it is introduced to the fractionation tower. This process is illustrated as shown in the following sketch in Figure 19.1. The next step in this process is to explain how and why the crude oil is separated into products?

Well, in the transformation of raw materials (*crude oil*); and in the presence of energy (*heat*) to produce finished products, *three modes of transfer* are encountered in this process. They are known as:

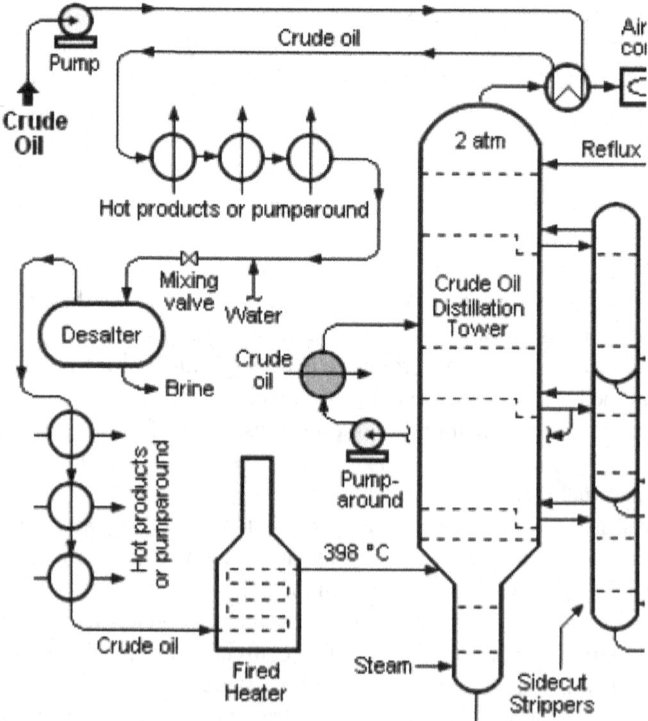

FIGURE 19.1 Illustrative example for a refining distillation operation.

1. Momentum Transfer (fluid flow), using a *pump.*
2. Heat Transfer of oil, using heat exchangers and a *furnace.*
3. Mass Transfer through the *distillation column* that leads to the separation of crude oil into different cuts (transfer is due to molecular diffusion of the components that separates the light from heavy). The *physical operations* (known as unit operations) and shown above in the example are: fluid flow, heat transfer, and distillation. *Unit operations deal chiefly with the transfer of energy and the transfer, separation, and conditioning of materials by physical means.*

At this stage, two basic questions arise:

1. What is the mechanism(s) underlying this process?
2. How and where it takes place?

The answer to the first question deals with theory of transfer or transport, as explained, within the boundaries of our system. For the second question, it is the combined effect of Momentum, Heat and Mass (MHM) that is responsible for the physical changes that took place in the distillation column to produce the finished products.

To make the story complete in the above example, it will be assumed, that the gasoline exit the distillation column is introduced into what is known as a "Reforming Unit", in order to obtain a *higher grade* gasoline. This reforming process represents a typical example of a *chemical conversion* or *chemical reaction* process-known as *Unit Process*—where the hydrocarbons undergo molecular changes and rearrangement leading to *high octane* gasoline. *Unit processes involve primarily the conversion of materials by means of chemical reactions.* Again, it should be pointed out that the three modes of transfer, MHM, take place as well for operations involving chemical reactions, or chemical changes.

In general, refineries rely on four major chemical processing operations in addition to the backbone physical operation of fractional distillation, in order to alter the ratios of the different fractions. These are normally called the Five Pillars of petroleum refining:

Pillar 1: Fractional distillation
Pillar 2: Cracking
Pillar 3: Unification (alkylation)
Pillar 4: Alteration (catalytic reforming)
Pillar 5: Hydroprocessing

19.1.3 Refinery Design as a Chemical Plant

The following one-page summary represents an avenue what is involved to design a refinery guided by the basic principles of plant design taught in chemical engineering.

Refinery design follows the outline of Figure 19.2 as shown:

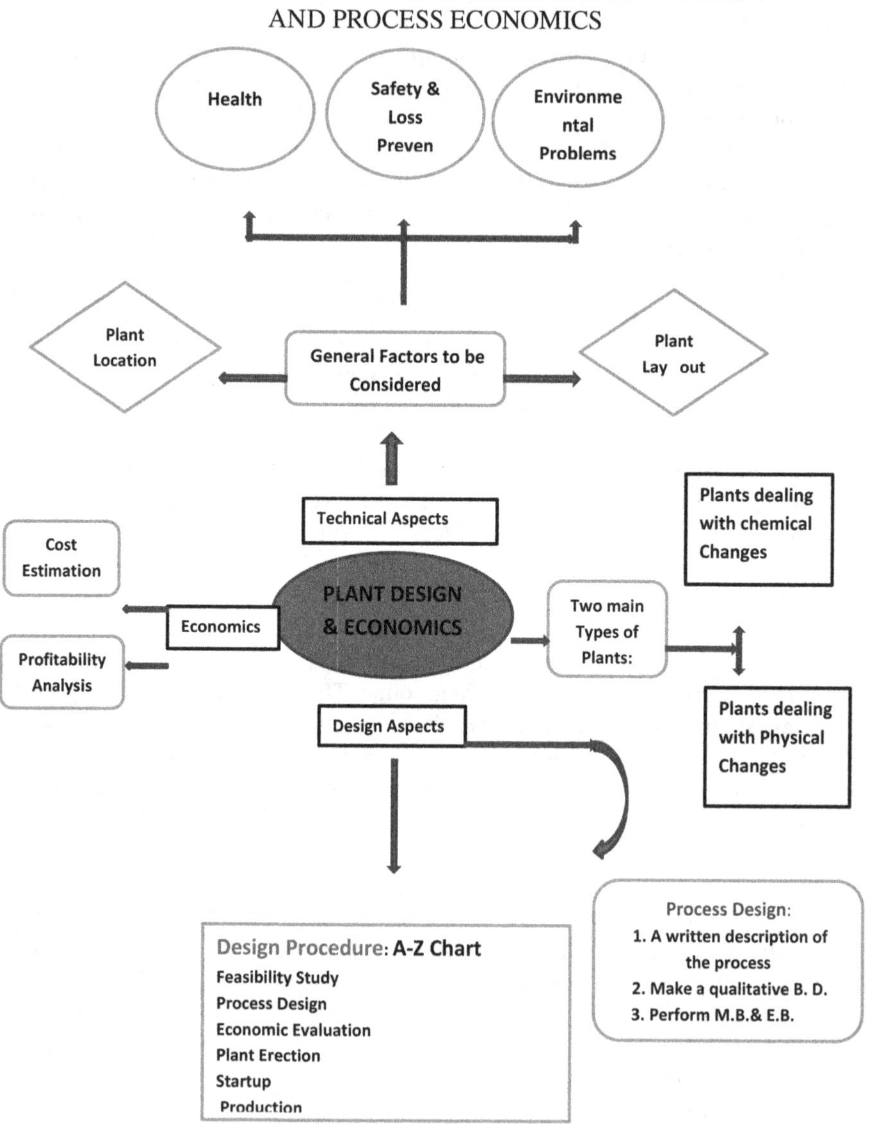

FIGURE 19.2 A design of a refinery follows the same aspects as plant design.

The design of a chemical plant (which is a refinery for our case) would normally go through the following steps:

- Inception of an idea (e.g., to produce a product).
- Find out if it is feasible to build a plant (technical and economic feasibility study).
- Carry out "Process Design" which involves three basic stages:
    a. Draw a *qualitative* block diagram based on a written description for the selected process.

   b. Carry out basic calculations using M.B. & E.B. (Material Balance & Energy Balance) to come up with a *quantitative* block diagram. Material balance is the basis of process design.
   c. Determine the size and capacity of equipment (equipment sizing).
- Do cost estimation for the capital investment of the plant.
- Carry out profitability analysis for the project.
- When it comes to computer applications, spread sheet software has become indispensable tool in plant design, because of the availability of personal computers, ease of use, and adaptability to many types of problems. On the other hand, many programs are available for the design of individual units of chemical process units. The Chemical Aids for Chemical Education (CACHE) Corporation makes available several programs mainly for educational use.

Apart from the engineering principles considered in the plant design, there are other important functions and items to be considered regarding safety, health, loss prevention, plant location, plant layout, and others.

A brief summary is given as follows:

- Health & Safety Hazards: One should consider toxicity of materials and frequency of exposure, fire, and explosion hazarads.
- Loss Prevention: HAZOP study.
- Environmental Protection and Pollution Control: This includes: air, water, solid wastes, thermal effects, noise effects, and others.
- For Plant Location: Primary factors and specific factors, both are to be considered.
- For Plant Layout: Optimum arrangement of equipment within a given area is a strategic factor.
- Plant Operation and Control: Designer should be aware of:
   a. Instrumentation
   b. Maintenance
   c. Utilities
   d. Structural design
   e. Storage
   f. Materials handling; pipes and pumps
   g. Patents aspects

Once crude oil is produced, exit oil well, it goes through the following treating operations before distillation (Figure 19.3).

If the mixture to be separated is a homogenous, single-phase solution, a second phase must generally be formed before separation takes place. This is an "*interphase*" operation, which involves the transfer of mass from one phase to another. This second phase is introduced by two methods:

a. By adding or removing energy: Energy Separating Agent (ESA), e.g., distillation.
b. By introducing a solvent: Mass Separating Agent, e.g., absorption.

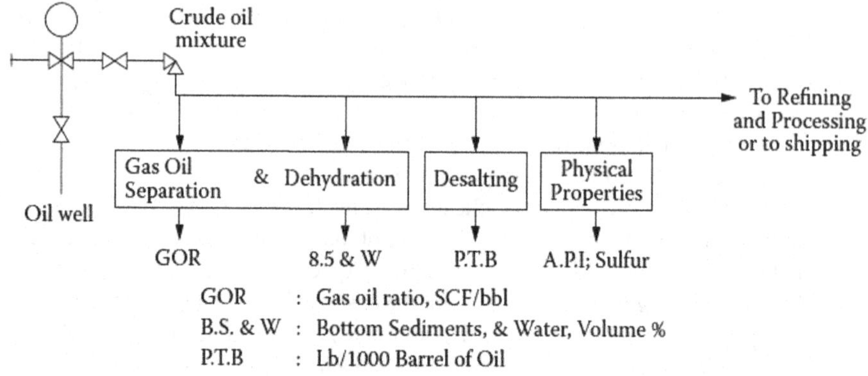

FIGURE 19.3 Surface petroleum operations for crude oil exit the well.

"Intraphase" separation, on the other hand, implies separation of components within a phase, such as diffusion through inert barriers or membranes. These are rate-governed operations.

Separation of components from a liquid mixture via distillation depends on the differences in boiling points of the individual components. Also, depending on the concentrations of the components present, the liquid mixture will have different boiling point characteristics. Therefore, distillation processes depends on the *vapor pressure* characteristics of liquid mixtures.

For separation to take place, say by distillation, the selection of an exploitable chemical or physical property difference is very important. Factors influencing this are:

a. The physical property itself.
b. The magnitude of the property difference.
c. The amount of material to be distilled.
d. The relative properties of different species

19.1.4 Different Types of Distillation Methods

1. Flash—Vapors are kept in intimate contact with the liquid.
2. Simple—Vapors are withdrawn as quickly as they are formed to be condensed by a condenser.

These methods are illustrated in Figure 19.4.

19.1.5 Design Aspects

There are two main factors that govern the design of equipment in diffusional operations:

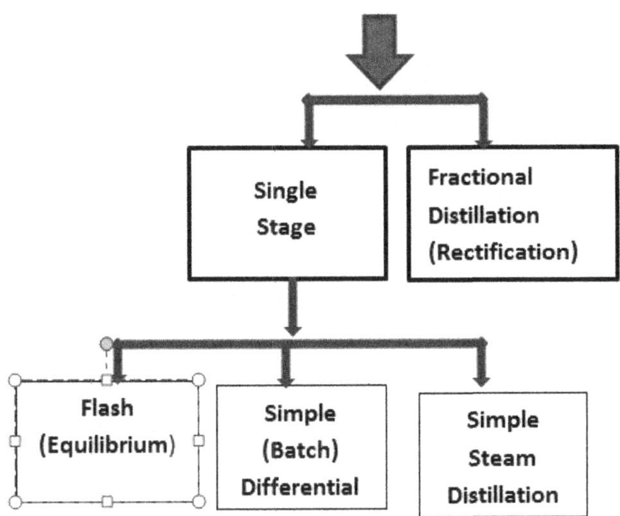

FIGURE 19.4 Classification of distillation method.

a. The thermodynamic equilibrium distribution of the components between the phases.
b. The rate of movement, diffusion rate, from one phase to the other.

Main factors to be considered in the design of finite-stage columns, other than calculating the number of the theoretical stages (plates) required for a given separation are the following:

a. Column diameter
b. Tray efficiency
c. Pressure drop across the tray

It should be pointed out that the number of plates in a column is a function of the degree of separation required, i.e.

$$\mathbf{N = f\,(separation)}$$

On the other hand, the diameter of a column is a function of the charge input to the column or capacity, i.e.

$$\mathbf{D = f(capacity)}$$

19.1.6 Operating Pressure

The primary physical separation process, which is used in almost every stage while processing the crude oil, is fractional distillation, as explained above. The distillation

TABLE 19.1
Three Systems of Oil Fractionation Wrt Operating Pressure

Operation features	Atmospheric distillation	Vacuum distillation	Pressure distillation
Application	Fractionation of crude oils	Fractionation of heavy residues (fuel oil)	Fractionation and/or separation of light hydrocarbons
Justification	Always, work near atmospheric pressure	To avoid thermal decomposition	To allow condensation of the overhead stream using cooling water
Extra equipment (as compared with atmospheric		Steam jet ejectors and condensors to produce and maintain vacuum	Stronger thickness for the vessel shell

operation can take place at atmospheric pressure, under vacuum, or under high-operating pressure. The three operations are common in the oil refining industry. For example, crude oil fractionation is always accomplished at atmospheric pressure (slightly higher), topped crude oil (fuel oil residue) is distilled under vacuum, while the *stabilization* of straight-run gasoline utilizes high-pressure fractionators or stabilizers. A comparison between these three systems of fractionation is shown in Table 19.1, which shows the technical merits and economic implications of each system.

19.1.6.1 Distillation Models

Distillation models are based on three pillars:

- Laws of conservation of mass and energy
- The concept of ideal stage
- Rault's law and Henery's law used (for ideal case) to describe the tendency of escape for vapor/liquid at equilibrium

19.1.7 Types of Refineries and Economic Analysis

Depending on the type of crude oil used, the processes selected, and the products needed, as well as the economic considerations involved, refineries can have different classifications, as shown in Figure 19.5. The products that dictate the design of a *fuel* refinery or conventional refinery are relatively few in number but are produced

CLASSIFICATIONS OF RREFINERIES

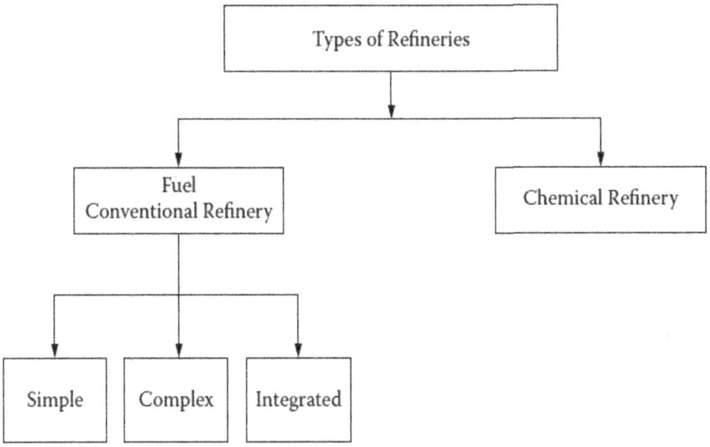

FIGURE 19.5 This illustratres these types of refinery.

in large quantities, such as gasoline, jet fuels, and diesel fuels. The number of products, however, increases with the degree of complexity of a fuel refinery, which varies from simple to complex or to fully integrated.

A simple refinery consists mainly of a crude oil atmospheric distillation unit, stabilization splitter unit, catalytic reforming plant, and product-treating facilities. Products are limited: LPG, gasoline, kerosene, gas oil, diesel oil, and fuel oil. A complex refinery will employ additional physical separation units (such as vacuum distillation) and a number of chemical conversion processes, including hydrocatalytic cracking, polymerization, alkylation, and others. The fully integrated refinery will provide other processes and operations necessary to produce practically all types of petroleum products, including lubrication oils, waxes, asphalts, and many others.

A *chemical* refinery, on the other hand, is a special case of the conventional oil refinery in which the emphasis is on manufacture of olefins and aromatics from crude oil. A chemical refinery can be defined as one that includes an olefin complex for the pyrolysis of petroleum fractions (e.g., C_2H_6 to C_2H_4). It must not produce motor gasolines; that is, it is a non-fuel-producing refinery. In other words, the purpose of chemical refining is to convert the whole crude oil directly into chemical feedstocks.

An example is the heavy oil cracking (HOC) process, in which the atmospheric residuum is catalytically cracked directly into lighter products. Chemical refining is an economically attractive venture for large chemical companies that can penetrate the market by selling large quantities of olefins and aromatics.

Economic analysis is used in refining to determine *the most economical refining operations*, to determine whether to use new or existing equipment, etc. Economic analysis, including cost analysis, is complicated in a refinery because an operation in a refinery with lower operating costs is not necessarily the most desirable procedure, and similarly, an operation giving higher yields, or production rates, is not necessarily a more economical one. A highest yield with lowest cost is what the refiner would like to achieve. Economic analysis is further complicated by the fact that several hundred different products may be produced from one basic raw material, crude oil.

There are also other complications. The basic crude may consist of a number of different crudes that have considerably different characteristics and different selling.

19.2 ECONOMIC BALANCE FOR DISTILLATION SCHEMES

19.2.1 Economic Balance

Economic balance in refining operations means that costs are balanced with revenue, inputs with outputs and crudes with refined products. The object is to find the combination of least cost with the “’greatest” contribution.

There are two corollaries of great significance to the oil refiner that follow from the principle of diminishing productivity: namely, the principle of variable proportion, and the principle of least-cost combination.

The principle of variable proportion enters into all decisions relative to combining economic factors (inputs) for full production. In chemistry, we know that elements combine in definite proportions. For instance, the combination of 2 atoms of hydrogen with 1 atom of oxygen will produce 1 molecule of water: $H_2 + O \dashrightarrow H_2O$. No other combination of hydrogen atoms and oxygen atoms will produce water. What is true in this instance is also true in all other chemical combinations, and in oil production as well. In other words, a law of definite proportions governs the combination of the various chemical elements and the various factors of production, such as amount of labor, materials consumed, and capital in a plant investment.

Economic balance applies to both physical operations (unit operations) and chemical conversion processes. It may involve a design problem or may address a processing operation or a separation step. In other words, economic balance may refer to the period before installation of equipment, in which case it consists of a study of costs and values received on *design* of equipment, or the period after installation of equipment, in which case it is a study of costs and values received on *processing operations*. The latter means on one hand an economic balancing of costs against optimum yield or optimum recovery, and on the other hand, elimination of as much waste as possible.

19.2.2 Economic Balance in Design

Design of equipment for process operations is a complex problem because of the many variables involved and the fact that broad generalizations about these variables cannot be made. Economic balance is not discussed in detail here, as much of it is beyond the scope of this book. A number of cases of economic balance in design, however, will be discussed.

- Economic balance in evaporation is a problem of determining the most economical number of effects to use in a multiple-effect evaporation operation. There is economy in increasing the amount of steam used because direct costs are reduced, but at the same time there is an increase in fixed costs when an increasing number of effects are used. So selection of which number of effects will balance direct costs is desirable.

- Economic balance in vessel design may involve specific design problems, such as heating and cooling, catalyst distribution, design of pressure vessels for minimum cost, etc.
- Economic balance in fluid flow involves the study of costs in which such direct costs as power costs for pressure drop and repairs, as well as fixed costs of pipe, fittings, and installation, are related to size of pipe. For example, power costs decrease as pipe size increases, and total costs are at a minimum point at some optimum pipe size.
- Economic balance in heat transfer requires an understanding of how fixed costs vary, with a selected common variable used as a basis for analysis. Variable costs must also be related to this same variable. Thus, both fixed costs and variable costs are required for economic balance.

In any study of either design or operations, only the variable cost often referred to as direct costs, which are affected by variations in operation is included.

The following case study stresses the role of an economic balance in design in many applications throughout the processing of crude oil, which may involve the transfer of material, heat, or mass with or without chemical conversions.

19.2.3 Case Study 1: Optimum Reflux Ratio

In designing a bubble plate distillation column, the design engineer must calculate:

1. The number of plates
2. The optimum reflux ratio
3. The diameter of the column

It is well established that if the reflux ratio is increased from its minimum value, R_m, the number of plates would be decreased to attain the same desired separation. This means lower fixed costs for the column. The other extreme limit for the reflux could be reached by further increase in R with corresponding decrease in the number of trays until the total reflux, R_t, is reached (case of minimum number of trays, N_m). Attention is now directed to the effect on the diameter of the column of increasing the reflux ratio, that is, increasing vapor load.

As R increases, the vapor load inside the column increases; consequently, the diameter of the column must be increased to attain the same vapor velocity. A point is reached where the increase in column diameter is more rapid than the decrease in the number of trays. Hence, the only way to determine the optimum conditions of reflux ratio that will result in the right number of trays for the corresponding column diameter is to use economic balance. For different variable reflux ratios, the corresponding annual fixed costs and operating costs must be combined and plotted versus the reflux ratio.

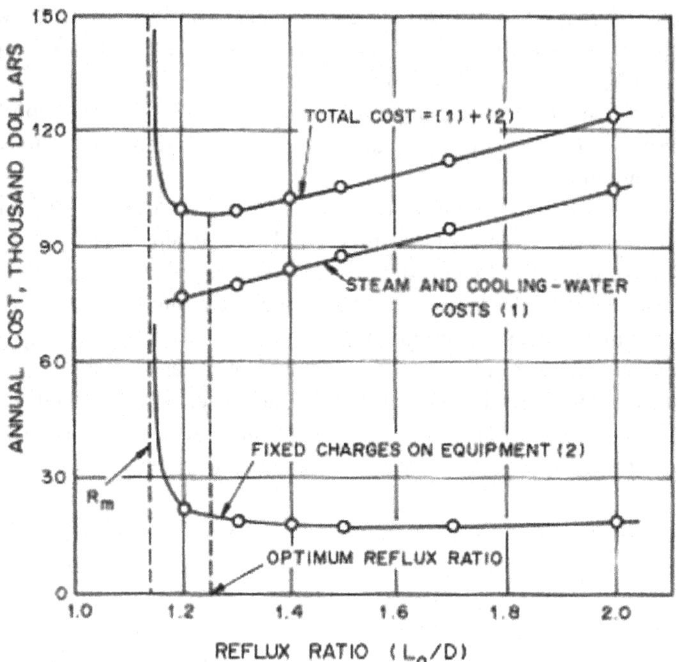

FIGURE 19.6 A plot of the total annual costs versus the reflux ratio.

(Source: Peters and Timmerhaus, Plant Design and Economics for Chemical Engineers, McGraw-Hill International, 1981.)

Annual fixed costs are defined as the annual depreciation costs for the column, the reboiler and the condenser, where the cost of a column for a given diameter equals the cost per plate of this particular diameter times the number of plates. Therefore, the operating cost equals the cost of the steam plus the cost of cooling water. Figure 19.6 illustrates how to obtain the optimum reflux ratio (a design parameter) by minimizing the total annual costs of the distillation column.

19.2.4 Case Study 2: Economic Balance in Yield and Recovery

Principles of economic balance must be applied to different processes in the oil refinery for the purpose of determining how variations in yield, as affected by design or operation will produce maximum profit. The effect of changing the crude feed and refined oil product compositions on the overall profit for a refinery process can best be illustrated, in most cases, as follow:

A typical study of economic balance in yield and recovery reveals that obtaining a higher-grade product from a fixed amount of given feed means an increase in variable costs because of costs of increased processing. The final refined oil product, of course, has a higher value, but for some product grades, the costs may equal the selling price, with the result that it becomes uneconomical to exceed that particular "'specification".

At some optimum grade of a product, however, a maximum gross profit, or difference between the sales dollars curve and the total costs curve, may be obtained per barrel of pure material (crude) in the feedstock.

In general, capacity is reduced as grade is increased, with the result that the maximum profit per barrel of pure material (crude) may not correspond to the maximum annual profit.

Although graphic analysis is the best procedure to use for such problems, there are also some useful mathematical relations. For example, if D is total refined product, F is total feed (crude), and Y is a conversion factor relating to feed (crude) and product (refined), then, under physical operations,

$$Y = D\text{, bbl of total refined product/F bbl of total crude feed}$$

or,

$$Y = \text{output/input}$$

or recovery in percent form. Also, if fixed costs are constant for a given process, then fixed costs will be constant for a given value of F or total feed (crude). However, as is usually the case, equipment costs will be higher for a higher-grade product, with the result that the annual fixed cost per unit of refined product increases.

For a given crude feed rate, raw material costs are constant, but refinery processing costs usually increase for a higher-grade product to give a variable cost curve that also increases.

The value of the finished product, like that of fixed costs and variable costs per unit of refined oil, will vary with the grade of product.

Figure 19.7 is a typical economic chart with curves illustrating economic balance curves in a refinery. Recovery, or ratio of output to input, in the oil refinery is greater than recovery in the oil fields. Note that to make a profit the refiner must stick to the product grades marked between A and B on Figure 19.7.

19.2.5 Selected Case Study 3: Crude Oil Desalter

The salt content of a Middle-Eastern crude oil (API gravity 24.2) was found to be 60 PTB. In order to ship and market this oil, it is necessary to install a desalting unit in the field, which will reduce the salt content to 15 PTB. This upgrading in the quality of oil in terms of an acceptable PTB could realize a possible saving of 0.1 $/bbl in the shipping cost of the oil.

Assume the following:

The crude oil desalter has a design capacity of 120,000 bbl/day.

The current capital investment of the desalting unit is estimated to be $3.0 million plus another $2.0 million for storage tanks and other facilities.

Service life of equipment is 10 years with negligible salvage value, while the operating factor = 0.95.

The total operating expenses of the desalter are estimated to be $10/1,000 bbl.

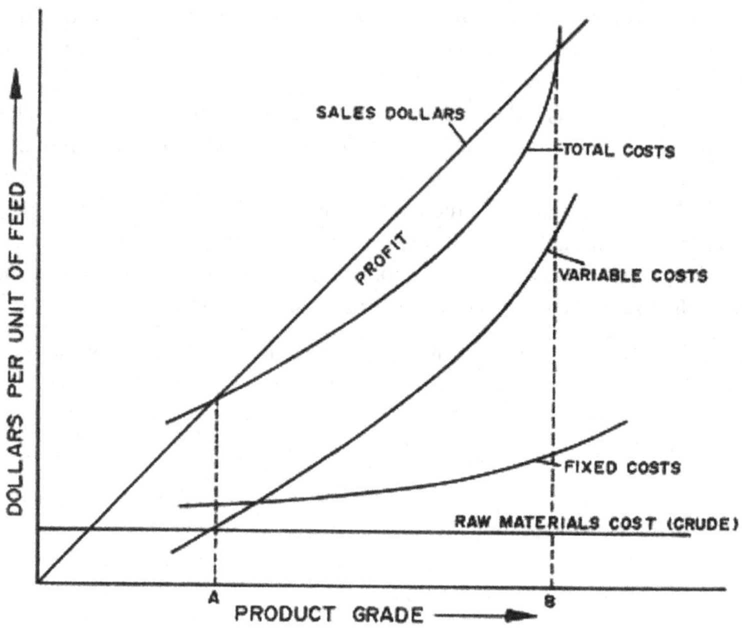

FIGURE 19.7 Economic level of refined oil production from a given feed of crude oil.

The annual maintenance expenses are 10% of the total capital investment. Evaluate the economic merits of the desalter by calculating, the ROI and payout period (P.P.).

SOLUTION

The total annual cost is the sum of the annual operating expenses plus annual depreciation costs. Assuming straight line depreciation:

$$\text{d, the annual depreciation cost} = 5\times106/10 = 0.5\times106 \text{ \$/year} \quad (19.1)$$

$$\begin{aligned}\text{The annual operating expenses} &= (10/1{,}000)(120{,}000)(365)(0.95)\\ &= 0.416\times106 \text{ \$/year}\end{aligned} \quad (19.2)$$

$$\begin{aligned}\text{The annual maintenance} &= (0.1)(5\times106)\\ &= 0.5\times106 \text{ \$/year}\end{aligned} \quad (19.3)$$

$$\text{Total annual costs} = (1)+(2)+(3)$$

$$\begin{aligned}\text{Annual savings} &= (0.1)(120{,}000)(365)(0.95)\\ &= 4.161\times10^{6} \text{ \$/year}\end{aligned}$$

$$\text{Net Savings} = (4.161 - 1.4161) \times 10^6$$
$$= 2.7449 \times 10^6 \,\$/\text{year}$$
$$\text{R.O.I} = \underline{\mathbf{54.9\%}}$$

19.3 SOLVED EXAMPLES AND COMPUTATIONS BY EXCEL

19.3.1 Theoretical Backgound

Thiel & Geddes method is used to calculate the number of trays. It involves the simultaneous solution of equilibrium relationships (VLE) and the operating line; where the operating line is used to compute the composition of one of the two streams passing each other for two consecutive plates, while the equilibrium relationship is used to compute the composition of either the vapor or liquid (in equilibrium) on the same plate.

19.3.2 Case Study 4: Plate-to-Plate Calculations (Case of Rectification Column)

This example is an oversimplified one for illustrative purposes.

Given:

1. Derivation of the Operating line: $\mathbf{y_{n+1} = [R/R + 1]\ X_n + [1/R + 1]\ X_D}$ Equation 19.1 R is reflux ratio (R.R.) = L/V, X_D is the composition of the overhead product.
2. Equilibrium Data: $\mathbf{X_n = Y_n/[Y_n + \alpha\ (1 - Y_n)]}$ Equation 19.2.

Statement of the Example:

Given: 40 mols/hr of feed (**vapor**) that contains 20% hexane and 80% octane entering the bottom plate, where D = 5 mols/hr, $X_D = 0.9$, R.R. = 7, $\alpha = 6$. Find the number of theoretical trays, N.

1st: Numerical Solution

Steps:

1. The liquid composition leaving the partial condenser (plate number 0) is in equilibrium with the vapor[top product] and is calculated by Equation 19.2. Hence, X_{reflux} (leaving plate 0) = 0.9/0.9 + 6(1-0.9) = 0.6.
2. Y_1 is calculated using Equation 19.1, substituting for R = 7, X = 0.6, and X_D = 0.9, we get $Y_1 = 0.637$.

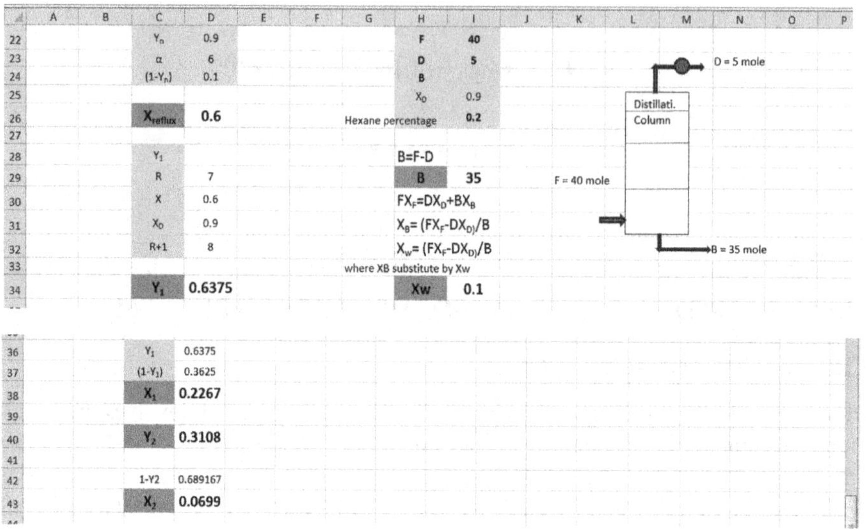

FIGURE 19.8 Solution of problem—Case study 4.

3. Get the equilibrium composition of the liquid on the same tray, $X_1 = 0.226$.
4. Again, using Equation 19.1, get $y_2 = 0.3$.
5. Next, get $X_2 = 0.066$, which is the bottom product leaving the column, call it x_w.

Finally, we make overall M.B., and C.M.B. (Component Material Balance): $(40)(0.2) = (5)(0.9) + (35)X_w$.

Solve for $X_w = 0.1$.

Therefore, 2 plates plus the condenser, make a total of **3 theoretical plates.**

2nd: Using Excel

19.3.3 Case Study 5: Plate-to-plate Calculations by Excel (*Case of Stripping Column*) Solution is Illustrated in Figure 19.8

A liquid mixture at the boiling point consists of 70 mole% Benzene and 30 mole% Toluene is fed to a stripping column. Pressure is taken 1 atm. Feed rate is 400 kg mole/hr. Stripping operation is carried out to achieve a bottom product W = 60 kg mole/hr. that contains no more than 2 mole% Benzene.

Solve the problem using Excel, in order to determine the number of theoretical trays, N required to obtain the desired specifications of the bottom product W. Use α_{AB} relative volatility for Benzene/Toluene, where $\alpha_{AB} = \mathbf{K}_{Benzene}/\mathbf{K}_{Toluene} = P^0{}_B/P^0{}_T = 2.45$

				BENZENE mol%	TOLUENE mol%	
F	FEED Lm	400 Kg mol/hr		70.0	30.0	100.0
W	BOTTOM	60 Kg mol/hr		10.0	90.0	100.0
D	TOP Vm	340 Kg mol/hr	T.M.B	80.6	19.4	100.0

Trayes	vapor Y	liquied X
1	0.806	0.629
2	0.757	0.560
3	0.677	0.461
4	0.560	0.342
5	0.420	0.228
6	0.286	0.140
7	0.183	0.084
8	0.116	0.051
9	0.078	0.033
10	0.057	0.024
11	0.046	0.019
12	0.040	0.017

Operating line constants	
Lm/Vm	1.1765
Constant	0.0176

Relative volatility	2.45

Design of distillation column
Plate-to-plate calculation case of striping column- binary system
By
GHADA EZZ ELDIN ALLAM
Under supervision of
DR. HUSSEIN ABDEL-AAL

FIGURE 19.9 Solution of problem—Case study 5.

SOLUTION

The number of trays required to reach a bottom product, exit the stripping column is found to be around 11 trays, as seen next; that corresponds to1.9 mole% benzene.

CASE STUDY: RECOVERY OF BUTANE USING LEAN OIL EXTRACTION

Objective:

This case study was presented as a senior project to students studying chemical engineering at KFUPM.

Case Study:

Associated natural gas is passed through an absorption unit to recover heavier hydrocarbons (butane plus), which can be sold for a value of \$7.5/gal. Calculations show that the minimum total cost for the recovery and the extraction of the butanes in the plant is estimated to be \$1.2/gal of *butane* recovered. Other additional costs for processing the absorbing oil used in the recovery are estimated to be \$27/million gal of the lean oil circulated.

The engineering group in the plant developed the following empirical relationship for the rate (R) of the absorber oil used as a function of the rate of butane produced (P):

$$\text{R, millions of gal/hr} = 0.004\ \text{P}^{1.3},\ \text{where P is in gal/hr}$$

1. Compute the optimum butane recovery P_o, and the optimum circulating oil rate, R_o applicable to this plant.
2. What is the value of P at which the process of recovery breaks even?

SOLUTION

$$\text{Profit} = \text{Income} - \text{Expenses}$$
$$\text{Income} = 7.5\ \$/\text{gal} * \text{P gal/hr} = 7.5\ \text{P}\ \$/\text{hr}$$
$$\text{Expenses} = 1.2\ \$/\text{m gal} * \text{P gal/hr} + 27\ \$/\text{m gal} * \text{R}$$
$$= 1.2\ \text{P} + 27\ (0.004\ \text{P}^{1.3}) = 1.2\ \text{P} + 0.108\ \text{P}^{1.3}\ \$/\text{hr}$$
$$\text{Profit} = 7.5\ \text{P} - 1.2\ \text{P} - 0.108\ \text{P}^{1.3} = 6.3\ \text{P} - 0.108\ \text{P}^{1.3}$$

d/dp(profit) = 6.3 – 1.3(0.108) $P^{0.3}$; setting this derivative equal to zero:

$$\text{P}_{\text{opt.}} = [6.3/1.3(0.108)]^{1/0.3} = \underline{320{,}647}\ \text{gal butane/hr}$$
$$\text{R}_{\text{opt.}} = 0.004(320{,}647)^{1.3} = \underline{57{,}530}\ \text{million gal oil/hr}$$

At the break-even point, profit = 0

$$6.3\ \text{P} - 0.108\ \text{P}^{1.3} = 0;\ \text{hence, P}^{0.3} = 6.3/0.108$$
$$\text{P}_{\text{B}} = (6.3/0.108)^{1/03} = \underline{768{,}777}\ \text{gal butane/hr}$$

CASE STUDY: UTILIZATION OF NATURAL GAS RECOVERED FROM GAS PLANT

This case study fits Chapter 19.

Objective:
To investigate the economics of utilizing natural gas as a fuel for heating crude oil.

Process Description:
Natural gas is recovered from gas–oil separator plant (GOSP) using an absorber de-ethanizer system, along with an amine treating unit and a gas dryer to have available desulfurized gas that can be used or sold as a fuel gas.

Given: The total cost for the recovery of this gas is estimated to be \$0.75/ MCF (Million Cubic Feet). It has been suggested to use this gas as a fuel for heating 5000 bbl/day of 40° API crude oil from 80°F to 250°F:

Find:

1. The cost of heating the crude oil using this gas.
2. Compare it with the cost of heating using fuel oil at \$2.2/MMBtu.
3. Do you recommend change in operation to use the fuel gas as a heating fuel instead of using the fuel oil?

SOLUTION

The heat duty required is calculated using the well-known equation: Q = m cp Δ T

$$= 127.7 \text{ MM Btu/day}$$

Assuming the heating value of the gas is 960 Btu/ft^3 and the heat efficiency is 60%; then the fuel gas consumption will be 221,700 ft^3/day.

$$\text{The cost of using this fuel gas for heating} = 2217000 \text{ ft3/day} \times \$0.75/\text{MCF}$$
$$= \$166.28/\text{day}$$
$$\text{The cost of using the fuel oil for heating} = [127.7 \text{ MM Btu} \times \$2.2/\text{MM Btu}]/0.6$$
$$= \$468.23/\text{day}$$

A daily savings in the cost of fuel of about \$300 is realized if the change to fuel gas takes place. One has to consider other economic factors in making this analysis. The capital cost involved in changing the burner system has to be considered.

CASE 2: HOW TO CONTROL THE CO_2 SPECS IN THE SWEET GAS? (DISCUSSION TYPE)

Methyl diethanolamine (MDEA) has become the amine molecule chosen to remove hydrogen sulfide, carbon dioxide, and other contaminants from hydrocarbon streams. Amine formulations based on MDEA can significantly reduce the costs of acid gas treating.

CASE STUDY: RECOVERY OF BUTANE USING LEAN OIL EXTRACTION

This case study fits Chapter 19.

Objective:
This case study was presented as a senior project to students studying chemical engineering at KFUPM, Dhahran, Saudi Arabia.

Case Study
Associated natural gas is passed through an absorption unit to recover heavier hydrocarbons (butane plus), which can be sold for a value of $7.5/gal. Calculations show that the minimum total cost for the recovery and the extraction of the butanes in the plant is estimated to be $1.2/gal of *butane* recovered. Other additional costs for processing the absorbing oil used in the recovery are estimated to be $27/million gal of the lean oil circulated.

The engineering group in the plant developed the following empirical relationship for the rate (*R*) of the absorber oil used as a function of the rate of butane produced (*P*):

$$\text{R, millions of gal/hr} = 0.004\ \text{P}^{1.3}, \text{ where P is in gal/hr}$$

1. Compute the optimum butane recovery P_o, and the optimum circulating oil rate, R_o applicable to this plant.
2. What is the value of *P* at which the process of recovery breaks even?

SOLUTION

$$\text{Profit} = \text{Income} - \text{Expenses}$$

$$\text{Income} = 7.5\ \$/\text{gal} * \text{P gal/hr} = 7.5\ \text{P}\ \$/\text{hr}$$

$$\text{Expenses} = 1.2\ \$/\text{m gal} * \text{P gal/hr} + 27\ \$/\text{m gal} * \text{R}$$

$$= 1.2\ \text{P} + 27\ (0.004\ \text{P}^{1.3}) = 1.2\ \text{P} + 0.108\ \text{P}^{1.3}\ \$/\text{hr}$$

$$\text{Profit} = 7.5\ \text{P} - 1.2\ \text{P} - 0.108\ \text{P}^{1.3} = 6.3\ \text{P} - 0.108\ \text{P}^{1.3}$$

d/dp(profit) = 6.3 - 1.3(0.108) $P^{0.3}$; setting this derivative equal to zero:

$$\text{P}_{\text{opt.}} = [6.3/1.3(0.108)]^{1/0.3} = \underline{320{,}647}\ \text{gal butane/hr}$$

$$\text{R}_{\text{opt.}} = 0.004(320{,}647)^{1.3} = \underline{57{,}530}\ \text{million gal oil/hr}$$

At the break-even point, profit = 0

$$6.3\ \text{P} - 0.108\ \text{P}^{1.3} = 0;\ \text{hence, } \text{P}^{0.3} = 6.3/0.108$$

$$\text{P}_{\text{B}} = (6.3/0.108)^{1/03} = \underline{768{,}777}\ \text{gal butane/hr}$$

20 Crude Oil Refining—Part 2

20.1 FLASH DISTILLATION

20.1.1 Introduction

Flash distillation (sometimes called "equilibrium distillation") is a single-stage separation technique. A liquid mixture feed is pumped through a heater to raise the temperature and enthalpy of the mixture. It then flows through a valve and the pressure is reduced, causing the liquid to partially vaporize.

Under the assumption of equilibrium conditions, and knowing the composition of the fluid stream coming into the separator and the working pressure and temperature conditions, we could apply our current knowledge of vapor/liquid/ equilibrium (flash calculations) and calculate the vapor and liquid fractions at each stage.

The problem of separating the gas from crude oil for well fluids (crude oil mixtures) breaks down to the well-known problem of flashing a partially vaporized feed mixture into two streams: vapor and liquid. In the first case, we use a gas–oil separator. In the second case, we use what we call a flashing column, as shown in Figure 20.1.

A "flash" is a single-stage distillation in which a feed is partially vaporized to give a vapor that is richer in the more volatile components. This is the case of a feed heated under pressure and flashed adiabatically across a valve to a lower pressure, the vapor being separated from the liquid residue in a flash drum. This is the case of "light liquids".

20.1.2 Conditions Necessary for Flashing

For flashing to take place, the feed has to be two-phase mixture, that is, it satisfies the following: $TBP < Tf < TDP$ (as indicated in Figure 20.2) or the sum of $ziKi$ for all components is greater than 1 and the sum of zi/Ki for all components is less than 1, where TBP, Tf, and TDP are the bubble point of the feed mixture, flash temperature, and dew point of the feed mixture, respectively; Zi and Ki are the feed composition and equilibrium constant, respectively, for component i.

The need to discuss flash calculation arises from the fact that it provides a tool to determine the relative amounts of the separation products V (gas) and L (oil) and their composition Yi and Xi, respectively.

The flash equation is derived by material balance calculations.

The following example illustrates the calculation procedure used in solving flash problems.

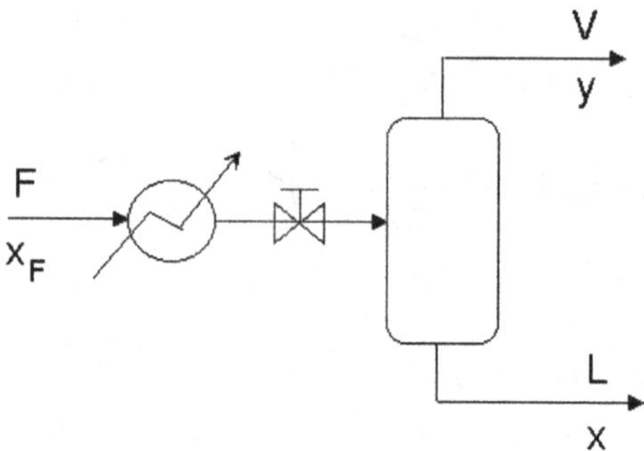

FIGURE 20.1 Flash column.

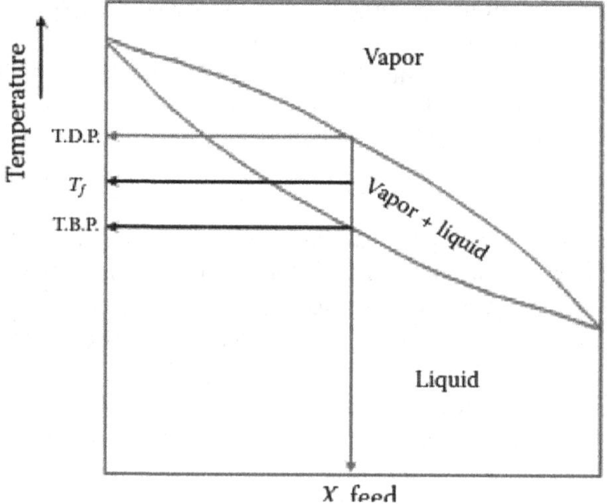

FIGURE 20.2 Conditions of flashing.

Example 20.1 (modified, after www.sciencedirect.com, topics, engineering)

Consider a gas with the following composition. This mixture is flashed at 1,000 psi. Determine the fraction of the feed-vaporized and composition of gas and liquid streams leaving the separator if the temperature of the separator is 150°F.

Component	Mole Fraction
C_1	0.70
C_2	0.07
C_3	0.03
C_4	0.05
C_5	0.05
C_6	0.1

SOLUTION

In this case we should determine the equilibrium ratio of each component at $T = 150°F$ and $P = 1,000$ psi as reported in the following table.

Component	Mole Fraction	P_c	T_c	ω_i	K_i
C_1	0.70	666.4	343.33	0.0104	3.1692
C_2	0.07	706.5	549.92	0.0979	1.1520
C_3	0.03	616.0	666.06	0.1522	0.5616
C_4	0.05	527.9	765.62	0.1852	0.2830
C_5	0.05	488.6	845.8	0.2280	0.1421
C_6	0.1	453	923	0.2500	0.0686

Then assume the vapor fraction $f = 0.5$. Calculation was carried out. The vapor fraction was found to be 0.78 after three iterations.

20.1.3 Pre-Flash Column

Pre-flash column is one of the major parts of a Crude Distillation Unit (CDU). The main purpose is to flash or vaporize the lighter (volatile) portion of the crude oil before it enters the furnace. This will reduce the feed rate and the energy load for the CDU.

The basic principle of this vaporization is the sudden decrease of pressure from around 12 bar to about 3 bar. This will cause an appreciable portion of the crude to get vaporized and is directed to the main distillation column bypassing the furnace. The remaining heavy portion is heated in the furnace and finally introduced in the main distillation column.

Item	Percent of "Other"	Percent of Total
Crude oil and blend stocks	–	85
Other operating costs		
Fuel oil, fuel gas	40.7	6.1
Electrical power	5.3	0.8
Maintenance	23.3	3.5
Operations	18	2.7
Catalysts and chemicals	12.7	1.9
Totals	**100**	**100**

Typical breakdown of refinery operating costs.

The advantage of having the pre-flash drum is to reduce the load in the furnace resulting in saving of fuels. Otherwise, we would have to heat the entire crude before introducing it to the crude distillation column. In this way it is possible to reduce the heat duty of the distillation unit and to have also an improvement of the hydraulic performance of the heat exchanger network.

Generally, every 30,000-barrel batch **takes** around 12–24 hours to undergo through analytical testing and pass quality control. A key stage is ultra-heating the **crude** to boiling point, with a distillation column used to separate the liquids and gases.

The problem of flash distillation is considered next. Flash calculations are very common, perhaps one of the most common chemical engineering calculations. They are a key component of simulation packages like Hysys, Aspen, etc.

20.1.3.1 Application

Let us use the feasibility study, as given next to calculate the savings in using a pre-flash column for a CDU processing 100,000 bbl/year, using the procedure outlined in Chapter 9 on Feasibility Study (Figure 20.3).

The cost of the pre-flash drum is assumed to be \$100,000, with a life time 5 years. This makes a TAC of \$20,000 per year, including an annual maintenance.

To determine the annual income or saving by installing the flash drum, bypassing the heating about 50% of vapor is flashed in the drum. This vapor is introduced directly to the main column for further rectification. This makes a saving in the energy, otherwise to be given to fractionate the vapor which is = $(q_1 - q_2)$.

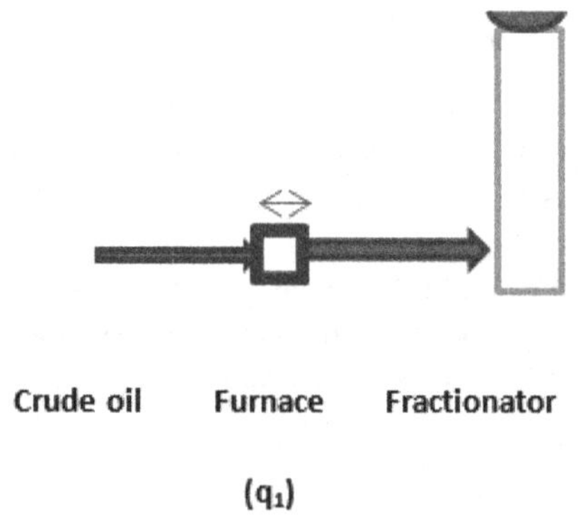

20.3.A- CDU without a preflash drum

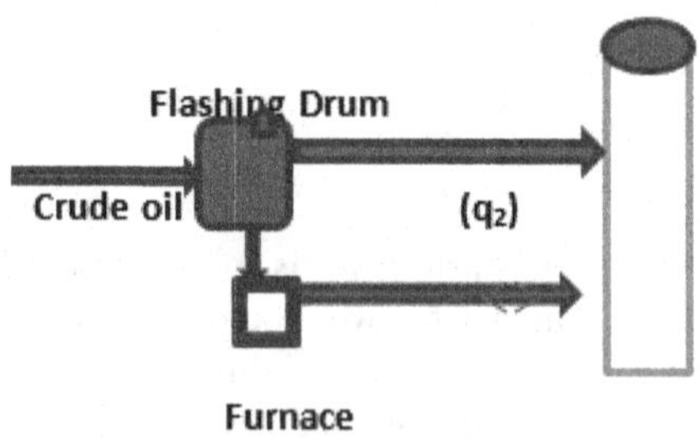

20.3.B- CDU with a preflash drum

FIGURE 20.3 Outline for a procedure for feasibility study.

Assuming the cost of refining oil is 4 \$/bbl (see foot note), the cost of refining a feed input of 100,000 bbl/year × 4 \$/bbl = 400,000 \$/year

Cost of electric power (energy) saved is about 0.8% of the total refinery operating cost. Since we are processing 50% only.

Therefore, saving (see footnote) = 400, 000 × 0.5 × 0.8 = \$160,000

$$\begin{aligned}\text{The net savings} &= \text{income} - \text{TAC}\\ &= 160{,}000 - 20{,}000\\ &= 140{,}000\ \$/\text{year}\end{aligned}$$

This makes an ARR = saving/TAC = (140,000/100,000) × 100 = 140%

This is a very attractive return to encourage the installation of flash drum in the CDU, which makes it a feasible project.

As in most questions the answer is, it depends on the type of crude oil, the configuration of the refinery, the size of the refinery, and the products produced. At a huge Texas refinery like Exxon Baytown, the cost is \$3–4 bbl. At a small California refinery like Torrance, the cost is \$10–12 per bbl.

20.2 CONVERSION OF CRUDE OIL INTO FUEL PRODUCTS AND OTHER VALUE-ADDED PRODUCTS

20.2.1 Introduction

This part discusses the various aspects of crude oil refining as a primary source of fuel and as a feedstock for petrochemicals. The main objective of chemical conversion in oil refining is to convert crude oils of various origins into valuable products having the qualities and quantities demanded by the market. Various refining processes based on chemical conversion such as thermal and catalytic processes as well as general properties of refined products are briefly reviewed.

Currently, the refining industry faces several challenges related to increasing demand for transportation fuels, stringent specifications of these products, crude oil availability, reduction of carbon emissions, and renewable fuels.

Refining does not lead to any chemical change instead entails in bringing only physical changes in crude. Its only in the downstream of refining process that some chemical processes are undertaken to improve the quality of petroleum products. First of all crude is a mixture of different hydrocarbon compounds (HC). In summary, petroleum refining implies changing crude oil into petroleum products for use as fuels for transportation, heating, and power generation.

The major unification process is called catalytic reforming and uses a catalyst (platinum, platinum-rhenium mix) to combine low weight naphtha into aromatics, which are used in making chemicals and in blending gasoline.

After distillation, heavy, lower-value distillation fractions can be processed further into lighter, higher-value products such as gasoline. This is where fractions from the distillation units are transformed into streams (intermediate components) that eventually become finished products.

Actual refinery operations are very complicated, but the basic functions of the refinery can be broken down into three categories of chemical processes:

- *Distillation* involves the separation of materials based on differences in their volatility. This is the first and most basic step in the refining process, and is the precursor to cracking and reforming.
- *Cracking* involves breaking up heavy molecules into lighter (and more valuable) hydrocarbons.
- *Reforming* involves changing the chemical nature of hydrocarbons to achieve desired physical properties (and also to increase the market value of those chemicals).

One can change one fraction into another by one of three methods shown here:

- Breaking large hydrocarbons into smaller pieces (cracking)
- Combining smaller pieces to make larger ones (unification)
- Rearranging various pieces to make desired hydrocarbons (alteration)

20.2.2 Cracking or Conversion Process

Cracking takes large hydrocarbons and breaks them into smaller ones. There are still many too heavy hydrocarbon molecules remaining after the separation process. To meet demand for lighter products, the heavy molecules are "cracked" into two or more lighter ones.

The conversion process, which is carried out at 500°C, is also known as catalytic cracking because it uses a substance called a catalyst to speed up the chemical reaction. This process converts 75% of the heavy products into gas, gasoline, and diesel. The yield can be increased further by adding hydrogen, a process called hydrocracking, or by using deep conversion to remove carbon.

The more complex the operation, the more it costs and the more energy it uses. The refining industry's ongoing objective is to find a balance between yield and the cost of conversion.

20.2.3 Major Chemical Conversion Processes

Processing options in a refinery can be classified into:

- **Light oil processing** prepares light distillates through rearrangement of molecules using isomerization and catalytic reforming or combination processes such as alkylation and polymerization.
- **Heavy oil processing** changes the size and/or structure of hydrocarbon molecules through thermal or catalytic cracking processes.
- **Treatment processes** involve a variety and combination of processes including hydrotreating, drying, solvent refining, and sweetening.

20.2.3.1 Light Oil Processing

(a) Catalytic Hydrotreating

Catalytic hydrotreating is used to remove about 90% of contaminants such as nitrogen, sulfur, oxygen, and metals from liquid petroleum fractions. These contaminants can have detrimental effects on the equipment and the quality of the finished product. Hydrotreating for sulfur or nitrogen removal is called hydrodesulfurization (HDS) or hydrodenitrogenation (HDN), respectively. World capacity for all types of hydrotreating currently stands at about 45.7 million b/d. Hydrotreating is used to pretreat catalytic reformer feeds, saturate aromatics in naphtha, desulfurize kerosene/jet, diesel, distillate aromatics saturation, and to pretreat catalytic cracker feeds. Hydrotreating processes differ depending upon the feedstock available and catalysts used. Mild hydrotreating is used to remove sulfur and saturate olefins. More severe hydrotreating removes nitrogen, additional sulfur, and saturates aromatics. In a typical catalytic hydrotreater, the feedstock is mixed with hydrogen, preheated in a fired heater (315–425°C) and then charged under pressure (up to 68 atm) through a fixed-bed catalytic reactor. In the reactor, sulfur and nitrogen compounds in the feed are converted into H_2S and NH_3. Hydrotreating catalysts contain cobalt or molybdenum oxides supported on alumina and less often nickel and tungsten.

(b) Catalytic Naphtha Reforming

The reforming process combines catalyst, hardware, and process to produce high-octane reformate for gasoline blending or BTX (benzene, toluene, and xylene) aromatics for petrochemical feedstocks. Reformers are also the source of much needed hydrogen for hydroprocessing operations. Naphtha reforming reactions comprise cracking, polymerization, dehydrogenation, and isomerization that take place simultaneously. UOP and Axens are the two major licensors and catalyst suppliers for catalytic naphtha reforming. There is a necessity of hydrotreating the naphtha feed to remove permanent reforming catalyst poisons and to reduce the temporary catalyst poisons to low levels. Currently, there are more than 700 reformers worldwide with a total capacity of about 11.5 million b/d [1]. About 40% of this capacity is located in North America followed by 20% each in West Europe and Asia-Pacific regions. Reforming processes are generally classified into semi-regenerative, cyclic, and continuous catalyst regenerative (CCR). Most grassroots reformers are designed with continuous catalyst regeneration. CCR is characterized by high catalyst activity with reduced catalyst requirements, more uniform reformate of higher aromatic content, and high hydrogen purity.

(c) Isomerization

Isomerization is an intermediate feed preparation-type process. There are more than 230 units worldwide with a processing capacity of 1.7 million bbl/d of light paraffins. Two types of units exist: C_4 isomerization and C_5/C_6 isomerization. A C_4 unit converts normal butane into isobutane, to provide additional feedstock for alkylation units, whereas a C_5/C_6 unit isomerizes mixtures of C_5/C_6 paraffins, saturates benzene, and removes naphtenes. Isomerization is similar to catalytic reforming in that the hydrocarbon molecules are rearranged, but unlike catalytic reforming, isomerization just converts normal paraffins to isoparaffins. The greater value of branched paraffins over straight paraffins is a result of their higher octane contribution. The extent of paraffin isomerization is limited by a temperature dependent thermodynamic equilibrium. For these reactions a more active catalyst permits a lower reaction temperature and that leads to higher equilibrium levels. Isomerization of paraffins takes place under medium pressure (typically 30 bar) in a hydrogen atmosphere.

(d) Alkylation

Alylation is the process that produces gasoline-range compounds from the combination of light C_3-C_5 olefins (mainly a mixture of propylene and butylene) with isobutene. The highly exothermic reaction is carried out in the presence of a strong acid catalyst, either sulfuric acid or hydrofluoric acid. World alkylation capacity is currently 2.1 million b/d [1]. The alkylate product is composed of a mixture of high-octane, branched-chain paraffinic hydrocarbons. Alkylate is a premium clean gasoline blending with octane number depending upon the type of feedstocks and operating conditions. Research efforts are directed toward the development of environmentally acceptable solid superacids capable of replacing HF and H_2SO_4.

(e) Polymerization and Dimerization

Catalytic polymerization and dimerization refer to the conversion of FCC light olefins such as ethylene, propylene, and butenes into higher-octane hydrocarbons for gasoline blending. The process combines two or more identical olefin molecules to form a single molecule with the same elements in the same proportions as the original molecules. World capacity of polymerization and dimerization processes is about 195,000 b/d [1]. In the catalytic process, the feedstock is either passed over a solid phosphoric acid catalyst on silica or comes in contact with liquid phosphoric acid, where an exothermic polymeric reaction occurs. Another process uses homogenous catalyst system of aluminum-alkyl and a nickel coordination complex. The hydrocarbon phase is separated, stabilized, and fractionated into liquefied petroleum gas (LPG) and oligomers or dimers.

20.2.3.2 Heavy Distillate Processing

(a) Fluid Catalytic Cracking (FCC)

Catalytic cracking is the largest refining process for gasoline production with global capacity of more than 14.4 million b/d [1]. The fluidized catalytic process (FCC) converts heavy feedstocks such as vacuum distillates, residues, deasphalted oil into lighter products rich in olefins and aromatics. FCC catalysts are typically solid acids of fine-particles especially zeolites (synthetic Y-Faujasite) with content generally in the range of 5–20 wt% while the balance is silica-alumina amorphous matrix. Additives to the FCC catalyst make no more than 5% of the catalyst and they are basically used to enhance octane, as metal passivator, SOx reducing agents, CO oxidation, enhance propylene and reduce gasoline sulfur. The FCC unit comprises a reaction section, product fractionation, and regeneration section. Typical operating temperatures of the FCC unit are from 500 to 550°C at low pressures. Hydrocarbon feed temperatures range from 260 to 425°C while regenerator exit temperatures for hot catalyst are 650 to 815°C. Since the FCC unit is a major source of olefins (for downstream alkylation unit or petrochemical feedstock), an unsaturated gas plant is generally considered a part of it.

(b) Catalytic Hydrocracking

Catalytic hydrocracking of heavy petroleum cuts is an important process for the production of gasoline, jet fuel, and light gas oils. The world capacity for hydrocarcking is about 5.5 million b/d [1]. The process employs high pressure, high temperature, a catalyst, and hydrogen. In contrast to FCC, the advantage of hydrocracking is that middle distillates, jet fuels, and gas oils of very good quality are provided. In general, hydrocracking is more effective in converting gas oils to lighter products, but it is more expensive to operate. Heavy aromatic feedstock is converted into lighter products under a wide range of very high pressures (70–140 atm) and fairly high temperatures (400–820°C) in the presence of hydrogen and special catalysts. Hydrocracking catalysts have bifunctional activity combining an acid function (halogenated aluminas, zeolites) and a hydrogenating function (one or more transition metals, such as Fe, Co, Ni, Ru, Pd, and Pt, or by a combination of Mo and W).

20.2.3.3 Residual Oil Processing

(a) Coking

About 90% of coke production comes from delayed coking. The process is one of the preferred thermal cracking schemes for residue upgrading in many refiners, mainly in the U.S. The process provides essentially complete rejection of metals and carbon while providing partial or complete conversion to naphtha and diesel. World capacity of coking units is 4.7 million b/d (about 54% of this capacity is in the U.S. refineries) and total coke production is about 172,000 t/d [1]. New cokers are designed to minimize coke and produce a heavy coker gas oil that is catalytically upgraded. The yield slate for a delayed coker can be varied to meet a refiner's objectives through the selection of operating parameters. Coke yield and the conversion of heavy coker gas oil are reduced, as the operating pressure and recycle are reduced and to a lesser extent as temperature is increased.

(b) Visbreaking

Visbreaking is a non-catalytic residue mild-conversion process with a world capacity of 3.8 million b/d [1]. The process is designed to reduce the viscosity of atmospheric or vacuum residues by thermal cracking. It produces 15–20% of atmospheric distillates with proportionate reduction in the production of residual fuel oil. Visbreaking reduces the quantity of cutter stock required to meet fuel oil specifications and, depending upon fuel oil sulfur specs, typically reduces the overall quantity of fuel oil produced by 20%. In general, visbreakers are typically used to process to vacuum residues. The process is available in two schemes: coil cracker and soaker cracker. The coil cracker operates at high temperatures during a short residence time of about 1 minute. The soaker cracker uses a soaking drum at 30–40°C at about 10–20 residence time.

(c) Residue Hydrotreating and RFCC

Refineries that have substantial capacity of coking, visbreaking, or deasphalting are faced with large quantities of visbreaker tar, asphalt, or coke, respectively. These residues have high viscosity and high organic sulfur content (4–7 wt%) with primary consequences reflected in the potential for sulfur emissions and the design requirements for sulfur removal system. Residue hydrotreating is another method for reducing high-sulfur residual fuel oil yields. Atmospheric and vacuum residue desulfurization units are commonly operated to desulfurize the residue as a preparatory measure for feeding low sulfur vacuum gas-oil feed to cracking units (FCC and hydrocrackers), low sulfur residue feed to delayed coker units, and low sulfur fuel oil to power stations. The processing units used for hydrotreating of resids are either a down-flow, trickle phase reactor system (fixed catalyst bed) or a liquid recycle and back mixing system (ebullating bed). Economics generally tend to limit residue hydrotreating applications to feedstocks containing less than 250 ppm nickel and vanadium.

Residue FCC (RFCC) is a well established approach for converting a significant portion of the heavier fractions of the crude barrel into a high-octane gasoline blending component. In addition to high gasoline yields, the RFCC unit also produces gaseous, distillate, and fuel oil-range products. The RFCC unit's product quality is directly affected by its feedstock quality. In particular, unlike hydrotreating, RFCC redistributes sulfur, but does not remove it from the products. Consequently, tightening product specifications have forced refiners to hydrotreat some, or all, of the RFCC's products. Similarly, in the future the SOx emissions from an RFCC may become more of an obstacle for residue conversion projects. For these reasons, a point can be reached where the RFCC's profitability can economically justify hydrotreating the RFCC's feedstock. Table 20.1 indicates these RFCC.

(d) Residue Gasification

The gasification of refinery residues into clean syngas provides an alternative route for the production of hydrogen and the generation of electricity in a combined turbine and steam cycle. Compared to steam-methane reforming, gasification of residues can

TABLE 20.1
Typical Yields (%) of Refineries Processing Selected Crude Oils—US Gulf Coast

Major Refined Product	Boiling Point, °C	West Texas Intermediate	Arabian Light	Arabian Heavy	Nigerian Bonny Light
Gasoline	10–200	48.1	38.9	36.8	44.9
Kerosene/Jet	200–260	8.1	8.2	6.7	7.8
Diesel	260–345	30.9	24.7	9.7	39.6
Fuel Oil	345+	9.8	23.7	41.6	4.5

be a viable process for refinery hydrogen production when natural gas price is in the range of $3.75–4.00/MMBtu. The largest application of syngas production is in the generation of electricity power by the integrated gasification combined cycle (IGCC) process. Electricity consumption in the modern conversion refinery is increasing and the need for additional power capacity is quite common, as is the need to replace old capacity. The IGCC plant consists of several steps: gasification section, gas desulfurization and combined cycle system.

(e) Aromatics Extraction

BTX aromatics are high-value petrochemical feedstocks produced by the catalytic naphtha reforming and extracted from the reformate stream. Whether or not other aromatics are recovered, it is sometimes necessary to remove benzene from reformate in order to meet mandated specifications on gasoline composition. Aromatics production in refineries reached 1.4 million b/d in 2011 [1]. Most new aromatic complexes are configured to maximize the yield benzene and paraxylene and sometimes orthoxylene. The solvents used in the extraction of aromatics include dimethylformamide (DMF), formylmorpholine (FM), dimethyl sulfoxide (DMSO), sulfolane, and ethylene glycols.

(f) Sulfur Recovery

Sulfur recovery converts hydrogen sulfide in sour gases and hydrocarbon streams to elemental sulfur. Total sulfur production in world refineries reached about 84,000 tons/d in 2011 compared to about 28,000 tons/d in 1996 corresponding to a yearly growing recovery rate of 14%. In other words, an average today's refinery recovers 1.0 kg sulfur from one processed barrel of crude oil compared to less than 0.4 kg sulfur recovered in 1996. This indicates the increasing severity of operations to meet stringent environmental requirements. The most widely used sulfur recovery system is the Claus process, which uses both thermal and catalytic-conversion reactions. A typical process produces elemental sulfur by burning hydrogen sulfide under controlled conditions. Knockout pots are used to remove water and hydrocarbons from feed gas streams. The gases are then exposed to a catalyst to recover additional sulfur. Sulfur vapor from burning and conversion is condensed and recovered.

20.3 REFINERY END PRODUCTS

20.3.1 INTRODUCTION

Petroleum refining implies changing crude oil into petroleum products for use as fuels for transportation, heating, and power generation. A base stock for making chemicals is recognized in the field of petrochemical industry. Refining breaks crude oil down into its various components, which are then selectively reconfigured into new products, as will be demonstrated.

The most widely used conversion method is called *cracking* because it uses heat, pressure, catalysts, and sometimes hydrogen to crack heavy hydrocarbon molecules into lighter ones. A cracking unit consists of one or more tall, thick-walled, rocket-shaped reactors and a network of furnaces, heat exchangers, and other vessels. Complex refineries may have one or more types of crackers, including fluid catalytic cracking units and hydrocracking/hydrocracker units.

Cracking is not the only form of crude oil conversion. Other refinery processes rearrange molecules to add value rather than splitting molecules. Normally, all refineries have three basic structures:

- Separation
- Conversion
- Treatment

The primary end-products produced in petroleum refining may be grouped into four categories: light distillates, middle distillates, heavy distillates, and others.

Petroleum refining implies changing crude oil into petroleum products for use as fuels for transportation, heating, and power generation. A base stock for making chemicals is recognized in the field of petrochemical industry. Refining breaks crude oil down into its various components, which are then selectively reconfigured into new products, as will be demonstrated.

Petroleum refineries process crude oil into many different petroleum products. The physical characteristics of crude oil determine how the refineries turn it into the highest value products.

Petroleum products are usually grouped into four categories: light distillates (LPG, gasoline, naphtha), middle distillates (kerosene, jet fuel, diesel), heavy distillates, and residuum (heavy fuel oil, lubricating oils, wax, asphalt). These require blending various feedstocks, mixing appropriate additives, providing short-term storage, and preparation for bulk loading to trucks, barges, product ships, and railcars. This classification is based on the way crude oil is distilled and separated into fractions [2].

The LPG constitutes the lowest boiling point (most volatile) product from a refinery and higher boiling fractions lead to most desirable distillate liquids, such as **gasoline**, jet fuel, diesel fuel, and fuel oil in the increasing order of boiling points, while asphalt is made from the residual.

20.3.2 Primary End-Products

Four categories are produced in petroleum refining: light distillates, middle distillates, heavy distillates, and others.

Light distillates:

- C1 and C2 components
- LPG
- Light naphtha
- Gasoline (petrol)
- Heavy naphtha

Middle distillates:

- Kerosene
- Automotive and rail-road diesel fuels
- Residential heating fuel
- Other light fuel oils

Heavy distillates:

- Heavy fuel oils
- Wax
- Lubricating oils
- Asphalt

Others:

- Coke (similar to coal)

Some of the major products from a typical refinery are:

- Propane: Used as a feedstock for ethylene cracking, or blended into LPG for uses as a fuel
- Butane: Used as a feedstock for ethylene cracking, or blended into LPG for uses as a fuel
- LPG: A blend of propane and butane used as fuel
- Light naphtha: Used as feedstock into ethylene crackers
- Gasoline: Used as a transportation fuel for passenger cars and light trucks
- Aviation gasoline: Used as an engine fuel in light aircraft
- Jet fuel: Used as a fuel for jet aircraft
- Kerosene fuel oil: Used as a residential cooking, heating, and lighting fuel
- Diesel: Used as a fuel for heavy-duty trucks, trains, and heavy equipment
- Industrial gasoil: Used as a furnace fuel in industrial plants and commercial/residential heating (heating oil)
- Residual fuel oil: Used as a fuel in power generation and for large ocean-going ships (bunker fuel)

Many refineries also produce specialty or non-fuel products such as:

- Asphalt: Used to pave roads and in the manufacture of building materials (e.g., roof shingles)
- Base oils: Used to make lubricating oils for use in industrial machinery and vehicle engines
- Propylene: Can be separated for sale to the petrochemicals industry
- Aromatics: Can be separated from reformate for sale to the petrochemicals industry

- Wax: Extracted from lubricating oil and either sold as a feedstock to specialty wax production (as slackwax) or treated at the refinery to a finished wax product
- Grease: Used as a solid lubricating oil, mostly in industrial uses
- White oil: A colorless, odorless, tasteless oil used by the food, cosmetics, and pharmaceuticals industries
- White spirit: Naphtha range material used as an industrial or household solvent
- Sulfur: A contaminant when present in other products, but once separated, it can be sold as a feedstock to the petrochemicals industry
- Pet coke: A by-product of the coking process that can be sold as a fuel for power plants and cement plants or to manufacture electrodes and anodes

20.3.3 Methodology

Petroleum refineries process crude oil into many different petroleum products. The physical characteristics of crude oil determine how the refineries turn it into the highest value products.

Petroleum products are usually grouped into four categories: light distillates (LPG, gasoline, naphtha), middle distillates (kerosene, jet fuel, diesel), heavy distillates, and residuum (heavy fuel oil, lubricating oils, wax, asphalt). These require blending various feedstocks, mixing appropriate additives, providing short-term storage, and preparation for bulk loading to trucks, barges, product ships, and railcars. This classification is based on the way crude oil is distilled and separated into fractions.

The LPG constitutes the lowest boiling point (most volatile) product from a refinery and higher boiling fractions lead to most desirable distillate liquids, such as **gasoline**, jet fuel, diesel fuel, and fuel oil in the increasing order of boiling points, while asphalt is made from the residual.

In the U.S. refineries, a principal focus is on the production of gasoline because of high demand. Diesel fuel is the principal refinery product in most other parts of the world.

The physical characteristics of crude oil determine how refineries process it. In simple terms, crude oils are classified by density and sulfur content. Less dense (lighter) crude oils generally have a higher share of light hydrocarbons. Refineries can produce high-value products such as gasoline, diesel fuel, and jet fuel from light crude oil with simple distillation. When refineries use simple distillation on denser (heavier) crude oils, it produces low-value products. Heavy crude oils require additional, more expensive processing to produce high-value products. Some crude oils also have a high sulfur content, which is an undesirable characteristic in both processing and product quality.

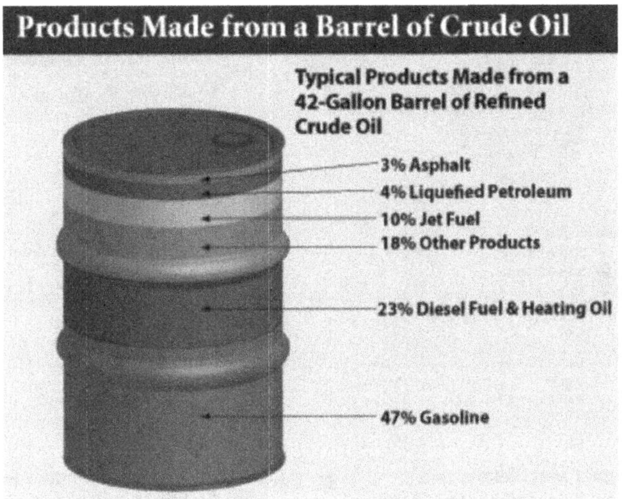

FIGURE 20.4 Refinery products.

20.3.3.1 Process Gain

The total volume of products refineries produce (output) is greater than the volume of crude oil that refineries process (input) because most of the products they make have a lower density than the crude oil they process. This increase in volume is called *processing gain*:

$$\text{Process gain} = \text{output} - \text{input}$$

The average processing gain at the U.S. refineries was about 6.5% in 2018. In 2018, the U.S. refineries produced an average of about 44.7 gallons of refined products for every 42-gallon barrel of crude oil they refined (Figure 20.4).

The U.S. refiner input versus output is shown next (Figure 20.5)

END OF CHAPTER: ECONOMIC EVALUATION AND APPLICATION

CASE STUDY 1: REFINERY COMPLEXITY INDEX

The concept of refinery complexity was introduced by W. Nelson back in the 1960s to quantify the relative cost of processing units that make up a refinery [4]. A refinery's complexity index indicates how complex it is in relation to a refinery that performs only crude distillation. Table 20.2 presents a list of complexity factors for refinery processes that are used in the calculation of refinery complexity index. A complexity factor of 1 was assigned to the atmospheric distillation unit and expressed the cost of all other units in terms of their cost relative to distillation.

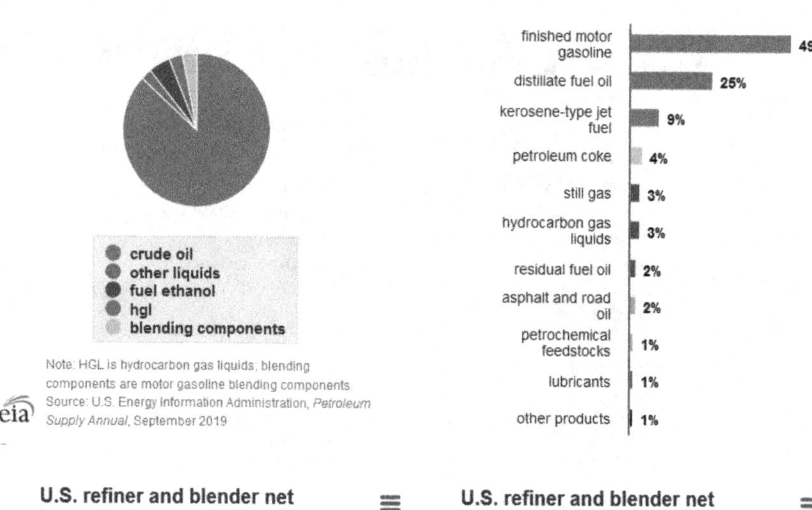

FIGURE 20.5 Input Vs Output.

Source: www.eia.gov › oil-and-petroleum-products › refining-c…

Source: Processing Gain—US Energy Information Administration-EIA

TABLE 20.2
Complexity Factors of Refinery Processes

Processing Unit	Complexity Factor
Atmospheric Distillation	1.0
Vacuum Distillation	2.0
Thermal Cracking	3.0
Delayed/Fluid Coking	6.0
Visbreaking	2.5
Catalytic Cracking (FCC)	6.0
Catalytic Reforming	5.0
Catalytic Hydrocracking	6.0
Catalytic Hydrorefining	3.0
Catalytic Hydrotreating	2.0
Alkylation	10.0
Aromatics, BTX	15.0
Isomerization	15.0
Polymerization	10.0
Lubes	6.0
Asphalt	1.5
Hydrogen Manufacturing, MMscfd	1.0
Oxygenates	10.0

TABLE 20.3
Complexity Index of Various Refineries

Refinery Type	Process	Complexity
Coking	Coking/resid upgrading to process medium/sour crude oil.	9
Cracking	Vacuum distillation and catalytic cracking to process light sour crude oil to produce light and middle distillates.	5
Hydroskimming	Atmospheric distillation, naphtha reforming, and desulfurization to process light sweet crude oil to produce gasoline.	2
Topping	Separate crude oil into refined products by atmospheric distillation, produce naphtha, but no gasoline.	1

For example, if a CDU of 100,000 b/d capacity costs $20 million to build, then the unit cost/daily barrel of throughput would be $200/b/d. If a 20,000 b/d catalytic reforming unit costs $20 million to build, then the unit cost is $1,000/b/d of throughput and the "complexity factor" of the catalytic reforming unit would be 1,000/200 = 5.

The complexity rating of a refinery is calculated by multiplying the complexity factor of each process by the percentage of crude oil it processes, then totaling these individual factors. This method accounts only for the refinery processing units of the Inside Battery Limits (ISBL) units, and not for off-sites and utilities. As an example, consider the case of a refinery with 400,000 b/d crude capacity and 140,000 b/d vacuum distillation capacity. The throughput of the vacuum tower relative to the crude distillation capacity is 35%. Given a vacuum unit complexity factor of 2, then the contribution of this unit to the overall complexity is: 2 × 0.35, or 0.7.

The complexity index can be generalized across any level of aggregation, such as a company, state, country, or region. In general, refineries can be classified as hydroskimming, cracking, and deep conversion, in order of both increasing complexity and cost. Table 20.3 compares the complexity indices of various types of refineries. A high-conversion coking refinery has a complexity index of 9 compared with a topping refinery that has a complexity index of 1. The most complex, deep conversion refinery is able to transform a wide variety of crudes, including the lower quality heavy sour crudes into the higher value products (e.g., gasoline, diesel). The ability to meet stringent product specifications, notably ultra low sulfur gasoline and diesel fuel, is also a characteristic of high complexity refineries.

CASE STUDY 2: REFINERY COST AND PROFITABILITY

As with any manufacturing plant, refinery costs are mainly associated with refinery construction cost (capital) and refinery operation cost (variable and fixed). In estimating construction cost, data are correlated with variables such as capacity, process units, complexity, location, and type of crude processed (light sweet, heavy sweet, light sour, or heavy sour crude). For complexity, the Nelson complexity index is generally used because it is publicly available on Oil & Gas Journal, which also publishes cost indices that can be used to estimate and update these costs using Nelson-Farrar Cost Index [4]. These include data on pumps and compressors, electrical machinery, internal-combustion engines, instruments, heat exchangers, chemical costs, materials component, fuel cost, labor cost, wages, chemical costs, maintenance, etc. One remark on the utilization of complexity factors in estimating cost is that they do not account for the impact of capacity on cost because the complexity factor is capacity-invariant, and trends in complexity factors change slowly (or not at all) over time.

Refiners may undertake capital investment for a variety of reasons, for example, expanding existing or creating new production facilities, implementing new or enhanced technology, or regulatory compliance. Facility expansion and new technology implementation are indicators that the industry expects increasing demand and economic growth. While there has not been a new refinery constructed in the last 30 years, there has been an increase in the U.S. refining capacity. At current situation, it is more cost effective to expand a refinery in the U.S. than to build new. API estimates it would cost at least $24,000 per daily barrel of oil process for a new refinery and $15,000 per daily barrel of oil process for the expansion of an existing refinery. Moreover, the permitting process for a new refinery could take at least 5–10 years.

Refinery gross profitability (margin) is a measure of the economics of a specific refinery. It is measured as the difference ($/b) between refinery's product income (total of barrels for each product multiplied by the price of each product) and the cost of raw materials (crude oil and other chemicals/catalysts). For example, if a refinery receives $120 from the sale of the products refined from a barrel of crude oil that costs $100/bbl, then the refinery gross margin is $20/bbl. The net (cash) margin is equal to the gross margin minus operating costs. Therefore, for this specific refinery that has an operating cost of $6/b, its net margin is $14/b. In many occasions, the measure of refinery profitability is complicated by the fact that the refinery produces several hundred different products from a mixture of different crudes that have various characteristics and different selling prices. Furthermore, it is becoming increasingly more difficult for refiners to determine which products are prime products and which are by-products [5,6].

Refining profitability varies according to competitive market demand for refined products. It may range between -$2/b, up to $20/b or more in refinery markets that have very limited spare capacity. In competitive markets the refinery margins change daily as the market prices of both crude oil and products change [7]. Under such conditions the refinery revenues (average margin × throughput) over the course of a year must be equal to or exceed its operating costs, depreciation, and taxes, plus a fair return on investment.

CASE STUDY 3: INTEGRATION AND ENVIRONMENTAL ISSUES

Refining-petrochemical integration is mainly carried out between a refinery and an aromatic complex, or between a refinery and olefins plants (steam crackers). While aromatics (paraxylene and benzene) are readily traded, olefins require further processing to polyolefins or other derivatives. Further integration issues suggests the utilization of FCC gasoline that is highly aromatic and naphthenic for aromatics production. Moreover, FCC units have long been a source of petrochemical propylene using special process designs and catalysts.

It is strongly believed that refining and petrochemical integration improves refining margins and overall profitability of the integrated venture. Industry experience has shown that refineries that are integrated with petrochemicals had greater savings in investment cost and operating costs. Other drivers for this integration include flexibility in upgrading low-value fuel streams to petrochemical feed and the utilization of hydrogen and C4 raffinate in refinery processing. The integration brings processing synergies that reduce the cost of production of both the fuels and petrochemical products.

Refiners, on the other hand, are faced with various environmental issues related to the changing specifications of refined products. In many locations, refinery configuration has changed substantially mainly due to the declining quality of crude oil supply and environmental regulations. Refiners are faced with huge investments to meet new stringent specifications for sulfur, aromatics, and olefins content. Gasoline sulfur reduction is centered around the FCC unit employing feed pretreatment or gasoline post-treatment. For diesel fuel, a sulfur content of less than 30 ppm or may be 15 ppm is needed and an increase in the cetane number as well as reduction in polyaromatics content. To fulfill all these requirements, refiners have either to revamp existing units or invest into new hydroprocessing and hydrogen production units. However, the need for more hydrogen may itself contribute to an increase of CO_2 emissions, which could stand at about 20% of total refineries emission by 2035 (14% in 2005) as natural gas steam reforming should be the dominant technology. In addition the upgrading of extra heavy crude will account for more than 15% of the refineries' emissions in 2035 (4% in 2005).

Most environmental concerns in waste gas are around the emissions of SO_x, NO_x, CO, hydrocarbons, and particulates. The oxides are present in flue gases from furnaces, boilers, and FCC regenerator. Tail gas treatment and selective catalytic reduction (SCR) units are being added to limit SO_2 and NO_x emissions. Water pollutants include oil, phenol, sulfur, ammonia, chlorides, and heavy metals. New biological processes can be used to convert H_2S or SO_x from gaseous and aqueous streams. Spent catalysts and sludges are also of concern to the refinery in reducing pollution.

CASE STUDY 4: REFINERY FCC REVAMPS

This is a case study of a Gulf coast refinery in which the conversion capability of the existing FCC unit was found to be limiting the refinery economics [9].

Changing feed quality, combined with feed rate increases, beyond the original design, were limiting the performance of the unit. Further changes in feed quality were proposed to increase the heavy syn-crude percentage processed by the refinery.

A team consisting of refinery personnel, UOP, IAG, and Andrews Consulting was assembled to evaluate the following refinery objectives:

- Increase production of more valuable liquid products
- Address catalyst circulation limits
- Maintain same level of coke yield
- Provide flexibility for future changes in feed quality

An economic analysis was performed based on installed cost estimates from IAG and yield estimates from UOP. The total installed cost estimate for the new regenerated catalyst standpipe, wye section, feed distributors, and upper riser was $5.9 MM. The unit profitability estimate based on the heavier feed and new yields was $4.2 MM per year for *a simple payback of less than 15 months.*

The post-revamp operation had an improved conversion and reduced coke yield per expectations. The following table shows the base case compared to the revamped operation:

Base Case		**Post-Revamp**
Feed Rate, BPD	48,000	47,450
Feed API	24.4	24.6
UOP K	11.75	11.69
Feed Con Carbon, wt%	0.3	0.2
Feed Steam, wt%	2.1	1.3
Cat/Oil	6.1	6.1
Yields, wt%		
C2 minus	2.9	2.8
C3s	5.5	6.3
C4s	9.6	9.9
Gasoline, 430°F TBP EP	46.0	48.8
LCO, 650°F TBP EP	17.8	18.3
MC Botts	14.7	9.7
Coke	4.4	4.2
Conversion	67.5	72

It has been stated that part of the success of the revamp in this case study was due to a focused team accountable for the goals and execution of the project.

21 Crude Oil Refining—Part 3: Background on Modeling and Computer Applications

21.1 INTRODUCTION

Learning how to solve problems is an important part of developing competency in science and engineering. It is worth noting that most engineering problems are based upon one of the following three underlying principles:

- Equilibrium, Force, Flux, and Chemical
- Conservation Laws: Energy and Mass
- Rate Phenomena

The solution of chemical engineering problems should be an integrated part of this text, since the principles of chemical engineering are introduced along with solving numerical problems.

In attempting to solve a problem, we will demonstrate how students can attack a problem. The next proposed procedure is an attempt to be followed:

- 1st Identify first, the type of problem at hand, using the following guidelines:

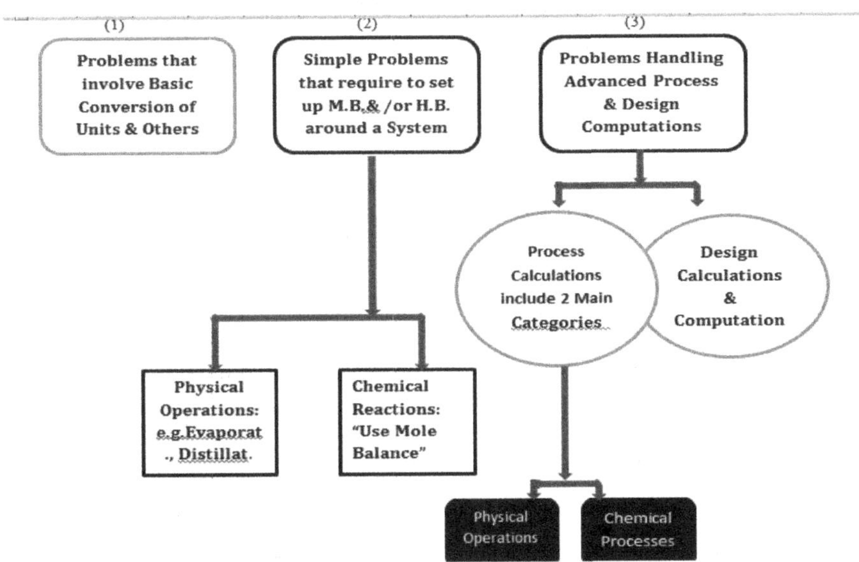

- 2nd Find out, for the problem as identified above, if you need additional help from the sources available at your fingertip (found in the text).

21.2 METHODOLOGY

One of our main objectives in this "primer" is to demonstrate how to solve chemical engineering problems that require numerical methods by using standard algorithms, such as MATLAB or Spreadsheets. To say it in simple words, is to device and evaluate numerical techniques for employing computers to solve problems in chemical engineering.

Problem solving using computers could be handled by using:

- Spreadsheets, such as Excel and
- A programming language such as MATLAB.

According to "Wikibooks"—"Introduction to chemical engineering"—a spreadsheet such as Excel is a program that lets you analyze moderately large amounts of data by placing each data point in a **cell** and then performing the same operation on groups of cells at once. *One of the advantages of spreadsheets is that data input and manipulation is relatively intuitive, and hence easier than doing the same tasks in MATLAB.*

Details on using Excel showing how to input and manipulate data and perform operations and others are fully-explained in many references found in the open literature.

Spreadsheet software has become indispensable tools in solving chemical engineering problems, because of the availability of personal computers, ease of use, and adaptability to many types of problems. Handheld calculations are encouraged as well to get a feeling for a numerical technique in solving a problem.

Problem-solving process using computers would normally involve the following four basic steps:

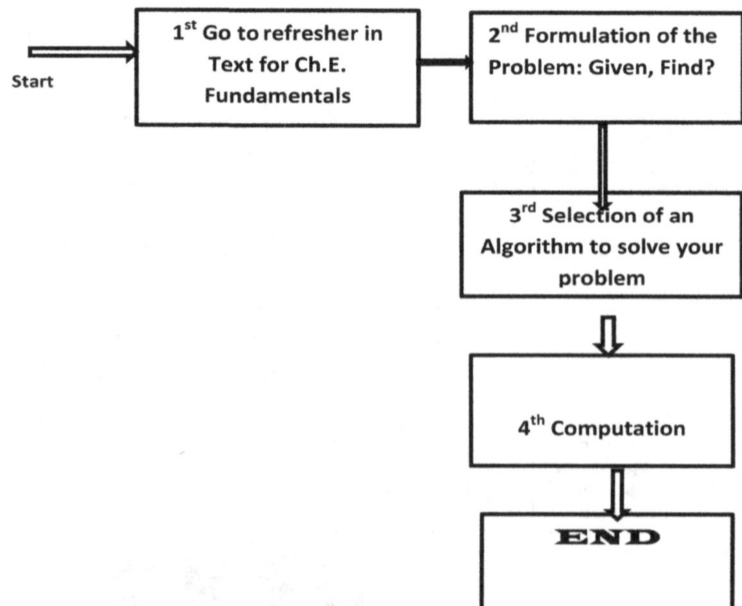

21.3 MODEL DEVELOPMENT AND MATHEMATICAL FORMULATION

The basic problems, that we face in sciences and engineering as far as mathematical modeling is concern, fall into three main categories:

- Equilibrium Problems: are recognized as steady state, where solution does not change with time.
- Eigenvalue Problems: are recognized as extensions of equilibrium problems in which critical values of certain parameters are to be determined in addition to the corresponding steady state configuration.
- Transient, Time-Varying, or Propagation Problems: are concerned with predicting the subsequent behavior of a system from knowledge of the initial stage.

Mathematical treatment involves four basic steps:

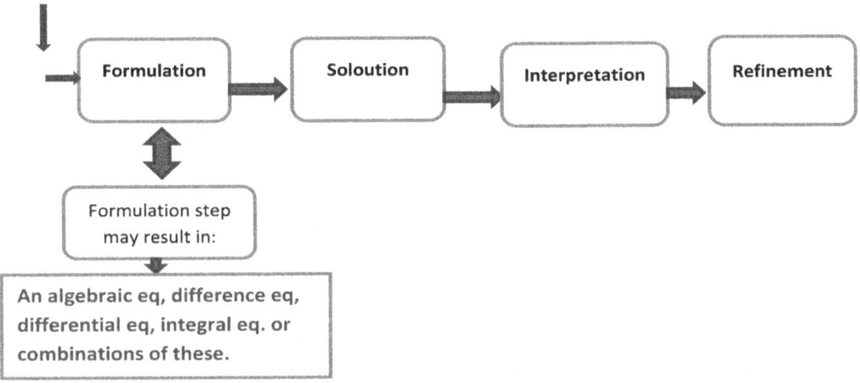

Modeling and simulation are principle approaches for quantitative description of chemical engineering processes and other disciplines in solving problems.

Quantitative process description is advantageous on two grounds:

- From a scientific point of view: It addresses the process mechanism study, which leads to the creation of a hypothesis about the process description. This is followed by a mathematical model.
- From an engineering aspect: It forms the basis of the engineering based on chemical process or a chemical plant. The Association for the Advancement of Modeling and Simulation Techniques in Enterprises (**AMSE**) gives the following definitions:
  a. The purpose of Modeling is *a schematic description of the processes and the systems.*

b. The Simulations are employments of the models for process investigations or process optimizations; without experiments with real systems. The introduction of interactive software packages brought about a major break-through in chemical engineering computations.

The solution of a problem is illustrated here:

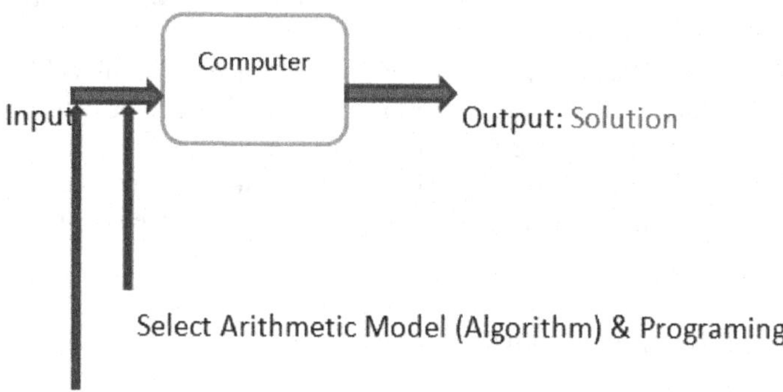

The following three steps are basically applied in this procedure:

1st Understanding the physical principles underlying the process involved in the problem in order to build a "*Conceptual*" model.

2nd Manipulation and formulation of these principles into a mathematical expression or a correlation: i.e., "A *Mathematical* model". This is achieved by a thorough analysis of the engineering problems at hand, which may involve two types:

a. Mathematical formulation (modeling) of engineering problems corresponding to specific physical situations such as momentum, heat, and mass transfer, chemical reactions, thermodynamics, and others.
b. Conversion of physical events and principles (e.g., a material balance), to mathematical model.

3rd In engineering practice, numerical values must be incorporated and a practical solution is obtained.

These steps are presented by Figure 21.1, which illustrates different options of numerical methods involved in solving problems.

21.4 APPLICATIONS

To demonstrate the above procedure, the problem of flash distillation is considered. Flash calculations are very common, perhaps one of the most common chemical engineering calculations. They are a key component of simulation packages like Hysys, Aspen, etc.

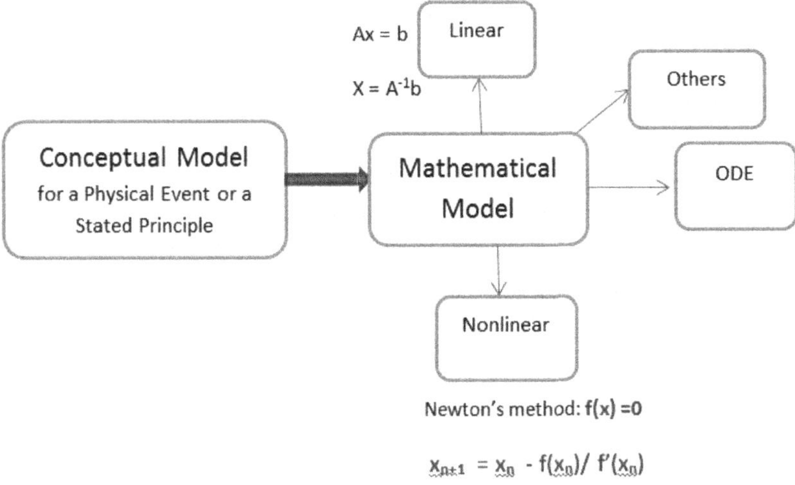

FIGURE 21.1 Formulation of mathematical model through numerical methods.

PROBLEM STATEMENT

Example 21.1

It is required to calculate the bubble point (B.P.) temperature, the dew point (D.P.) temperature, the flow rates of the streams leaving the *flash distillation column* as well as their composition. *mathematical formulation to determine the B.P. is done 1st, through the following analysis.*

Definition: *The B.P. temperature is physically defined as the lowest temperature at which the first bubble comes out as vapor, when the liquid is slowly heated at constant pressure.* Mathematically, at the B.P. the following relationships hold: (a) $\Sigma\, y_i = 1.0$, and (b) $\Sigma\, pi = P_T$

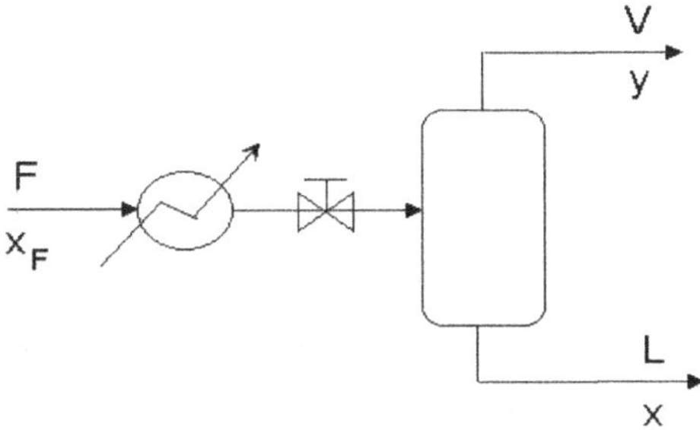

Now, mathematical formulation is pursued through these two equivalent definitions for the B.P. temperature. But let us first, present the following fundamental relationships:

- Rault's law: For a gas-liquid mixture, the partial pressure of component "i" in the liquid phase is given by:

$$p_i = P_i^0 x_i \tag{21.1}$$

- Dalton's law: In the vapor phase, where the vapor is in equilibrium with the liquid, the partial pressure of component, i, is:

$$p_i = P_T y_i \tag{21.2}$$

Take the sum of both sides of Equation (21.2):

$$\sum pi = PT$$

Where: p_i = partial pressure of component i
P_i^0 = vapor pressure of pure component i
P_T = total pressure

Equating (21.1) and (21.2), we obtain:

$$y_i / x_i = P_i^0 / P_T = K_i \tag{21.3}$$

- Antoine's equations: The *Antoine's equation* is a simple three-parameter fit to experimental vapor pressures measured over a restricted temperature range:

$$\text{Log}\,P^0 = A - \frac{B}{T+C}$$

Where: A, B, and C are "Antoine coefficients" that vary from substance to substance, P is the vapor pressure of the pure component.

1st: Determination of B.P. using the definition given in terms of the sum of y_i, by the equation (a): $\sum y_i = 1.0$, as shown in Figure 21.2. Equation (21.4) given next represents this case, where $[\sum y_i]$ is rewritten in terms of the x_i's ($y_i = k_i x_i$):

$$f\left(T_{assu.}\right) = \left[\sum k_i x_i\right] - 1, \text{ goes to zero at } T_{assu} = T_{B.P.} \tag{21.4}$$

2nd Determination of B.P. using the definition in terms of the sum of p_i, as per the equation (b): $\sum p_i = P_T$, as shown in Figure 21.3.

At the B.P., the sum of partial pressure of the components should be equal to the total pressure on the system, P_T.

or: $f(T_{assu})$. = $\{[\sum p_i] - P_T\}$, goes to zero at $T_{assu} = T_{B.P.}$

Using Equation (21.1) to replace p_i:

$$f\left(T_{assu}\right) = \left\{\left[\sum x_i P^0\right] - P_T\right\}, \text{ goes to zero at } T_{assu} = T_{B.P.} \tag{21.5}$$

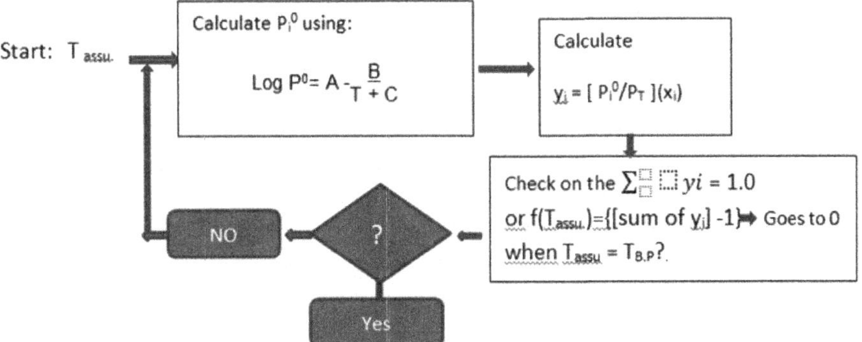

FIGURE 21.2 Determination of $T_{B.P.}$ using equation (a).

Solution of the above nonlinear algebraic equations for the B.P, (21.4) and (21.5), using MATLAB, is presented as shown in the following examples.

Similarly, *the D.P.* temperature is physically defined *as the temperature at which the first liquid drop would form when the temperature of a mixture of vapors is slowly decreased (cooled) at a specified constant* pressure.

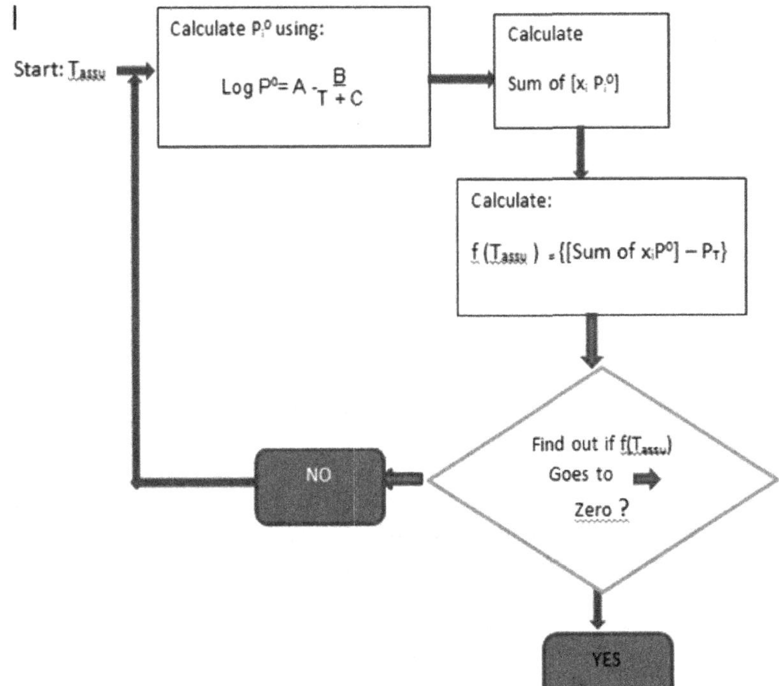

FIGURE 21.3 Determination of $T_{B.P.}$ Using Equation (b).

END OF CHAPTER SOLVED EXAMPLES

Example 21.1

For a three-component mixture, the following information is available:

Component Number	K_i	Composition (x_i): Mole Fraction
1	$K_1 = (0.01\ T)/P$	1/3
2	$K_2 = (0.02\ T)/p$	1/3
3	$K_3 = (0.03\ T)/P$	1/3

Compute the B.P. temperature, $T_{B.P.}$ at the specified pressure of 1 atm by using Newton's method. Take the first assumed value for T_n be equal to 100°F.

MANUAL SOLUTION

Assuming T_1 = 100°F and P = 1 atm. The following calculation is carried out for the 1st trial:

Component	x_i	K_i @ 1 atm & T = 100°F	$(K_i)(x_i)$	$[dK_i/dT]_{Tn=100}$	$(x_i)(dk_i/dT)$
1	1/3	1	1/3	0.01	(0.01)/3
2	1/3	2	2/3	0.02	(0.02)/3
3	1/3	3	3/3	0.03	(0.03)/3
Σ	**1.0**		**6/2 = 3**		**0.06/3 = 0.02**

From the above results, it follows that:

$$f(100) = \left[\sum K_i x_i\right] - 1.0 = 2 - 1 = 1, \text{and}$$

$$f'(100) = \sum x_i dK_i/dT = 0.02$$

where, f' stands for the first derivative of the [$\Sigma\ x_i\ dK_i$] w r t T.

Applying Newton's formula: $T_2 = T_1 - \{f(T_1)/f'(T_1)\}$

$$= 100 - (1/0.02)$$

$$= \mathbf{50°F}$$

Carry on one more trial to check the final answer.

Solution by Excel

T_1	Component	X_i	K_i	X_iK_i	$[dK_i/dT]_{Tn}$	$(x_i)(dk_i/dT)$
100	1	1/3	1	1/3	0.01	1/300
	2	1/3	2	2/3	0.02	1/150
	3	1/3	3	1	0.03	1/100
Σ	6	1	6	2	0.06	0.02

$f(100) = [\sum K_iX_i] - 1$	1	
$f'(100) = \sum X_i\, dK_i/dT$	0.02	
$T_2 = T_1 - [f(T_1)/f'(T_1)]$	50	°F

Example 21.2

An equimolar vapor mixture of benzene and ethylbenzene is kept at 100°C.

Calculate the pressure at which the first drop of liquid will form, and its composition.

SOLUTION

This is a dew pressure calculation. Antoine's constants are: (% *P* in kPa and *t* in °C)

$$A = [13.8858,\ 14.0045];$$
$$B = [2788.51,\ 3279.47];$$
$$C = [220.79,\ 213.201];$$

T and the vapor mole fractions are introduced:

$$T = 100;$$
$$Y = [0.5, 1-0.5]$$
$$P_{sat} = \exp\,[A - B./(T + C)]$$
$$= 180.0377 \quad 34.2488$$

The total pressure and the liquid mole fraction are calculated:

$$P = 1/\text{sum}\,(y./P_{sat})$$
$$x = P^{*}y./Psat$$
$$P = 57.5498$$
$$x = 0.1598 \quad 0.8402$$

Example 21.3

[Numerical solution of linear equations using MATLAB]

Case of distillation column: [Solution of *n* algebraic equations in *n* unknowns].

A stream containing 35.0 wt% benzene (B), 50.0% toluene (T), and the balance xylene (X) is fed to a distillation column. The overhead product from the column

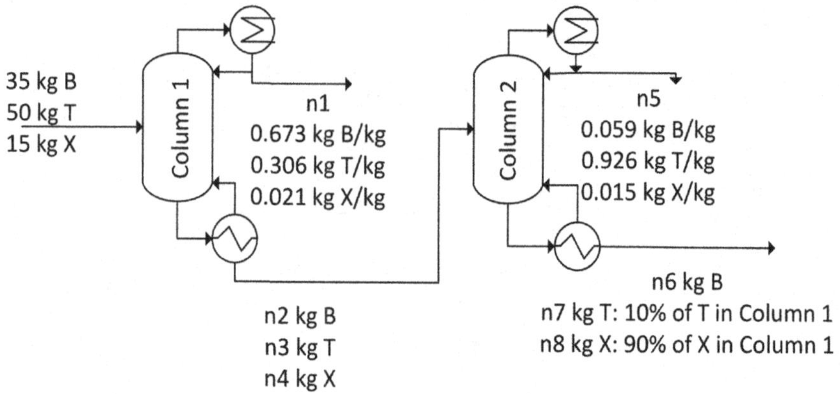

FIGURE 21.4 Distillation system comprising two columns in series.

contains 67.3 wt% benzene and 30.6% toluene. The bottoms product is fed to a second column. The overhead product from the second column contains 5.9 wt% benzene and 92.6% toluene. Of the toluene fed to the process, 10.0% is recovered in the bottoms product from the second column, and 90.0% of the xylene fed to the process is recovered in the same stream. This is shown in Figure 21.4.

Column 1 Balances	B:	35.0 = 0.673n1 + n2	(1)
	T:	50.0 = 0.306n1 + n3	(2)
	X:	15.0 = 0.021n1 + n4	(3)
Column 2 Balances	B:	n2 = 0.059n5 + n6	(4)
	T:	n3 = 0.926n5 + n7	(5)
	X:	n4 = 0.015n5 + n8	(6)
	10% T recovery:	n7 = 0.100(50.0) = 5.00	(7)
	93.3% x recovery:	n8 = 0.933(15.0) = 14.0	(8)

A solver tool can be used to solve the MATLAB ®equations simultaneously

[n1 n2 n3 n4 n5 n6 n7 n8] = solve ('35 = 0.673*n1 + n2','50 = 0.306*n2 + n3','15 = 0.021*n1 + n4','n2 = 0.059*n5 + n6','n3 = 0.926*n5 + n7','n4 = 0.015*n5 + n8','n7 = 5','n8 = 14')

Example 21.4

Calculate the temperature and composition of a vapor in equilibrium with a liquid that is 40.0 mole% benzene, 60.0 mole% toluene at 1 atm. Is the calculated temperature a bubble-point or D.P. temperature?

SOLUTION

Raooult's Law

$$P = x_A p_A^*(T_{bp}) + x_B p_B^*(T_{bp}) + \cdots$$

Let A = benzene and B = toluene.

$$f(T_{bp}) = 0.400p*(T_{bp}) + 0.600p*(T_{bp}) - 760 \text{ mm Hg} = 0$$

The solution procedure is to choose a temperature, evaluate P^*_A and P^*_B for that temperature from the Antoine's equation, evaluate $f(T_{bp})$ from the above equation, and repeat the calculations until a temperature is found for which $f(T_{bp})$ is sufficiently close to 0.

Solve using initial guess (100°C)

$$0.40 \times 10^{6.89272 - \frac{1203.5311}{T+219.888}} + 0.6 \times 10^{6.95805 - \frac{1346.773}{T+219.693}} - 760 = 0$$

```
>> T = fzero(@(T)0.40*10^(6.89272-1203.5311/
(T+219.888))+0.6*10^(6.95805-1346.773/(T+219.693))-760,100)

T = 95.1460

>> pA = 0.40*10^(6.89272-1203.5311/(T+219.888))

pA = 472.5616

>> pB = 0.6*10^(6.95805-1346.773/(T+219.693))

pB = 287.4384

>> yA = pA/760

yA = 0.6218

>> yb = pB/760

yb = 0.3782
```

The D.P. pressure, which relates to condensation brought about by increasing system pressure at constant temperature can be determined by solving the following equation for P:

$$\frac{y_A P}{p_A^*(T_{dp})} + \frac{y_B P}{p_B^*(T_{dp})} + \cdots = 1$$

$$\frac{0.1*760}{10^{6.89272 - \frac{1203.5311}{T+219.888}}} + \frac{0.1*760}{10^{6.95805 - \frac{1346.773}{T+219.693}}} - 1 = 0$$

Solve using initial guess

```
Tdp = fzero(@(T)(0.1*760)/10^(6.89272-1203.5311/
    (T+219.888))+(0.1*760)/10^(6.95805-1346.773/(T+219.693))-1,50)
```

Tdp = 52.4354 Final Answer

B = 200. C = 900. W = 100

Example 21.5

For the following distillation column calculate the values of F1, F3, and F4? As indicated in Figure 21.5

F1,F3,F4] = solve('.2*F1+250 = .5*F3+.2*F4','.3*F1+250 = .3*F3+.4*F4','.
5*F1 = .2*F3+.4*F4') F1 = 1000
F3 = 500
F4 = 1000

Example 21.6

TXY Diagram for Benzene/Toluene mixture Using Excel

Find: The goal is to create a Txy phase diagram for mixtures of benzene and toluene, where *T* is temperature, *x* is the mole fraction of benzene in the liquid, and *y* is the mole fraction of benzene in the vapor. A horizontal line drawn for a given *T* gives the compositions of liquid and vapor in equilibrium at that *T*. Such diagrams are very useful for distillation calculations.

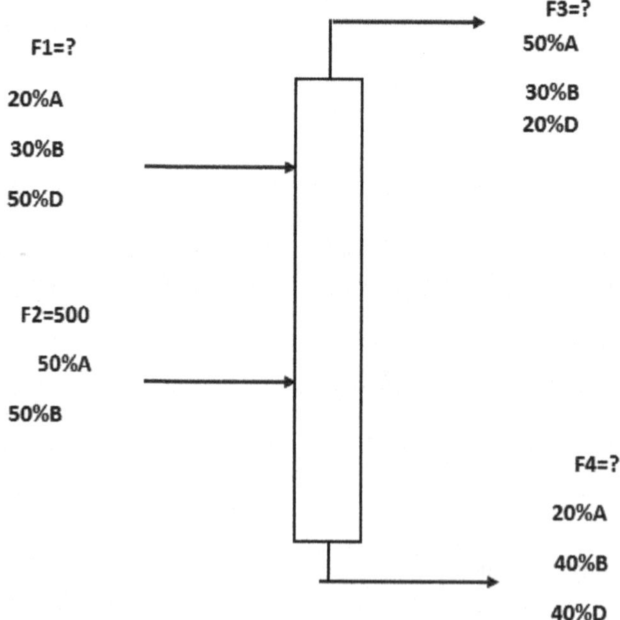

FIGURE 21.5 Calculation of flow rates for a given distillation column.

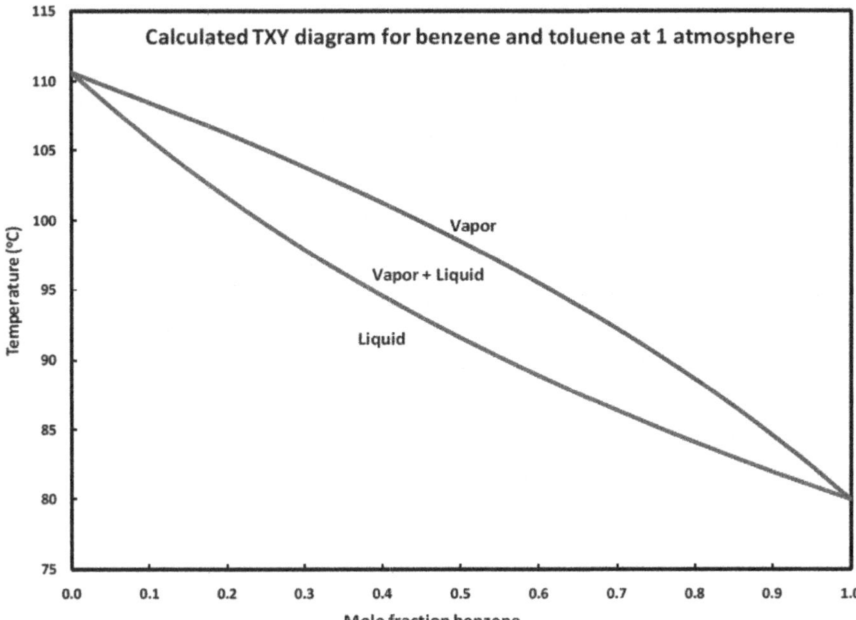

FIGURE 21.6 Calculation of TXY diagram for B/T mixture.

Approach: Using these equations, you will create a TXY graph for benzene-toluene mixtures at 1 atm as shown in Figure 21.6. This graph will show the mole fraction *ya* of benzene in the vapor corresponding to equilibrium with liquid of mole fraction *xa* at temperature *T* required to give a total pressure of 760 mm Hg.

Proposed Procedure

$$\log_{10} p_a^* = 6.814 - \frac{1090}{197.1 + T} \tag{1}$$

$$\log_{10} p_b^* = 7.136 - \frac{1457}{231.8 + T} \tag{2}$$

$$p_a = x p_a^* \tag{3}$$

$$p_b = (1 - x) p_b^* \tag{4}$$

$$y = \frac{p_a}{p_a + p_b} \tag{5}$$

1. Create an Excel spreadsheet giving *T* and *y* versus *x* for *P* = 760 Torr (mm Hg), then execute the following steps:
   a. Enter the values for the Antoine's constants from Equations (21.1) and (21.2).
   b. In Column A enter values for *x* from 0 to 1 by increments of 0.1.

c. Leave Column B blank for values of *T*.
d. In Column C calculate pa* using the Antoine's equation, the constants for benzene, and the temperature in Column B.
e. In Column D calculate pb* using the Antoine's equation, the constants for toluene, and the temperature in Column B.
f. In Column E calculate P = pa + pb using Columns C & D and Equations (21.3) and (21.4).
g. In Column F use Equation (21.5) to calculate values of *y*.
h. Use Goal Seek row-by-row to calculate *T* required to give *P* = 760 mm Hg. This will automatically fill in the correct values in all columns.

2. Use the results in the spreadsheet to create a graph, with *x* and *y* on the horizontal (X) axis and *T* on the vertical (Y) axis.
3. Format the graph.

Results will be as shown below:

	A	B	C	D	E	F
3	**For total pressure P = 760 Torr (1 atm)**					
4						
5	**Antoine equation coefficients**					
6	**Compound**	A	B	C		
7	Benzene	6.814	1090	197.1		
8	Toluene	7.136	1457	231.8		
9					P	
10	x	T (°C)	p_a* (Torr)	p_b* (Torr)	(p_a + p_b)	y
11	0.0	110.6	1869.2	760.0	760.0	0.000
12	0.1	105.8	1644.0	661.8	760.0	0.216
13	0.2	101.6	1463.6	584.1	760.0	0.385
14	0.3	97.9	1316.6	521.5	760.0	0.520
15	0.4	94.6	1194.9	470.1	760.0	0.629
16	0.5	91.6	1092.7	427.3	760.0	0.719
17	0.6	88.9	1005.8	391.3	760.0	0.794
18	0.7	86.4	931.2	360.5	760.0	0.858
19	0.8	84.1	866.5	334.0	760.0	0.912
20	0.9	82.0	809.9	311.0	760.0	0.959
21	1.0	80.0	760.0	290.8	760.0	1.000

Source: ***www.clarkson.edu/~wwilcox/ES100/xl-tut2.pdf*** **Clarkson University Oct 14, 2014**

Solution of Example 21.6—TXY Diagram for Benzene/Toluene mixture Using Excel

22 Oil and Gas Transportation

22.1 INTRODUCTION

This chapter is devoted to highlight the application and the use of different methods of transferring crude oil and natural gas. Oil is normally transported by one of *four options*:

- Tankers
- Pipelines
- Railroad
- Tank cars and tank trucks

Oil and gas pipelines act as veritable arteries inside the Earth. Using extensive steel and plastic pipes, they transport gas and oil throughout the planet. Pipeline—the most commonly used form of oil transportation is through **oil** pipelines. Pipelines are typically used to move crude oil from the wellhead to gathering and processing facilities and from there to refineries and tanker loading facilities.

Supply-end pipelines and railroads carry crude oil from production areas to a loading terminal at a port. Tankers then carry the crude oil directly to demand-side pipelines that connect to the refineries that convert the raw material into useful products. Most crude oil is transported by pipelines on land and by tankers across the seas. Moving natural gas, on the other hand, requires a network of pipelines from the production wells to the processing plants and to the final consumers.

Price, cost, and investment issues in transportation garner intense interest boosting prosperity. Oil and gas they transported contributed $81 billion to our GDP through exports. A recent study by Angevine Economic Consulting Ltd. estimates the total GDP contribution of the pipeline industry over the next 30 years is $175 billion. Pipelines power prosperity.

22.2 OVERVIEW

Tankers, railroads, and pipelines are proven, efficient and economical means of connecting petroleum supply and demand. Supply-end pipelines and railroads carry crude oil from production areas to a loading terminal at a port. Tankers then carry the crude oil directly to demand-side pipelines that connect to the refineries that convert the raw material into useful products.

With the advances in exploration and production, great stripes are achieved to locate and recover a supply of oil and natural gas from major reserves across the globe. At the same time, demand for petroleum-based products has grown in every corner of the world. Transportation therefore is indispensible and vital to ensure a reliable and affordable flow of petroleum we all count on to fuel our cars, heat our homes, and improve the quality of our lives.

22.3 PIPING SYSTEM AND PUMPS

Pipelines along with pumps are needed as an efficient means of transporting crude oil, hydrocarbon products, natural gas, and other important fossil fuels, quickly, safely, and smoothly. Pipelines are pipes, usually underground, that transport and distribute fluids. When discussing pipelines in an energy context, the fluids are usually either oil, oil products, or natural gas. Petroleum pipelines transport crude oil or natural gas liquids, and there are three main types of petroleum pipelines involved in this process. Pipelines need to be constantly and reliably operated and monitored in order to ensure maximum operating efficiency, safe transportation, and minimal downtimes, and to maintain environmental and quality standards. Powerful pumps, on the other hand, are needed for oil transport of crude oil within the oil field and for the delivery of oil to terminal points.

The role of pipelines and pumps in oil field operations is demonstrated as follows:

- Gathering systems in the oil field
- Crude oil delivery network
- Sizing of pipeline and selection of wall thickness
- Other aspects of piping
- Classification and types of pumps

Cross-country pipelines are globally recognized as the safest, cost-effective, energy-efficient, and environment-friendly mode for transportation of crude oil and petroleum products. … The pipeline was jointly dedicated to nation by Hon'ble Prime Ministers of India and Nepal.

Pipelines are the second most important form of oil and gas transportation. Their uses are more complex than uses of tankers, which by their nature only move crude oil or products and gas from or to a rather limited number of points on the oceans or navigable rivers. Pipelines, however, are used for gathering systems in oil fields, for moving the crude oil to refineries, marine terminals, and often for moving refined products from refineries to local distribution points.

Market demand growth can, of course, outstrip a pipeline's basic ability to handle the demanded volumes. The first way to solve this problem is to increase the speed with which the oil passes along the line by adding pumping stations. But since pipeline friction increases geometrically with the speed of flow, at some point it becomes economical to add more pipes. This process is called "looping", and it consists of laying another pipeline alongside the existing one. In summary, pipelines serve a vital function in the transportation of both oil and natural gas.

There are two main categories of pipelines used to transport energy products:

- Petroleum pipelines and
- Natural gas pipelines.

Petroleum pipelines transport crude oil or natural gas liquids, and there are three main types of petroleum pipelines involved in this process: gathering systems, crude oil pipeline systems, and refined products pipelines systems. The gathering pipeline systems gather the crude oil or natural gas liquid from the production wells. It is then transported with the crude oil pipeline system to a refinery. Once the petroleum is refined into products such as gasoline or kerosene, it is transported via the refined products pipeline systems to storage or distribution stations.

Natural gas pipelines transport natural gas from stationary facilities such as gas wells or import/export facilities, and deliver to a variety of locations, such as homes or directly to other export facilities. This process also involves three different types of pipelines: gathering systems, transmission systems, and distribution systems. Similar to the petroleum gathering systems, the natural gas gathering pipeline system gathers the raw material from production wells. It is then transported with large lines of transmission pipelines that move natural gas from facilities to ports, refiners, and cities across the country. Lastly, the distribution systems consist of a network that distributes the product to homes and businesses. The two types of distribution systems are the main distribution line, which are larger lines that move products close to cities, and the service distribution lines, which are smaller lines that connect main lines into homes and businesses.

22.3.1 Gathering Systems in Oil Fields

The value of a pipeline is in its economy of operation and in its consistency of operation. Today, there is great diversity in size of pipe used to carry crude oil refined oil products and natural gas ranging from 6 in. to as much as 36 in., and in some cases in the Middle East, even 48 in. piping. Lines are single or multiple, laid on top of the surface or buried in the ground, with booster pumps spaced anywhere from approximately every 25 miles to as much as 200 miles apart.

Pipeline costs vary, of course, with capacity, the character of the terrain which the lines will traverse and the type of product which the line is intended to carry, that is, its function. In general, there are three types of pipeline:

1. Those which run from the oil field to loading ports and are complementary to ocean transport. Without these, there would be no transport by tankers at all, so they are not competitive with transport by tankers.
2. Those long-distance pipelines which naturally shorten the alternative sea route. They can be competitive with ocean transport tankers if tanker rates are high. But in times of low tanker rates, such pipelines are not competitive with transport by tankers. A good example of this type of pipeline is Tapline, the 1,100-mile pipeline from Ras Tanura in Saudi Arabia through four countries to Sidon, Lebanon. Transport by Tapline saves approximately 3,300 miles each way of ocean transport, and also saved Suez tolls when the Suez Canal was open. At this writing the Suez Canal has just reopened.
3. Those pipelines which transport oil from ports of discharge to inland refineries located in industrial areas, remote from a seaport. They can be competitive with domestic railroad and motor carriers. Examples of this type of pipeline are the pipelines of Rotterdam on the Rhine and Wilhemshaven on the Ruhr.

TABLE 22.1
Assumed Values for Velocity for Different Fluids

Type of Fluid	Reasonable Assumed *u* (ft/s)
Water or fluid similar to water	3–10
Low pressure steam (25 psig)	50–100
High pressure steam (>100 psig)	100–200

To determine the fluid velocity in a pipe, the rule of thumb, the economic velocity for turblant flow, is used, as reported by Peters and Timmerhaus (4th ed) and shown in Table 22.1

22.3.2 Pumps

A fluid moves through a pipe or a conduit by increasing the pressure of the fluid using a pump that supplies the driving force for flow. In doing so, power must be provided to the pump. There are six basic means that cause the transfer of fluid flow: gravity, displacement, centrifugal force, electromagnetic force, transfer of momentum, and mechanical impulse.

Excluding gravity, *centrifugal force* is the means most commonly used today in pumping fluids. Centrifugal force is applied by using a *centrifugal pump or compressor*, of which the basic function of each is the same: To produce kinetic energy (K.E.) by the action of centrifugal force, and then converting this K.E. into pressure energy (P.E.) by efficient reduction of the velocity of the flowing fluid.

Fluid flow in pipes applying centrifugal devices have in general the following basic advantages and features:

- Fluid discharge is relatively free from pulsations.
- No limitation on throughput capacity of the operating pump.
- Discharge pressure is a function of the fluid density, i.e., $P = f(\ell f)$.
- To provide efficient performance in a simple way with low first cost.

Pumps can be classified into three major groups according to the method they use to move the fluid: direct lift, displacement, and gravity pumps. Pumps can also be classified by their method of displacement as positive displacement pumps, impulse pumps, velocity pumps, and gravity pumps. Pumps operate by a reciprocating or rotary mechanism. Mechanical pumps may be submerged in the fluid they are pumping or placed external to the fluid. A concise summary for the comparison between different types of pumps is in Table 22.2.

Pumps are used for many different applications. Understanding which pump type one needs for his application is very important. For the oil and gas industry, some basic features are listed next:

TABLE 22.2
Comparison between Different Types of Pumups

Type of Pump	Features
Centrifugal	Most common, high capacity, discharge lines can be shut off (safe) to handle liquids with solids.
Reciprocating	Low capacity and high head, can handle viscous fluids, used to discharge bitumen (asphalt) in vacuum distillation columns
Rotary positive-displacement	Combination of rotary motion and positive displacement, used in gas pumps, screw pumps, and metering pumps
Air displacement	Nonmechanical, air-lift type, used for "acid eggs" and jet pumps

- Pumps should handle the fluids with low shear and least damage to droplet sizes causing no emulsions for the effective separation of water from oil.
- Pumps should be self-priming and experience no gas locking.
- The requirement of having low net positive suction head (NPSH) is an advantage. This is advantageous for vessel-emptying applications such as closed drain drums or flare knockout drums or any applications encountering high-vapor pressure liquids.
- Pumps should handle multiphase fluids. construction site, pond, mine shaft, or any other area.

Fire pumps—A type of centrifugal pump used for firefighting. They are generally horizontal split case, end suction, or vertical turbine.

22.3.3 Global Overview for Pipelines

Globally, North America has the highest oil and gas pipelines length of 834,152.5 km (with start years up to 2023), of which, crude oil pipelines constitute 154,200.9 km, petroleum products pipelines constitute 103,106.3 km, natural gas pipelines constitute 495,555.3 km, and NGL pipelines constitute 81,290.0 km. The region's share in the global transmission pipeline length is 41.0%.

World's longest pipelines: Natural gas

- West-East Gas Pipeline: 8,707 km. …
- GASUN, Brazil: 4,989 km. …
- Yamal-Europe Pipeline: 4,196 km. …
- Trans-Saharan Pipeline: 4,127 km. …
- Eastern Siberia-Pacific Ocean Oil Pipeline: 4,857 km. …
- Druzhba Pipeline: 4,000 km. …
- Keystone Pipeline: 3,456 km. …
- Kazakhstan-China Pipeline: 2,798 km.

Source: From Wikipedia, the free encyclopedia

Africa

- Chad–Cameroon pipeline – Chad–Cameroon
- Sudeth Pipeline – South Sudan–Ethiopia (under construction)
- Transnet Pipelines – South Africa
- Sumed Pipeline – Egypt
- Tazama Pipeline – Tanzania–Zambia
- Nembe Creek Trunk Line – Nigeria
- CPMZ-Mozambique-Zimbabwe Pipeline company–Mozambique
- [KMPP] Khartoum-Madani Petroleum Products Pipeline–[inside SUDAN]
- Kenya pipeline–Kenya

Europe

1. Balkan area – Southeast Europe Pipelines (includes Albania, Bosnia and Herzegovina, Bulgaria, Greece, Hungary, Romania, Serbia, Slovakia, Slovenia, FYR Macedonia, and Turkey)
2. France and Belgium Pipelines
3. Germay, Netherlands, and Czech Republic Pipelines
4. Italy Switzerland, and Austria Pipelines
5. Norway, Sweden, and Denmark Pipelines
6. Russia and former Soviet states Pipelines (includes Russia, Kazahstan, Lithuania, Turkmenistan, Ukraine, Uzbekistan, Azerbaijan, Georgia, Belarus, Latvia, Estonia, and Tajikistan)
7. Spain and Portugal Pipelines
8. United Kingdom and Ireland Pipelines

Source: Europe Pipelines map - Crude Oil (petroleum) pipelines … *theodora.com › pipelines › europe_oil_gas_and_produ.*

As shown in Figure 22.1

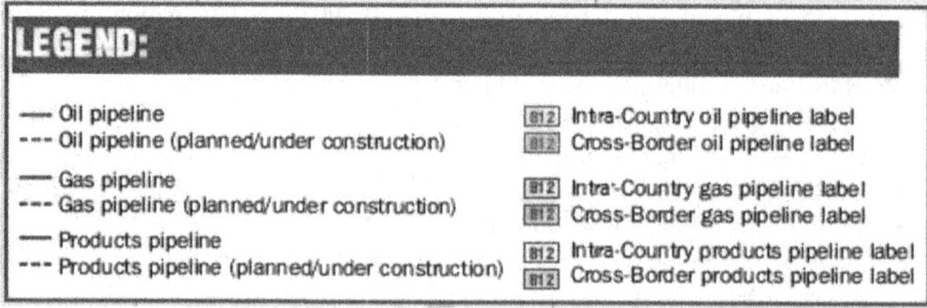

- Adria Oil Pipeline
- AMBO Pipeline
- Baltic Pipeline System
- Brent System

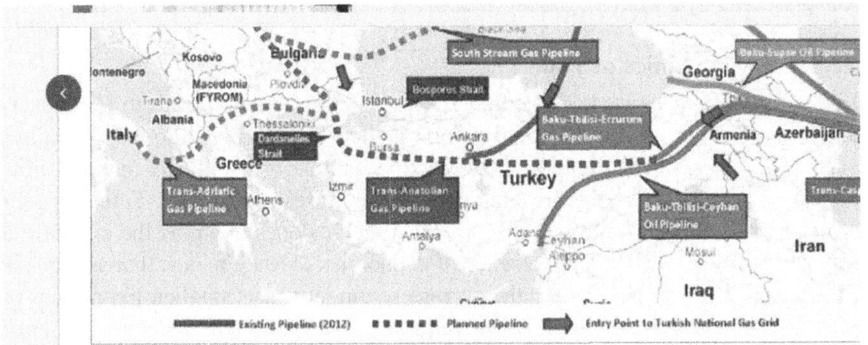

FIGURE 22.1 Existing and planned oil and natural gas pipelines to Europe.

- Burgas-Alexandroupoli Pipeline
- CLH Pipelines – Spain
- Druzhba Pipeline
- Forties Pipeline System
- Grozny-Tuapse Pipeline
- Ninian Pipeline
- Odessa-Brody Pipeline
- Pan-European Pipeline
- Transalpine Pipeline
- South European Pipeline
- TRAPIL – France

Oil and Gas Pipelines Industry, Global, Trunk/Transmission Pipeline Length by Region are indicated in Figure 22.2, Sept 2019.

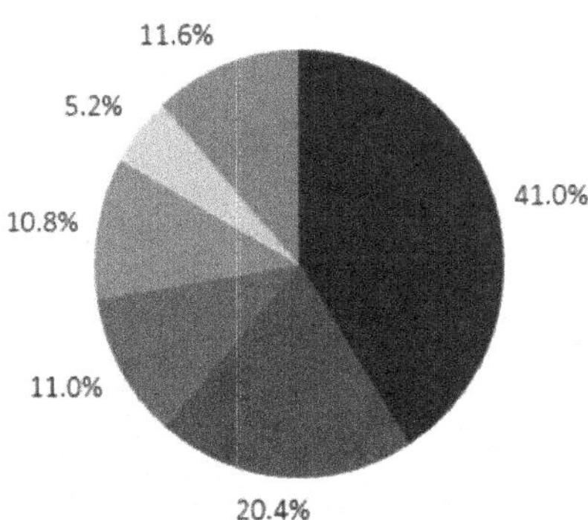

FIGURE 22.2 Oil and Gas Pipelines Industry, Global, Trunk/Transmission Pipeline Length by Region, Sept 2019.

Source: Midstream Analytics, GlobalData Oil and Gas © GlobalData

22.3.4 Pipeline Economics

22.3.4.1 Economics of Scale

Economic of scale exists because the larger scale of production leads to lower average costs. The cost of the materials for producing a pipe is related to the circumference of the pipe and its length. However, the volume of chemicals that can flow through a pipe is determined by the cross-section area of the pipe. Economies of scale refer to a long run average cost curve, which slopes down as the size of the transport firm increases. The presence of economies of scale means that as the size of the transport firm gets larger, the average or unit cost gets smaller. Economics of scale is calculated by dividing the percentage change in cost with percentage change in output. A cost elasticity value of less than 1 means that economies of scale exists. Economies of scale exist when increase in output is expected to result in a decrease in unit cost while keeping the input costs constant.

Once a firm has determined the least costly production technology, it can consider the optimal scale of production, or quantity of output to produce. Many industries experience economies of scale. Economics of scale refers to the situation where, as the quantity of output goes up, the cost per unit goes down. Figure 22.3 illustrates the idea of economies of scale, showing the average cost of producing an item falling as the quantity of output rises.

A doubling of the cost of producing the pipe allows the chemical firm to process four times as much material. This pattern is a major reason for economies of scale in chemical production, which uses a large quantity of pipes. Of course, economies of scale in a chemical plant are more complex than this simple calculation suggests. But the chemical engineers who design these plants have long used what they call the "six-tenths rule", a rule of thumb which holds that increasing the quantity produced in a chemical plant by a certain percentage will increase total cost by only six-tenths as much (Table 22.3).

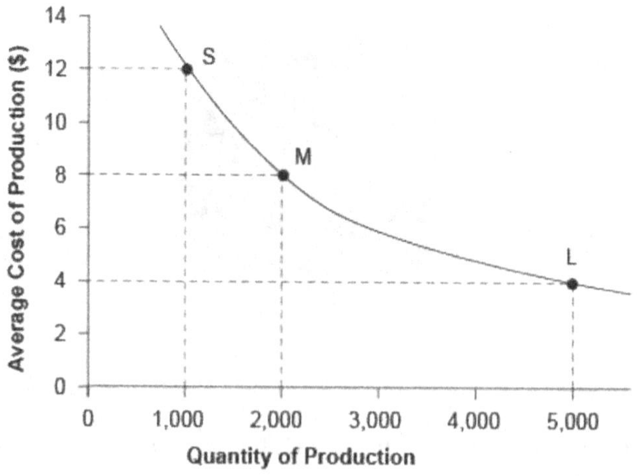

FIGURE 22.3 Economies of scale for a small factory.

TABLE 22.3
Comparing Pipes: Economies of Scale in the Chemical Industry

	Circumference ($2\pi r$)	Area (πr^2)
4-inch pipe	12.5 inches	12.5 square inches
8-inch pipe	25.1 inches	50.2 square inches
16-inch pipe	50.2 inches	201.1 square inches

Source Economies of Scale: Economies of Scale | Microeconomics - Reading courses.lumenlearning.com › chapter › economies-of-scale.

22.3.5 Economic Balance in Piping and Optimum Pipe Diameter

When pumping of a specified quantity of oil over a given distance is to be undertaken a decision has to be made as to

1. whether to use a large-diameter pipe with a small pressure drop, or
2. Whether to use a smaller-diameter pipe with a greater pressure drop. The first alternative involves a higher capital cost with lower running costs; the second, a lower capital cost with higher running costs specifically because of the need for more pumps.

So, it is necessary to arrive at an economic balance between the two alternatives. Unfortunately, there are no hard and fast rules or formulas to use; every case is different. Costs of actual pumping equipment undoubtedly must be considered, but the area in which the pipes will "run" is also important. For instance, to obtain the same pumping effort in the desert as opposed to a populated area could involve much higher costs in the form of providing outside services and even creating a small, self-contained township.

In the flow of oil in pipes, the fixed charges are the cost of the pipe, all fittings and installation. All these fixed costs can be related to pipe size to give an approximate mathematical expression for the sum of the fixed charges.

In the same way, direct costs, or variable costs, comprising mostly the costs of power for pressure drop plus costs of minor items such as repairs and maintenance, can be related to pipe size. For a given flow, the power cost decreases as the pipe size increases. Thus direct costs decrease with pipe size. And total costs, which include fixed charges, reach a minimum at some optimum pipe size. This factor can be expressed roughly in a series of simplified equations which express relations in terms of weight rate of flow and fluid density, then weight (or mass) rate of flow and annual cost per foot for most cases of turbulent flow.

To summarize, in choosing the inside diameter of pipe to be used, either in the oil field or in a refinery, selection should generally be based on costs of piping versus costs of pumping. Small-diameter pipe, which usually involves quicker drops in pressure than large-diameter pipe and therefore must be supplemented with more pumping equipment when laid for long distances, costs less than large-diameter pipe, but cost of pumping can add considerably to total cost of transferring a given amount of oil. Conversely, large-diameter pipe will have a fixed capital charge, even though pumping costs are minimized since natural pressure drops are less than with small-diameter pipe. Thus, an economic balance is desirable.

Example 22.1

This is an example of the principle of economic balance as applied to piping involving two alternatives. One alternative is the use of a large-diameter pipe with a small pressure drop; the other alternative is a small-diameter pipe with a greater pressure drop and more pumps. Pumps and pump room installation are considered part of the investment in pipelines.

Assume that the requirement is to transfer 100,000 bbl/day of crude oil for a distance of 200 miles by pipe. In order to arrive at the optimum conditions where total annual costs will be minimized; the fixed costs, or installation costs, and corresponding operating costs for the pipeline for different diameters must be determined and the optimization technique then applied. This is illustrated as follows:

1st: Calculate the fixed charges (installation costs) of piping and pumps and their installation. For a distance of 200 miles and for such a quantity of oil 100,000 bbl/day, the number of pump stations varies between two and three.

In order to convert the total fixed costs to an annual basis, a payout time has to be assumed. This is taken to be 5 years, plus 5% annual maintenance. Therefore, the annual "fixed charges" are 0.20 + 0.05 = 0.25% of the total fixed costs.

2nd: Operating expenses should include the following:

1. Labor, supervision, and salaries
2. Electrical power consumed

Using the above data and taking into consideration the pressure drop (P.D.) for each diameter of pump, one can estimate the number of stations needed and the brake horsepower used in pumping the oil. The ultimate solution leading to the optimum diameter is found from the graph shown in Figure 22.4.

Mathematically speaking, one can obtain the economic pipe/diameter for a pipeline using the optimization techniques.

Figure 22.5 illustrates the transport of oil by pipelines which run into millions of pipe feet and tonnage per oil field, as well as per refinery. From each individual wellhead in an oil field, the crude oil is collected in small-diameter gathering pipelines, which then converge on a collecting center. At the collecting center, the crude oil passes through gas separators, where gas is "linerated" from the crude oil. Usually, there are a number of collecting centers in different parts of the oil field. Figure 22.6 indicates this function; while it is pictorially shown in Figure 22.7.

From the collecting center, pipes of extremely large diameter lead the crude oil to a tank farm, a center or group of large circular enclosed storage tanks. From here, the crude is conveyed either to a refinery or to storage tanks at terminals for overseas delivery by sea tankers or long-distant pipeline. Large-diameter pipe

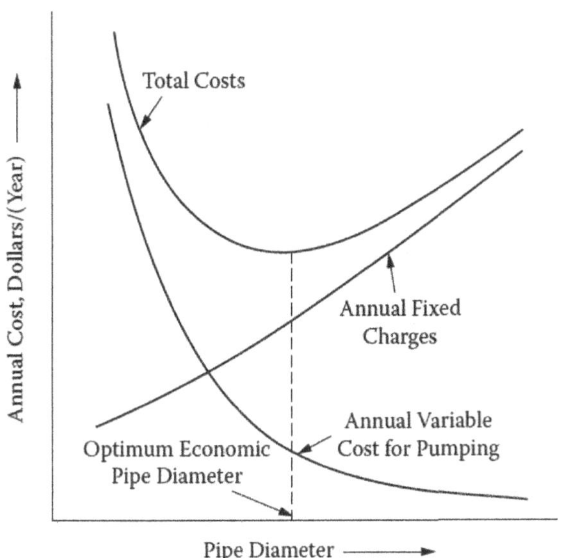

FIGURE 22.4 Optimum pipe diameter.

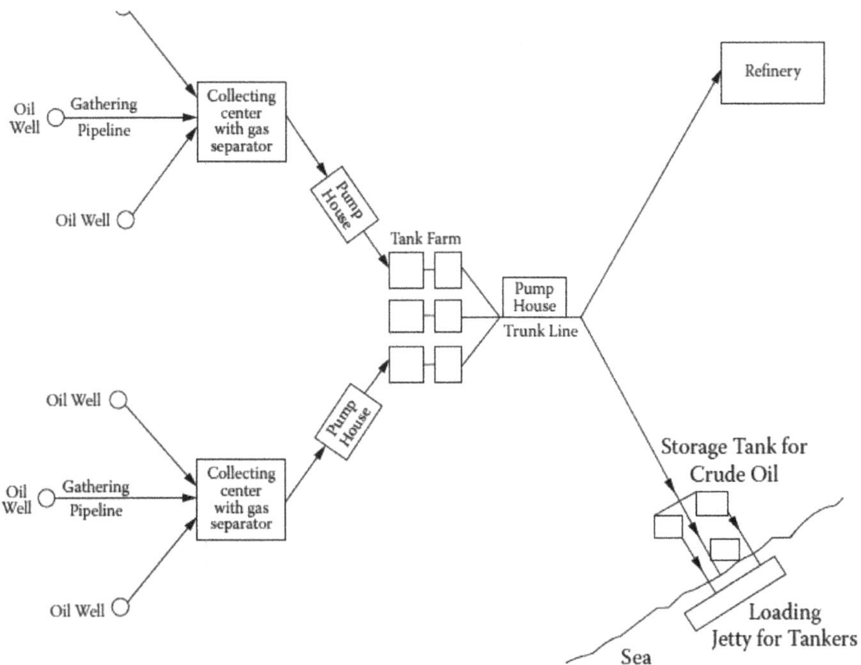

FIGURE 22.5 Transport of oil by pipes.

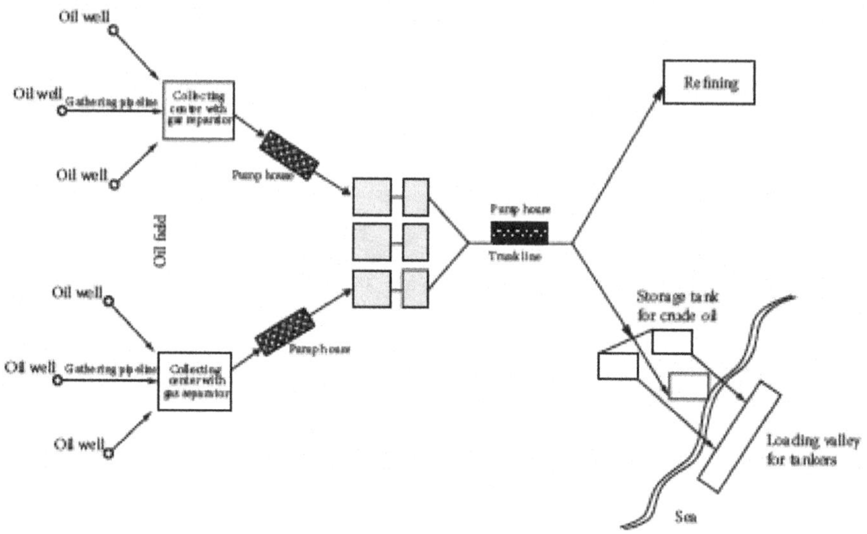

FIGURE 22.6 Net work for the delivery of oil from an oil Field.

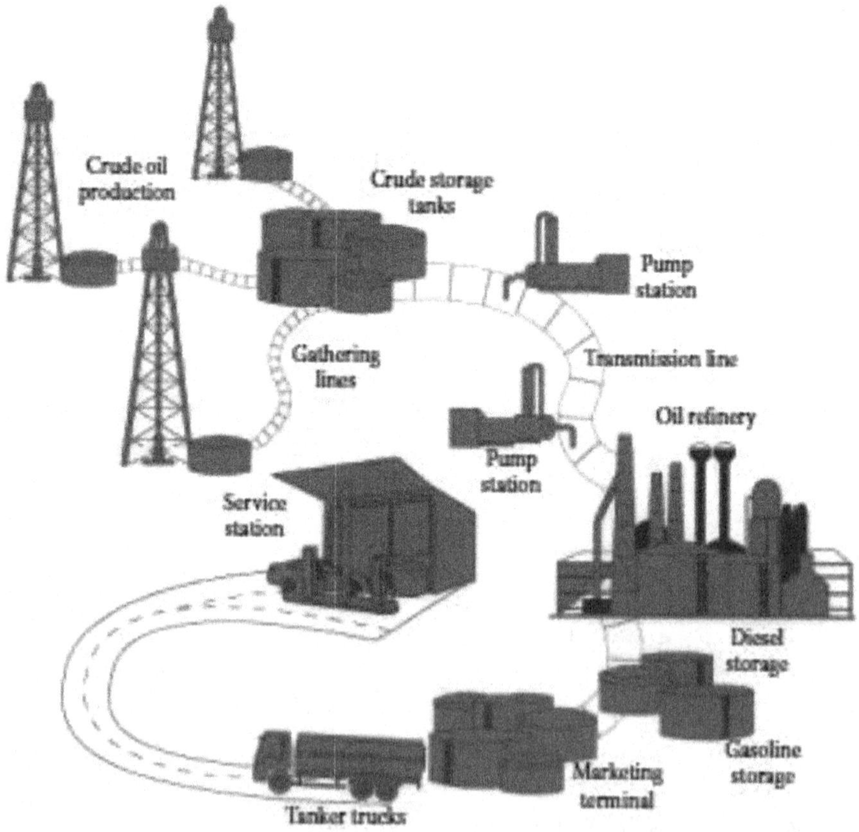

FIGURE 22.7 Network from wellheads to terminal points, from Canadian energy pipeline association, http//:www.cepa.com/about-pipelines/types-of-pipelines/liquids-pipelines.).

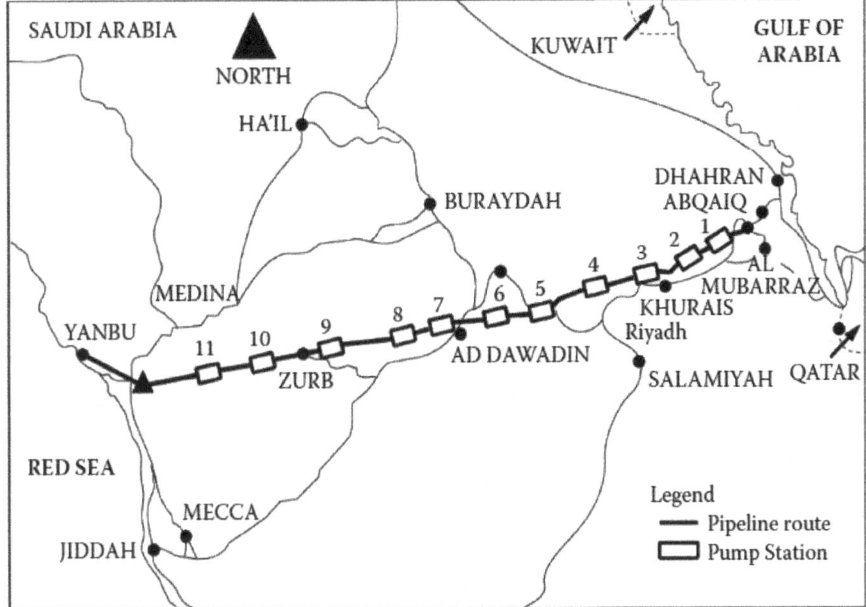

FIGURE 22.8 East-West pipeline of Saudi Arabia.

is used where volume is large, where it is practical and where long distances are involved, for the greater the diameter of the pipe, the less is the fall in pressure and thus the fewer pumping stations required. For example, the East-West pipeline of Saudi Aramco, which is known as the (Petroline), is presented in Figure 22.8. The 1,200 km, and 48 in. pipeline transports nearly 50% of Aramco's total crude oil output to Saudi refineries on the Red sea and more than 2.3 mbd crude export via Yanbu terminal.

Producing oil fields commonly have a number of small diameter gathering lines that gather crude oil from the wells and move it to central gathering facilities called oil batteries. In general, there are four types of pipelines that are in common usage:

- Oil field gathering pipelines; their function in the oil field is of great impact on production operations.
- Larger diameter feeder pipelines transport the crude oil from the oil field to loading ports and nearby refineries.
- Long-distance pipelines, which naturally shorten the alternative sea route.
- Pipelines that transport oil from ports of discharge to inland refineries, located in industrial areas remote from a seaport. These are called transmission pipelines.

Pipelines used to carry crude oil and petroleum products differ a great deal in size, ranging from 2 in. to as much as 36 in. diameter. In some cases, even 48 in. piping is used. As far as the design of an oil gathering system, flowlines and trunklines make a combination of different schemes.

22.3.6 Sizing Pipelines

In the design of a pipe, one should be aware of two fundamental concepts:

- The diameter of a pipe is a function of the flow rate of the fluid: $D = f(Q)$;
- The thickness of a pipe is a function of the working pressure inside the pipe: $t = f(p)$.

By sizing, we mean to determine the pipe diameter first. An engineer in charge must specify the diameter of pipe that will be used in a given piping system. Normally, the economic factor must be considered in determining the optimum pipe diameter.

To calculate pipe diameter for noncompressible fluids, one can apply the well-known equation:

$$Q = u.\text{A (cross section area of pip l e)}$$
$$= u.(\pi/4)d^2$$

Pipe diameter, *d*, is readily calculated from this equation for a specified flow rate, *Q* (bbl/hr) and for an assumed fluid velocity, *u* (ft/sec).

SOLVED EXAMPLES

Example 22.2

This example illustrates determination of the optimum pipe (D_{opt}) through optimization of the total annual cost. Assume the following formulas:

$$\text{Annual operating cost} = F_1(l/D_{pipe})$$
$$\text{Annual fixed costs} = F_2(D_{pipe})$$

where F_1 and F_2 are some defined functions of the diameter D of the pipe.

The total annual costs for transferring oil will be = $F_1 + F_2$ = Total costs

Optimum economic diameter of the pipeline is reached when the total annual costs are at the minimum, that is, taking the derivative of the total annual cost w.r.t. the pipe diameter, *D*.

Therefore,

$$d/dD_{pipe}[(\text{total costs}) = d/dD((F_1\ 1/D_{pipe}) + F2(D))]$$

and letting this product equal zero, solving for the value of $D = D_{opt}$.

To illustrate the principle of $D = D_{opt}$ in a simplified manner, take F_1 and F_2 as linear functions of some constants:

$$F_1(1/d) = a/D + b \text{ and } F_2(D) = cD + d$$

where a, b, c are constants to be defined.

$$\text{The total annual costs} = l/D + b + CD + d$$

and

$$d(T.C.)/dD = -a/D^2 + C = 0.$$

This gives

$$a/D^2 = c$$

Hence,

$$D_{opt} = (a/c)^{1/2}$$

The exact equation for predicting D_{opt} for turbulent flow for incompressible fluids inside steel pipes of constant diameter is given by the equation:

$$D_{opt} = 2.2\ W^{0.45} / \ell^{0.32}$$

where:

$D > 1''$
W = thousands of pounds mass flowing per hour
ℓ = density, or lb-mass/ft³

Then, to calculate D_{opt}, if we are considering the transfer of 500,000 bbl/day of oil of an average API of 33° (with $\ell = 53.70$ lb/ft³) across a distance of 1,000 miles, we have:

1st: Calculation for D_{opt}:

$$W = 500{,}000 \text{ bbl/day} \times 300 \text{ lb/bbl} \times 1/24 = 6.25 \text{ lb/hr}$$

$$D_{opt} = 2.2\ (6.25)^{0.45}/(53.7)^{0.32}$$

$$= 31 \text{ inches}$$

Therefore, 31 in. is the optimum economic pipe diameter in this particular case.

22.3.7 Construction Costs of Pipelines

There are four categories of pipeline construction costs; material, labor, miscellaneous, and right-of-way (ROW). The cost per category, expressed as a percentage of total construction costs, tends to vary by both location and year. Materials may include line pipe, pipe coating, and cathodic protection.

The average cost-per-mile for the projects rarely shows consistent trends related to either length or geographic area. In general, however, the cost-per-mile within a given diameter decreases as the number of miles rises, suggesting that fewer and longer pipelines are more cost efficient. Lines built nearer populated areas tend to have higher unit costs.

The Table 22.4 shows the results of some pipelines in a survey of projects documented by the Global Fossil Infrastructure Tracker. The survey is based on a diverse collection of projects worldwide. It shows the wide range in costs per km, which can likely be attributed primarily to differences between offshore and onshore projects, the inclusion or exclusion of additional infrastructure such as drilling platforms or pressurization stations, and regional costs differences. For that reason, these results should be viewed as non-conclusive.

Estimated average pipeline investment for any amount of piping involves millions of dollars. Size of pipe in diameter, length of the line in distances of miles and feet traveled and type of pipe used all contribute to total investment in pipelines.

The following example illustrates how the immense costs of a pipeline could be recovered quickly by pumping crude oil.

TABLE 22.4
Investment Per Km, US $, for a Number of Pipeline Contructors

Wiki	Name	Owner	Status	Estimated Investment (US$)	Lengthkm	Inv per km (US$)
http://bit.ly /2mzAq9t	Delfin Offshore Pipeline	Fairwood Peninsula Energy Corporation, Golar LNG	Proposed	8,000,000,000	50	160,000,000
http://bit.ly /2PCaLu9	Israel Cyprus Gas Pipeline	Energean	Proposed	7,000,000,000	200	35,000,000
http://bit.ly /2Rz8ZHM	Sakhalin-Hokkaido Gas Pipeline	Gazprom, Japanese Pipeline Development Organization (JPDO)	Proposed	50,000,000,000	1500	33,333,333
https://bit.ly /2Pj6M5J	Liza Gas Pipeline	ExxonMobil	Proposed	4,400,000,000	190	23,157,895
http://bit.ly /2PmmZHc	Saddle West Pipeline	TransCanada	Proposed	655,000,000	29	22,586,207
http://bit.ly /2gqy1bb	Alaska LNG Pipeline (AKLNG)	Alaska Gasline Development Corp (AGDC)	Proposed	55,000,000,000	2760	19,927,536
http://bit.ly						

For proposed onshore US gas pipeline projects in 2015-16, the average cost was $7.65 million/mile, up from both the 2014-15 average cost of $5.2 million/mile and the 2013-14 average cost of $6.6 million/mile.

Source Oil and Gas Pipeline Construction Costs - Global Energy Monitor ***www.gem.wiki › Oil_and_Gas_Pipeline_Construction_...***

Example 22.3

If the investment cost of pipeline in flat terrain is taken to be \$900,000/mile and the pipeline is 1,000 miles, while the rate of pumping crude oil is assumed to be 500,000 bbl/day, calculate the total capital investment of the pipeline and compare this figure with the gross revenue per year received by selling the oil at \$80/bbl.

$$\text{The capital investment of the pipeline} = 900{,}000\ \$/\text{mile} \times 1000\ \text{miles}$$

$$= \$9 \times 10^{8}$$

$$\text{The annual revenue of sales (gross)} = 500{,}000\ \text{bbl/day} \times 350\ \text{day/yr} \times 80\ \$/\text{bbl}$$

$$= \$1.4 \times 10^{10}$$

As far as the crude oil pipeline capacities are concerned, each pipeline must be considered an individual problem. Generally speaking, the economic capacity of each of the various diameters of pipelines as well as the usual spacing between pump stations (booster pumps) lies between the limits given in Table 20.12.

When moving oil and oil products, such operating costs as the following, based on a per-ton mile basis, will be important:

1. Construction costs of pipeline and equipment
2. Amortization of investment
3. Interest on invested capital
4. Energy costs for operating pumping stations, etc.
5. Personnel and maintenance costs
6. Royalties to governments of countries crossed by the pipeline

Finally, it has to be noted that these large sizes of pipe are costly to ship, because the space they occupy, relative to their weight, is high, and therefore freight costs are up. To reduce freight costs, it has become the practice today to design these large pipelines for equal quantities of two slightly different sizes of pipe, so that they can be "nested" for shipment; for example, one length of 20 in. pipe is placed inside each length of 22 in. pipe.

22.4 OIL TANKERS

Primarily the transportation of bulk liquids began in the year of the late 19th century when the discovery and expedition of oils began. At that time, tankers emerged as the main mode of transportation to carry bulk liquids from the refineries to the global market. On the way, as different energy products emerged, the need for a different type of tankers came into the real picture.

There are two basic types of oil tankers: crude tankers and product tankers. Crude tankers move large quantities of unrefined crude oil from its point of extraction to refineries. For example, moving crude oil from oil wells in a producing country to refineries in another country.

Today's cutting-edge tankers are the product of a commitment to safety combined with the power of computer-assisted design. As a result, the new ships traveling the seas are stronger, more maneuverable, and more durable than their predecessors. Oil tankers are the one of the best ways to transport extremely large quantities of oil. These tankers traverse the oceans and vast waterways of the world with millions of gallons of oil and liquefied natural gas.

Examples of commercial tankers:

The commercial oil tanker *AbQaiq*, in ballast

	Class overview
Name:	Oil tanker
Subclasses:	Handysize, Panamax, Aframax, Suezmax, Very Large Crude Carrier (VLCC), Ultra Large Crude Carrier (ULCC)
Built:	c. 1963–present
	General characteristics
Type:	Tank ship
Capacity:	up to 550,000 DWT
Notes:	Rear house, full hull, midships pipeline
	Source: Wikipedia, the free encyclopedia

An oil tanker, also known as a petroleum tanker, is a ship designed for the bulk transport of oil or its products.

There are two basic types of oil tankers:

- Crude tankers and
- Product tankers

Crude tankers move large quantities of unrefined crude oil from its point of extraction to refineries. For example, moving crude oil from oil wells in a producing country to refineries in another country.

Product tankers, generally much smaller, are designed to move refined products from refineries to points near consuming markets. For example, moving gasoline from refineries in Europe to consumer markets in Nigeria and other West African nations.

Oil tankers are often classified by their size as well as their occupation. The size classes range from inland or coastal tankers of a few thousand metric tons of deadweight (DWT) to the mammoth ultra large crude carriers (ULCCs) of 550,000 DWT. Tankers move approximately 2.0 billion metric tons (2.2 billion short tons) of oil every year. Second only to pipelines in terms of efficiency, the average cost of transport of crude oil by tanker amounts to only US$5 to $8 per cubic metre ($0.02 to $0.03 per US gallon).

Some specialized types of oil tankers have evolved. One of these is the naval replenishment oiler, a tanker which can fuel a moving vessel. Combination ore-bulk-oil carriers and permanently moored floating storage units are two other variations on the standard oil tanker design. Oil tankers have been involved in a number of damaging and high-profile oil spills. As a result, they are subject to stringent design and operational regulations.

Source: From Wikipedia, the free encyclopedia

22.5 RAILROAD TANK CARS

Before the development of pipeline systems, transportation of oil by railroad tank cars was by a wide margin the most important method of moving oil from its point of production through refining and to its point of final consumption. This dominance was initially a function of the fact that railroads were extensively developed in most areas at least a half century before the economic use of motor trucks and the road networks that were established to serve more local markets than could be reached by rail transport. The other factor that contributes to the importance of railroad tank cars in today's markets is that on a ton/kilometer basis, rail transport is generally between two and three times as efficient as oil and oil product movement by truck. This is partly because railroad tank cars are significantly larger than even the biggest tank trucks and thus enjoy greater economies of scale, and partly because each tank truck needs a driver while an entire trainload of perhaps a hundred cars requires only two or three employees. Roadbed costs also tend to be less, and required maintenance is not as expensive as the tank truck alternative requirements.

The relative economies of the three land-based transportation systems—pipelines, railroad tank cars, and tank trucks—can be illustrated by the way Iraq moved its crude oil to world markets during the Iraq/Iran War. Being essentially barred from using tankers in the Gulf by Iran's control of the Shatt El-Arab waterway, and with its pipelines to the Mediterranean Sea blocked by political action by Syria, Iraq turned principally to a pipeline across Saudi Arabia to Yanbu on the Red Sea, secondarily to a rail link with Turkey and finally to the most expensive mode of all, tank trucks by road to Turkey and to the Gulf of Aqaba through Jordan. With the war over, Iraq established limited tanker access through the Gulf, implemented an expansion of its pipelines across Saudi Arabia to Yanbu, discontinued its long-haul truck movements across Jordan and phased out its truck movements to Turkey, in that order.

Railroad tank cars remain in many parts of the world, in both the industrialized and developing countries, an important mode of transportation. Many small markets do not economically justify building pipelines to serve them, but are still large enough and close enough to rail connections to make rail the main method of basic oil product transportation. This means that tank trucks only have to do short hauls to get the oil to its final consumers.

22.6 TANK TRUCKS

Tank trucks tend to be very much oriented to specific local consumer markets. All gasoline and diesel service stations, for example, are supplied by tank trucks, as are all home heating oil customers. Rail transportation systems are not flexible enough to reach many small or medium-sized consumers of even commercial and industrial oil products. Large fuel users, such as electric utilities or steel plants, are likely to be supplied by individual pipelines from local refineries, barges if they are on the waterfront and railroad tank cars if they are both not available to water and too far away to justify a product pipeline. Heavy fuel oil is also sometimes too viscous to pump at ambient temperatures and thus requires heated delivery systems, whether pipelines, railroad tank cars, or tank trucks; this involves added capital and operating costs and is a significant factor in heavy fuel oil's competitive position with coal.

Tank trucks, because of their flexibility, are also involved fairly extensively on the crude oil supply side, particularly in North America but also in other countries where field size and flow rates do not justify pipeline gathering systems. Oil from small wells is pumped into small tanks at the well sites; these are regularly emptied and the oil trucked to the nearest refinery, rail connection, or pipeline access point. In the United States, for example, about 3% of total oil production, from well over half of the country's wells, is handled in this fashion.

The inefficiencies of this system, relative to the gathering costs of major oil fields, are such as to make such production barely inframarginal. This was why in the 1985-1986 decline in world oil prices about 500 barrels per day of U.S. producing capacity was shut down. Had the transport costs of bringing the output of many wells to market not been so high, it is likely that these cutbacks would have been substantially lower.

A significant exception to the generalization that most final consumers are served by tank trucks is the airline sector. Because of the volumes involved and the need to maintain product purity as well as consistent availability, most airports are served by pipelines from local refineries or distribution points. Again, relative economics are the dominant factor. But in these cases, the importance of assured supply and tight product specifications as to quality are enough to justify a market premium of perhaps U.S. two cents per gallon, or 85 cents per barrel. (Final delivery for the last few hundred meters, however, is by tank truck into the aircraft fuel tanks.)

22.7 RISKS AND HAZARDS ASSOCIATED WITH PIPELINES

The pipeline will be 24–30 inches in diameter. It will carry over 300,000 barrels of oil a day with a volatility of 32. For natural gas pipelines, the greatest risk is associated with fires or explosions caused by ignition of the natural gas. This can cause significant property damage and injuries or death. Additionally, the release of natural gas, primarily methane which is a very potent greenhouse gas, contributes to climate changeeleases of products carried through pipelines can impact the environment and may result in injuries or fatalities as well as property damage. The risk associated with pipelines varies depending on a number of factors such as the product being transported in the pipeline, size and operating pressure of the pipeline, as well as the population and natural resources near the pipeline.

Appendix A: Conversion Factors

TABLE A.1
Alphabetical Conversion Tables

To Convert From	Do This
Atmospheres to inches of mercury @32°F (Atm to inHg32	(atm) * 29.9213 = (inHg32)
Atmospheres to inches of mercury @60°F (Atm to inHg60)	(atm) * 30.0058 = (inHg60)
Atmospheres to millibars (atm to mb)	(atm) * 1013.25 = (mb)
Atmospheres to pascals (atm to Pa)	(atm) * 101325 = (Pa)
Atmospheres to pounds/square inch (atm to lb/in^2)	(atm) * 14.696 = (lb/in^2)
Centimeters to feet (cm to ft)	(cm) * 0.032808399 = (ft)
Centimeters to inches (cm to in)	(cm) * 0.39370079 = (in)
Centimeters to meters (cm to m)	(cm) * 0.01 = (m)
Centimeters to millimeters (cm to mm)	(cm) * 10 = (mm)
Degrees to radians (deg to rad)	(deg) * 0.01745329 = (rad)
Degrees Celsius to degrees Fahrenheit (C to F)	[(C) * 1.8] + 32 = (F)
Degrees Celsius to degrees Kelvin (C to K)	(C) + 273.15 = (K)
Degrees Celsius to degrees Rankine (C to R)	[(C) * 1.8] + 491.67 = (R)
Degrees Fahrenheit to degrees Celsius (F to C)	[(F) – 32)] * 0.555556 = (C)
Degrees Fahrenheit to degrees Kelvin (F to K)	[(F) * 0.555556] + 255.37 = (K)
Degrees Fahrenheit to degrees Rankine (F to R)	(F) + 459.67 = (R)
Degrees Kelvin to degrees Celsius (K to C)	(K) – 273.15 = (C)
Degrees Kelvin to degrees Fahrenheit (K to F)	[(K) – 255.37] * 1.8 = (F)
Degrees Kelvin to degrees Rankine (K to R)	(K) * 1.8 = (R)
Degrees Rankine to degrees Celsius (R to C)	[(R) – 491.67] * 0.555556 = (C)
Degrees Rankine to degrees Fahrenheit (R to F)	(R) – 459.67 = (F)
Degrees Rankine to degrees Kelvin (R to K)	(R) * 0.555556 = (K)
Feet to centimeters (ft to cm)	(ft) * 30.48 = (cm)
Feet to meters (ft to m)	(ft) * 0.3048 = (ft to m)
Feet to miles (ft to mi)	(ft) * 0.000189393 = (mi)
Feet/minute to meters/second (ft/min to m/s)	(ft/min) * 0.00508 = (m/s)
Feet/minute to miles/hour (ft/min to mph)	(ft/min) * 0.01136363 = (mph)
Feet/second to kilometers/hour (ft/s to kph)	(ft/s) * 1.09728 = (kph)
Feet/second to knots (ft/s to kt)	(ft/s) * 0.5924838 = (kt)

(Continued)

TABLE A.1
Alphabetical Conversion Tables *(Continued)*

To Convert From	Do This
Feet/second to meters/second (ft/s to m/s)	(ft/s) * 0.3048 = (m/s)
Feet/second to miles/hour (ft/s to mph)	(ft/s) * 0.681818 = (mph)
Grams/cubic centimeter to pounds/cubic foot (gm/cm^3 to lb/ft^3)	(gm/cm^3) * 62.427961 = (lb/ft^3)
Grams/cubic meter to pounds/cubic foot (gm/m^3 to lb/ft^3)	(gm/m^3) * 0.000062427961 = (lb/ft^3)
Hectopascals to millibars (hPa to mb)	Nothing, they are equivalent units
Inches to centimeters (in to cm)	(in) * 2.54 = (cm)
Inches to millimeters (in to mm)	(in) * 25.4 = (mm)
Inches of mercury @32°F to atmospheres (inHg32 to atm)	(inHg32) * 0.0334211 = (atm)
Inches of mercury @32°F to millibars (inHg32 to mb)	(inHg32) * 33.8639 = (mb)
Inches of mercury @32°F to pounds/square inch (inHg32 to lb/in^2)	(inHg32) * 0.49115 = (lb/in^2)
Inches of mercury @60°F to atmospheres (inHg60 to atm)	(inHg60) * 0.0333269 = (atm)
Inches of mercury @60°F to millibars (inHg60 to mb)	(inHg60) * 33.7685 = (mb)
Inches of mercury @60°F to pounds/square inch (inHg60 to lb/in^2)	(inHg60) * 0.48977 = (lb/in^2)
Kilograms/cubic meters to pounds/cubic foot (kg/m^3 to lb/ft^3)	(kg/m^3) * 0.062427961 = (lb/ft^3)
Kilograms/cubic meters to slugs/cubic foot (kg/m^3 to $slug/ft^3$)	(kg/m^3) * 0.001940323 = ($slug/ft^3$)
Kilometers to meters (km to m)	(km) * 1000 = (m)
Kilometers to miles (km to mi)	(km) * 0.62137119 = (mi)
Kilometers to nautical miles (km to nmi)	(km) * 0.5399568 = (nmi)
Kilometers/hour to feet/second (kph to ft/s)	(kph) * 0.91134 = (ft/s)
Kilometers/hour to knots (kph to kt)	(kph) * 0.5399568 = (kt)
Kilometers/hour to meters/second (kph to m/s)	(kph) * 0.277777 = (m/s)
Kilometers/hour to miles/hour (kph to mph)	(kph) * 0.62137119 = (mph)
Kilopascals to millibars (kPa to mb)	(kPa) * 10 = (mb)
Knots to feet/second (kt to ft/s)	(kt) * 1.6878099 = (ft/s)
Knots to kilometers/hour (kt to kph)	(kt) * 1.852 = (kph)
Knots to meters/second (kt to m/s)	(kt) * 0.514444 = (m/s)
Knots to miles/hour (kt to mph)	(kt) * 1.1507794 = (mph)
Knots to nautical miles/hour (kt to nmph)	Nothing, they are equivalent units
Langleys/minute to watts/square meter (ly/min to W/m^2)	(ly/min) * 698.339 = (W/m^2)

(Continued)

TABLE A.1
Alphabetical Conversion Tables *(Continued)*

To Convert From	Do This
Watts/square meter to langleys/minute (W/m^2 to ly/min)	(W/m^2) * 0.00143197 = (ly/min)
Meters to centimeters (m to cm)	(m) * 100 = (cm)
Meters to feet (m to ft)	(m) * 3.2808399 = (ft)
Meters to kilometers (m to km)	(m) * 0.001 = (km)
Meters to miles (m to mi)	(m) * 0.00062137119 = (mi)
Meters/second to feet/minute (m/s to ft/min)	(m/s) * 196.85039 = (ft/min)
Meters/second to feet/second (m/s to ft/s)	(m/s) * 3.2808399 = (ft/s)
Meters/second to kilometers/hour (m/s to kph)	(m/s) * 3.6 = (kph)
Meters/second to knots (m/s to kt)	(m/s) * 1.943846 = (kt)
Meters/second to miles/hour (m/s to mph)	(m/s) * 2.2369363 = (mph)
Miles to feet (mi to ft)	(mi) * 5280 = (ft)
Miles to kilometers (mi to km)	(mi) * 1.609344 = (km)
Miles to meters (mi to m)	(mi) * 1609.344 = (m)
Miles/hour to feet/minute (mph to ft/min)	(mph) * 88 = (ft/min)
Miles/hour to feet/second (mph to ft/s)	(mph) * 1.466666 = (ft/s)
Miles/hour to kilometers/hour (mph to kph)	(mph) * 1.609344 = (kph)
Miles/hour to knots (mph to kt)	(mph) * 0.86897624 = (kt)
Miles/hour to meters/second (mph to m/s)	(mph) * 0.44704 = (m/s)
Millibars to atmospheres (mb to atm)	(mb) * 0.000986923 = (atm)
Millibars to hectopascals (mb to hPa)	Nothing, they are equivalent units
Millibars to inches of mercury @32°F (mb to inHg32)	(mb) * 0.02953 = (inHg32)
Millibars to inches of mercury @60°F (mb to inHg60)	(mb) * 0.02961 = (inHg60)
Millibars to kilopascals (mb to kPa)	(mb) * 0.1 = (kPa)
Millibars to millimeters of mercury @32°F (mb to mm Hg)	(mb) * 0.75006 = (mm Hg)
Millibars to millimeters of mercury @60°F (mb to mm Hg)	(mb) * 0.75218 = (mm Hg)
Millibars to newtons/square meter (mb to N/m^2)	(mb) * 100 = (N/m^2)
Millibars to pascals (mb to Pa)	(mb) * 100 = (Pa)
Millibars to pounds/square foot (mb to lb/ft^2)	(mb) * 2.088543 = (lb/ft^2)
Millibars to pounds/square inch (mb to lb/in^2)	(mb) * 0.0145038 = (lb/in^2)
Millimeters to centimeters (mm to cm)	(mm) * 0.1 = (cm)
Millimeters to inches (mm to in)	(mm) * 0.039370078 = (in)
Millimeters of mercury @32°F to millibars (mm Hg to mb)	(mm Hg) * 1.33322 = (mb)
Millimeters of mercury @60°F to millibars (mm Hg to mb)	(mm Hg) * 1.32947 = (mb)
Nautical miles to kilometers (nmi to km)	(nmi) * 1.852 = (km)

(Continued)

TABLE A.1
Alphabetical Conversion Tables *(Continued)*

To Convert From	Do This
Nautical miles to statute miles (nmi to mi)	(nmi) * 1.1507794 = (mi)
Nautical miles/hour to knots (nmph to kt)	Nothing, they are equivalent units
Newtons/square meter to millibars (N/m^2 to mb)	(N/m^2) * 0.01 = (mb)
Pascals to atmospheres (Pa to atm)	(Pa) * 0.000009869 = (atm)
Pascals to millibars (Pa to mb)	(Pa) * 0.01 = (mb)
Pounds/cubic foot to grams/cubic centimeter (lb/ft^3 to gm/cm^3)	(lb/ft^3) * 0.016018463 = (gm/cm^3)
Pounds/cubic foot to grams/cubic meter (lb/ft^3 to gm/m^3)	(lb/ft^3) * 16018.46327 = (gm/m^3)
Pounds/cubic foot to kilograms/cubic meter (lb/ft^3 to kg/m^3)	(lb/ft^3) * 16.018463 = (kg/m^3)
Pounds/square foot to millibars (lb/ft^2 to mb)	(lb/ft^2) * 0.478803 = (mb)
Pounds/square inch to atmospheres (lb/in^2 to atm)	(lb/in^2) * 0.068046 = (atm)
Pounds/square inch to inches of mercury @32°F (lb/in^2 to inHg32)	(lb/in^2) * 2.03602 = (inHg32)
Pounds/square inch to inches of mercury @60°F (lb/in^2 to inHg60)	(lb/in^2) * 2.04177 = (inHg60)
Pounds/square inch to millibars (lb/in^2 to mb)	(lb/in^2) * 68.9474483 = (mb)
Radians to degrees (rad to deg)	(rad) * 57.29577951 = (deg)
Slugs/cubic foot to kilograms/cubic meter ($slug/ft^3$ to kg/m^3)	($slug/ft^3$) * 515.378 = (kg/m^3)
Statute miles to nautical miles (mi to nmi)	(mi) * 0.86897624 = (nmi)

Note: Follow simple formulas to make **conversions** in speed, pressure, and various units; for example, MPH to M/S or C to F.

Source: CSGNetwork, Conversion Factors Table, www.csgnetwork.com/convfactorstable.html.

TABLE A.2
Metric Tons to Barrels (Crude Oil)

Abu Dhabi	7.624	Denmark	7.650
Algeria	7.661	Ecuador	7.580
Argentina	7.196	France	7.287
Austria	6.974	Bahrain	7.335
Albania	6.672	Brazil	7.315
Angola	7.206	Bulgaria	7.300
Australia	7.775	Canada	7.428
Bolivia	8.086	China	7.300
Brunei	7.334	Congo	7.478
Burma	7.464	Czechoslovakia	6.782
Chile	7.802	Dubai	7.295
Colombia	7.054	Egypt	7.240
Cuba	6.652	Gabon	7.245
Germany	7.223	Hungary	7.630
India	7.441	Indonesia	7.348
Iran	7.370	Iraq	7.453
Japan	7.352	Italy	6.813
Libya	7.615	Kuwait	7.281
Mexico	7.104	Malaysia	7.709
Morocco	7.602	Mongolia	7.300
Neutral Zone	6.825	Netherlands	6.816
New Zealand	8.043	New Guinea	7.468
Norway	7.444	Nigeria	7.410
Pakistan	7.308	Oman	7.390
Poland	7.419	Peru	7.517
Romania	7.453	Qatar	7.573
Senegal	7.535	Saudi Arabia	7.338
Spain	7.287	Sharjah	7.650
Taiwan	7.419	Syria	6.940
Tunisia	7.709	Trinidad	6.989
United Arab Emirates	7.522	Turkey	7.161
United States	7.418	United Kingdom	7.279
Zaire	7.206	USSR	7.350
Venezuela	7.005	Yugoslavia	7.407

TABLE A.3
Metric Tons to Barrels (Products)

Refined	Products		Other Products	
Aviation gasoline	8.90	Grease		6.30
Motor gasoline	8.50	Paraffin	oil, pure	7.14
White spirits	8.50	Paraffin	wax	7.87
Kerosene	7.75	Petrolatum		7.87
Jet fuel	8.00	Asphalt and road oil		6.06
Distillate gas and diesel oil	7.46	Petroleum coke		5.50
Residual fuel oil	6.66	Bitumen		6.06
Lubricating oil	7.00	LPG		11.60
		Miscellaneous products		7.00

TABLE A.4
Crude Oil Measure[a]

From	Tons	Long Tons	To Barrels	Gallons (Imperial)	Gallons (U.S.)	Tons/ Year
			Multiply by			
Tons (metric)	1	0.984	7.33	256	308	
Long tons	1.016	1	7.45	261	313	
Barrels	0.136	0.134	1	35	42	
Gallons (Imperial)	0.00391	0.0383	0.0286	1	1.201	
Gallons (U.S.)	0.00325	0.00319	0.0238	0.833	1	
Barrels/day						49.8

[a] Based on average Arabian light (33.5 API gravity).

TABLE A.5
Refined Product Measures

To Convert:	Barrels to Tons	Tons to Barrels	Barrels/Day to Tons/Year	Tons/Year to Barrels/Day
		Multiply by		
Motor spirit	0.118	8.45	43.2	0.0232
Kerosene	0.128	7.80	46.8	0.0214
Gas oil/diesel	0.133	7.50	48.7	0.0205
Fuel oil	0.149	6.70	54.5	0.0184

TABLE A.6
Calorific Equivalent

	One Million Tonnes of Oil Approximately Equals
Heat Units	
In Btu's	40×10^{12}
In therms	397×10^{6}
In teracalories	10,000
Solid Fuels[a]	
In tonnes of coal	1.5×10^{6}
In tonnes of lignite	3×10^{6}
Natural Gas[b]	
In cubic meters	1.111×10^{9}
In cubic feet	39.2×10^{9}

[a] Calorific values of coal and lignite, as produced.

[b] 1 cubic foot = 1,000 Btu; 1 cubic meter = 9,000 Kcal.

TABLE A.7
Natural Gas/LNG/LPG

Natural Gas	LNG	LPG
(One billion cubic meters equals approximately 35.3×10^{9} cubic feet)	**(One million tonnes equals approximately 0.05 TCF [gas])**	**(One million tonnes equals approximately 11.8×10^{6} barrels of LPG)**
0.89×10^{6} tonnes of crude oil	1.23×10^{6} tonnes of crude oil	1.1×10^{6} tonnes of crude oil
0.8×10^{6} tonnes of LPG	1.1×10^{6} tonnes of LPG	1.25×10^{9} cubic meters (gas)
0.725×10^{6} tonnes of LNG	1.4×10^{9} cubic meters (gas)	0.91×10^{6} tonnes of LNG
1.35×10^{6} tonnes of coal	1.9×10^{6} tonnes of coal	1.7×10^{6} tonnes of coal
36×10^{12} British Thermal Units (Btu)	52×10^{12} Btu	47×10^{12} Btu
38×10^{15} joules (38 PJ)	55 PJ	50 PJ

Notes: Tonnes, metric tons; TCF, trillion cubic feet; Mtoe, million tonnes crude oil equivalent; Mtpa, million tonnes per annum; 1 trillion, 1 million million (10^{12}); 1 billion, 1 thousand million (10^{9}); mmscfd, million cubic feet per day; mmbtu, million British Thermal Units; PJ, petajoules (10^{15} joules).

Appendix B: Compound Interest Factors

B.1 COMPOUND INTEREST TABLES (USING MS EXCEL)

Using MS Excel a compound interest table could be established to calculate compound interest factors for different interest rates and time periods. An example is cited next for $i = 10\%$ and $n = 1$ to 50 years.

10.00%								**10.00%**
n	*F/P*	*P/F*	*A/F*	*A/P*	*F/A*	*P/A*	*A/G*	*P/G*
1	1.1000	0.9091	1.0000	1.1000	1.0000	0.9091	0.0000	0.0000
2	1.2100	0.8264	0.4762	0.5762	2.1000	1.7355	0.4762	0.8264
3	1.3310	0.7513	0.3021	0.4021	3.3100	2.4869	0.9366	2.3291
4	1.4641	0.6830	0.2155	0.3155	4.6410	3.1699	1.3812	4.3781
5	1.6105	0.6209	0.1638	0.2638	6.1051	3.7908	1.8101	6.8618
6	1.7716	0.5645	0.1296	0.2296	7.7156	4.3553	2.2236	9.6842
7	1.9487	0.5132	0.1054	0.2054	9.4872	4.8684	2.6216	12.7631
8	2.1436	0.4665	0.0874	0.1874	11.4359	5.3349	3.0045	16.0287
9	2.3579	0.4241	0.0736	0.1736	13.5795	5.7590	3.3724	19.4215
10	2.5937	0.3855	0.0627	0.1627	15.9374	6.1446	3.7255	22.8913
11	2.8531	0.3505	0.0540	0.1540	18.5312	6.4951	4.0641	26.3963
12	3.1384	0.3186	0.0468	0.1468	21.3843	6.8137	4.3884	29.9012
13	3.4523	0.2897	0.0408	0.1408	24.5227	7.1034	4.6988	33.3772
14	3.7975	0.2633	0.0357	0.1357	27.9750	7.3667	4.9955	36.8005
15	4.1772	0.2394	0.0315	0.1315	31.7725	7.6061	5.2789	40.1520
16	4.5950	0.2176	0.0278	0.1278	35.9497	7.8237	5.5493	43.4164
17	5.0545	0.1978	0.0247	0.1247	40.5447	8.0216	5.8071	46.5819
18	5.5599	0.1799	0.0219	0.1219	45.5992	8.2014	6.0526	49.6395
19	6.1159	0.1635	0.0195	0.1195	51.1591	8.3649	6.2861	52.5827
20	6.7275	0.1486	0.0175	0.1175	57.2750	8.5136	6.5081	55.4069
21	7.4002	0.1351	0.0156	0.1156	64.0025	8.6487	6.7189	58.1095
22	8.1403	0.1228	0.0140	0.1140	71.4027	8.7715	6.9189	60.6893
23	8.9543	0.1117	0.0126	0.1126	79.5430	8.8832	7.1085	63.1462
24	9.8497	0.1015	0.0113	0.1113	88.4973	8.9847	7.2881	65.4813
25	10.8347	0.0923	0.0102	0.1102	98.3471	9.0770	7.4580	67.6964
26	11.9182	0.0839	0.0092	0.1092	109.1818	9.1609	7.6186	69.7940
27	13.1100	0.0763	0.0083	0.1083	121.0999	9.2372	7.7704	71.7773
28	14.4210	0.0693	0.0075	0.1075	134.2099	9.3066	7.9137	73.6495
29	15.8631	0.0630	0.0067	0.1067	148.6309	9.3696	8.0489	75.4146
30	17.4494	0.0573	0.0061	0.1061	164.4940	9.4269	8.1762	77.0766
31	19.1943	0.0521	0.0055	0.1055	181.9434	9.4790	8.2962	78.6395

(Continued)

10.00%								10.00%
n	F/P	P/F	A/F	A/P	F/A	P/A	A/G	P/G
32	21.1138	0.0474	0.0050	0.1050	201.1378	9.5264	8.4091	80.1078
33	23.2252	0.0431	0.0045	0.1045	222.2515	9.5694	8.5152	81.4856
34	25.5477	0.0391	0.0041	0.1041	245.4767	9.6086	8.6149	82.7773
35	28.1024	0.0356	0.0037	0.1037	271.0244	9.6442	8.7086	83.9872
40	45.2593	0.0221	0.0023	0.1023	442.5926	9.7791	9.0962	88.9525
45	72.8905	0.0137	0.0014	0.1014	718.9048	9.8628	9.3740	92.4544
50	117.3909	0.0085	0.0009	0.1009	1163.9085	9.9148	9.5704	94.8889

B.2 STANDARD COMPOUND INTEREST TABLES

½%

	To find *F*, given *P*: $(1+i)^n$	To find *P*, given *F*: $\frac{1}{(1+i)^n}$	To find *A*, given *F*: $\frac{i}{(1+i)^n-1}$	To find *A*, given *P*: $\frac{i(1+i)^n}{(1+i)^n-1}$	To find *F*, given *A*: $\frac{(1+i)^n-1}{i}$	To find *P*, given *A*: $\frac{(1+i)^n-1}{i(1+i)^n}$	
n	$(f/p)_n^{1/2}$	$(p/f)_n^{1/2}$	$(a/f)_n^{1/2}$	$(a/p)_n^{1/2}$	$(f/a)_n^{1/2}$	$(p/a)_n^{1/2}$	n
1	1.005	0.9950	1.00000	1.00500	1.000	0.995	1
2	1.010	0.9901	0.49875	0.50375	2.005	1.985	2
3	1.015	0.9851	0.33167	0.33667	3.015	2.970	3
4	1.020	0.9802	0.24183	0.25313	4.030	3.950	4
5	1.025	0.9754	0.19801	0.20301	5.050	4.926	5
6	1.030	0.9705	0.16460	0.16960	6.076	5.896	6
7	1.036	0.9657	0.14073	0.14573	7.106	6.862	7
8	1.041	0.9609	0.12283	0.12783	8.141	7.823	8
9	1.046	0.9561	0.10891	0.11391	9.182	8.779	9
10	1.051	0.9513	0.09777	0.10277	10.288	9.730	10
11	1.056	0.9466	0.08866	0.09366	11.279	10.677	11
12	1.062	0.9419	0.08107	0.08607	12.336	11.619	12
13	1.067	0.9372	0.07464	0.07964	13.397	12.556	13
14	1.072	0.9326	0.06914	0.07414	14.464	13.489	14
15	1.078	0.9279	0.06436	0.06936	15.537	14.417	15
16	1.083	0.9233	0.06019	0.06519	16.614	15.340	16
17	1.088	0.9187	0.05615	0.06151	17.697	16.259	17
18	1.094	0.9141	0.05323	0.05823	18.786	17.173	18
19	1.099	0.9096	0.05030	0.05530	19.880	18.082	19
20	1.105	0.9051	0.04767	0.05267	20.979	18.987	20
21	1.110	0.9006	0.04528	0.05028	22.084	19.888	21
22	1.116	0.8961	0.04311	0.04811	23.194	20.784	22
23	1.122	0.8916	0.04113	0.04613	24.310	21.676	23
24	1.127	0.8872	0.03932	0.04432	25.432	22.563	24

½%

n	To find F, given P: $(1+i)^n$ $(f/p)_n^{1/2}$	To find P, given F: $\frac{1}{(1+i)^n}$ $(p/f)_n^{1/2}$	To find A, given F: $\frac{i}{(1+i)^n-1}$ $(a/f)_n^{1/2}$	To find A, given P: $\frac{i(1+i)^n}{(1+i)^n-1}$ $(a/p)_n^{1/2}$	To find F, given A: $\frac{(1+i)^n-1}{i}$ $(f/a)_n^{1/2}$	To find P, given A: $\frac{(1+i)^n-1}{i(1+i)^n}$ $(p/a)_n^{1/2}$	n
25	1.133	0.8828	0.03767	0.04265	26.559	23.446	25
26	1.138	0.8784	0.03611	0.04111	27.692	24.324	26
27	1.144	0.8740	0.03469	0.03969	28.830	25.198	27
28	1.150	0.8697	0.03336	0.03836	29.975	26.068	28
29	1.156	0.8653	0.03213	0.03713	31.124	26.933	29
30	1.161	0.8610	0.03098	0.03598	32.280	27.794	30
31	1.167	0.8567	0.02990	0.03490	33.441	28.651	31
32	1.173	0.8525	0.02889	0.03389	34.609	29.503	32
33	1.179	0.8482	0.02795	0.03295	35.782	30.352	33
34	1.185	0.8440	0.02706	0.03206	36.961	31.196	34
35	1.191	0.8398	0.02622	0.03122	38.145	32.035	35
40	1.221	0.8191	0.02265	0.02765	44.159	36.172	40
45	1.252	0.7990	0.01987	0.02487	50.324	40.207	45
50	1.283	0.7793	0.01765	0.02265	56.645	44.143	50
55	1.316	0.7601	0.01548	0.02084	63.126	47.981	55
60	1.349	0.7414	0.01433	0.01933	69.770	51.726	60
65	1.383	0.7231	0.01306	0.01806	76.582	55.377	65
70	1.418	0.7053	0.01197	0.01697	83.566	58.939	70
75	1.454	0.6879	0.01102	0.01602	90.727	62.414	75
80	1.490	0.6710	0.01020	0.01520	98.068	65.802	80
85	1.528	0.6545	0.00947	0.01447	105.594	69.108	85
90	1.567	0.6383	0.00883	0.01383	113.311	72.331	90
95	1.606	0.6226	0.00825	0.01325	121.222	75.476	95
100	1.647	0.6073	0.00773	0.01273	129.334	78.543	100

1%

n	To find F, given P: $(1+i)^n$ $(f/p)_n^{1/2}$	To find P, given F: $\frac{1}{(1+i)^n}$ $(p/f)_n^{1/2}$	To find A, given F: $\frac{i}{(1+i)^n-1}$ $(a/f)_n^{1/2}$	To find A, given P: $\frac{i(1+i)^n}{(1+i)^n-1}$ $(a/p)_n^{1/2}$	To find F, given A: $\frac{(1+i)^n-1}{i}$ $(f/a)_n^{1/2}$	To find P, given A: $\frac{(1+i)^n-1}{i(1+i)^n}$ $(p/a)_n^{1/2}$	n
1	1.010	0.9901	1.00000	1.01000	1.000	0.990	1
2	1.020	0.9803	0.49751	0.50751	2.010	1.970	2
3	1.030	0.9706	0.33002	0.34002	3.030	2.941	3
4	1.041	0.9610	0.24628	0.25628	4.060	3.902	4

(Continued)

1%

	To find F, given P: $(1+i)^n$	To find P, given F: $\frac{1}{(1+i)^n}$	To find A, given F: $\frac{i}{(1+i)^n-1}$	To find A, given P: $\frac{i(1+i)^n}{(1+i)^n-1}$	To find F, given A: $\frac{(1+i)^n-1}{i}$	To find P, given A: $\frac{(1+i)^n-1}{i(1+i)^n}$	
n	$(f/p)_n^{1/2}$	$(p/f)_n^{1/2}$	$(a/f)_n^{1/2}$	$(a/p)_n^{1/2}$	$(f/a)_n^{1/2}$	$(p/a)_n^{1/2}$	n
5	1.051	0.9515	0.19604	0.20604	5.101	4.853	5
6	1.062	0.9420	0.16255	0.17255	6.152	5.795	6
7	1.072	0.9327	0.13863	0.14863	7.214	6.728	7
8	1.083	0.9235	0.12069	0.13069	8.286	7.652	8
9	1.094	0.9143	0.10674	0.11674	9.369	8.566	9
10	1.105	0.9053	0.09558	0.10558	10.462	9.471	10
11	1.116	0.8963	0.08645	0.09645	11.567	10.368	11
12	1.127	0.8874	0.07885	0.08885	12.683	11.255	12
13	1.138	0.8787	0.07241	0.08241	13.809	12.134	13
14	1.149	0.8700	0.06690	0.07690	14.947	13.004	14
15	1.161	0.8613	0.06212	0.07212	16.097	13.865	15
16	1.173	0.8528	0.05794	0.06794	17.258	14.718	16
17	1.184	0.8444	0.05426	0.06426	18.430	15.562	17
18	1.196	0.8360	0.05098	0.06098	19.615	16.398	18
19	1.208	0.8277	0.04805	0.05805	20.811	17.226	19
20	1.220	0.8195	0.04542	0.05542	22.019	18.046	20
21	1.232	0.8114	0.04303	0.05303	23.239	18.857	21
22	1.245	0.8034	0.04086	0.05086	24.472	19.660	22
23	1.257	0.7954	0.03889	0.04889	25.716	20.456	23
24	1.270	0.7876	0.03707	0.04707	26.973	21.243	24
25	1.282	0.7798	0.03541	0.04541	28.243	22.023	25
26	1.295	0.7720	0.03387	0.04387	29.526	22.795	26
27	1.308	0.7644	0.03245	0.04245	30.821	23.560	27
28	1.321	0.7568	0.03112	0.04112	32.129	24.316	28
29	1.335	0.7493	0.02990	0.03990	33.450	25.066	29
30	1.348	0.7419	0.02875	0.03875	34.785	25.808	30
31	1.361	0.7346	0.02768	0.03768	36.133	26.542	31
32	1.375	0.7273	0.02667	0.03667	37.494	27.270	32
33	1.391	0.7201	0.02573	0.03573	38.869	27.990	33
34	1.403	0.7130	0.02484	0.03484	40.258	28.703	34
35	1.417	0.7059	0.02400	0.03400	41.660	29.409	35
40	1.489	0.6717	0.02046	0.03046	48.886	32.835	40
45	1.565	0.6391	0.01771	0.02771	56.481	36.095	45
50	1.645	0.6080	0.01551	0.02551	64.463	39.196	50
55	1.729	0.5785	0.01373	0.02373	72.852	42.147	55
60	1.817	0.5504	0.01224	0.02224	81.670	44.955	60
65	1.909	0.5237	0.01100	0.02100	90.937	47.627	65
70	2.007	0.4983	0.00993	0.01993	100.676	50.169	70

1%

n	To find F, given P: $(1+i)^n$ $(f/p)_n^{1/2}$	To find P, given F: $\frac{1}{(1+i)^n}$ $(p/f)_n^{1/2}$	To find A, given F: $\frac{i}{(1+i)^n-1}$ $(a/f)_n^{1/2}$	To find A, given P: $\frac{i(1+i)^n}{(1+i)^n-1}$ $(a/p)_n^{1/2}$	To find F, given A: $\frac{(1+i)^n-1}{i}$ $(f/a)_n^{1/2}$	To find P, given A: $\frac{(1+i)^n-1}{i(1+i)^n}$ $(p/a)_n^{1/2}$	n
75	2.109	0.4741	0.00902	0.01902	110.913	52.587	75
80	2.217	0.4511	0.00822	0.01822	121.672	54.888	80
85	2.330	0.4292	0.00752	0.01752	132.979	57.078	85
90	2.449	0.4084	0.00690	0.01690	144.863	59.161	90
95	2.574	0.3886	0.00636	0.01636	157.354	61.143	95
100	2.705	0.3697	0.00587	0.01587	170.481	63.029	100

1½%

n	To find F, given P: $(1+i)^n$ $(f/p)_n^{11/2}$	To find P, given F: $\frac{1}{(1+i)^n}$ $(p/f)_n^{1\,1/2}$	To find A, given F: $\frac{i}{(1+i)^n-1}$ $(a/f)_n^{1\,1/2}$	To find A, given P: $\frac{i(1+i)^n}{(1+i)^n-1}$ $(a/p)_n^{1\,1/2}$	To find F, given A: $\frac{(1+i)^n-1}{i}$ $(f/a)_n^{1\,1/2}$	To find P, given A: $\frac{(1+i)^n-1}{i(1+i)^n}$ $(p/a)_n^{1\,1/2}$	n
1	1.015	0.9852	1.00000	1.01500	1.000	0.985	1
2	1.030	0.9707	0.49629	0.51128	2.015	1.956	2
3	1.046	0.9563	0.32838	0.34338	3.045	2.912	3
4	1.061	0.9422	0.24444	0.25944	4.091	3.854	4
5	1.077	0.9283	0.19409	0.20909	5.152	4.783	5
6	1.093	0.9145	0.16053	0.17553	6.230	5.697	6
7	1.110	0.9010	0.13656	0.15156	7.323	6.598	7
8	1.126	0.8877	0.11858	0.13358	8.433	7.486	8
9	1.143	0.8746	0.10461	0.11961	9.559	8.361	9
10	1.161	0.8617	0.09343	0.10843	10.703	9.222	10
11	1.178	0.8489	0.08429	0.09930	11.863	10.071	11
12	1.196	0.8364	0.07668	0.09168	13.041	10.908	12
13	1.214	0.8240	0.07024	0.08524	14.237	11.732	13
14	1.232	0.8118	0.06472	0.07972	15.450	12.543	14
15	1.250	0.7999	0.05994	0.07494	16.682	13.343	15
16	1.269	0.7880	0.05577	0.07077	17.932	14.131	16
17	1.288	0.7764	0.05208	0.06708	19.201	14.908	17
18	1.307	0.7649	0.04881	0.06381	20.489	15.673	18
19	1.327	0.7536	0.04588	0.06086	21.797	16.426	19
20	1.347	0.7425	0.04325	0.05825	23.124	17.169	20
21	1.367	0.7315	0.04087	0.05587	24.471	17.900	21
22	1.388	0.7207	0.03870	0.05370	25.838	19.621	22

(Continued)

1½%

	To find F, given P: $(1+i)^n$	To find P, given F: $\frac{1}{(1+i)^n}$	To find A, given F: $\frac{i}{(1+i)^n-1}$	To find A, given P: $\frac{i(1+i)^n}{(1+i)^n-1}$	To find F, given A: $\frac{(1+i)^n-1}{i}$	To find P, given A: $\frac{(1+i)^n-1}{i(1+i)^n}$	
n	$(f/p)_n^{11/2}$	$(p/f)_n^{1\,1/2}$	$(a/f)_n^{1\,1/2}$	$(a/p)_n^{1\,1/2}$	$(f/a)_n^{1\,1/2}$	$(p/a)_n^{1\,1/2}$	n
23	1.408	0.7100	0.03673	0.05173	27.225	19.331	23
24	1.430	0.6995	0.03492	0.04992	28.634	20.030	24
25	1.451	0.6892	0.03325	0.04826	30.063	20.720	25
26	1.473	0.6790	0.03173	0.04673	31.514	21.399	26
27	1.495	0.6690	0.03032	0.04532	32.987	22.068	27
28	1.517	0.6591	0.02900	0.04400	34.481	22.727	28
29	1.540	0.6494	0.02778	0.04278	35.999	23.376	29
30	1.563	0.6396	0.02664	0.04164	37.539	24.016	30
31	1.587	0.6303	0.02557	0.04057	39.102	24.646	31
32	1.610	0.6210	0.02458	0.03958	40.688	25.267	32
33	1.634	0.6118	0.02364	0.03864	42.229	25.879	33
34	1.659	0.6028	0.02276	0.03776	43.933	26.482	34
35	1.684	0.5939	0.02193	0.03693	45.592	27.076	35
40	1.814	0.5513	0.01834	0.03343	54.268	29.916	40
45	1.954	0.5117	0.01572	0.03072	63.614	32.552	45
50	2.105	0.4750	0.01357	0.02857	73.683	35.000	50
55	2.268	0.4409	0.01183	0.02683	84.530	37.271	55
60	2.443	0.4093	0.01039	0.02539	96.215	39.380	60
65	2.632	0.3799	0.00919	0.02419	108.803	41.338	65
70	2.835	0.3527	0.00817	0.02317	122.364	43.155	70
75	3.055	0.3274	0.00730	0.02230	136.973	44.842	75
80	3.291	0.3039	0.00655	0.02155	152.711	46.407	80
85	3.545	0.2821	0.00589	0.02089	169.665	47.861	85
90	3.819	0.2619	0.00532	0.02032	187.930	49.210	90
95	4.114	0.2431	0.00482	0.01982	207.606	50.462	95
100	4.432	0.2256	0.00437	0.01937	228.803	51.625	100

2%

	To find F, given P: $(1+i)^n$	To find P, given F: $\frac{1}{(1+i)^n}$	To find A, given F: $\frac{i}{(1+i)^n-1}$	To find A, given P: $\frac{i(1+i)^n}{(1+i)^n-1}$	To find F, given A: $\frac{(1+i)^n-1}{i}$	To find P, given A: $\frac{(1+i)^n-1}{i(1+i)^n}$	
n	$(f/p)_n^2$	$(p/f)_n^2$	$(a/f)_n^2$	$(a/p)_n^2$	$(f/a)_n^2$	$(p/a)_n^2$	n
1	1.020	0.9804	1.00000	1.02000	1.000	0.980	1
2	1.040	0.9612	0.49505	0.51505	2.020	1.942	2
3	1.061	0.9423	0.32675	0.34675	3.060	2.884	3

2%

n	To find F, given P: $(1+i)^n$	To find P, given F: $\frac{1}{(1+i)^n}$	To find A, given F: $\frac{i}{(1+i)^n-1}$	To find A, given P: $\frac{i(1+i)^n}{(1+i)^n-1}$	To find F, given A: $\frac{(1+i)^n-1}{i}$	To find P, given A: $\frac{(1+i)^n-1}{i(1+i)^n}$	n
	$(f/p)^2_n$	$(p/f)^2_n$	$(a/f)^2_n$	$(a/p)^2_n$	$(f/a)^2_n$	$(p/a)^2_n$	
4	1.082	0.9238	0.24262	0.26262	4.122	3.808	4
5	1.104	0.9057	0.19216	0.21216	5.204	4.713	5
6	1.126	0.8880	0.15853	0.17853	6.308	5.601	6
7	1.149	0.8706	0.13451	0.15451	7.434	6.472	7
8	1.172	0.8535	0.11651	0.13651	8.583	7.325	8
9	1.195	0.8368	0.10252	0.12252	9.755	8.162	9
10	1.219	0.8203	0.09133	0.11133	10.950	8.983	10
11	1.243	0.8043	0.08216	0.10218	12.169	9.787	11
12	1.268	0.7885	0.07456	0.09456	13.412	10.575	12
13	1.294	0.7730	0.06812	0.08812	14.680	11.348	13
14	1.319	0.7579	0.06260	0.08260	15.974	12.106	14
15	1.346	0.7430	0.05783	0.07783	17.293	12.849	15
16	1.373	0.7284	0.05365	0.07365	18.639	13.578	16
17	1.400	0.7142	0.04997	0.06997	20.012	14.292	17
18	1.428	0.7002	0.04670	0.06670	21.412	14.992	18
19	1.457	0.6864	0.04378	0.06378	22.841	15.678	19
20	1.486	0.6730	0.04116	0.06116	24.297	16.351	20
21	1.516	0.6598	0.03878	0.05878	25.783	17.011	21
22	1.546	0.6468	0.03663	0.05663	27.299	17.658	22
23	1.577	0.6342	0.03467	0.05467	28.845	18.292	23
24	1.608	0.6217	0.03287	0.05287	30.422	18.914	24
25	1.641	0.6095	0.03122	0.05122	32.030	19.523	25
26	1.673	0.5976	0.02970	0.04970	33.671	20.121	26
27	1.707	0.5859	0.02829	0.04829	35.344	20.707	27
28	1.741	0.5744	0.02699	0.04699	37.051	21.281	28
29	1.776	0.5631	0.02578	0.04578	38.792	21.844	29
30	1.811	0.5521	0.02465	0.04465	40.568	22.396	30
31	1.848	0.5412	0.02360	0.04360	42.379	22.938	31
32	1.885	0.5306	0.02261	0.04261	44.227	23.468	32
33	1.922	0.5202	0.02169	0.04169	46.112	23.989	33
34	1.961	0.5100	0.02082	0.04082	48.034	24.499	34
35	2.000	0.5000	0.02000	0.04000	49.994	24.999	35
40	2.208	0.4529	0.01656	0.03656	60.402	27.355	40
45	2.438	0.4102	0.01391	0.03391	71.893	29.490	45
50	2.692	0.3715	0.01182	0.03182	84.579	31.424	50
55	2.972	0.3365	0.01014	0.03014	98.587	33.175	55

(Continued)

2%

n	To find F, given P: $(1+i)^n$ $(f/p)^2_n$	To find P, given F: $\frac{1}{(1+i)^n}$ $(p/f)^2_n$	To find A, given F: $\frac{i}{(1+i)^n-1}$ $(a/f)^2_n$	To find A, given P: $\frac{i(1+i)^n}{(1+i)^n-1}$ $(a/p)^2_n$	To find F, given A: $\frac{(1+i)^n-1}{i}$ $(f/a)^2_n$	To find P, given A: $\frac{(1+i)^n-1}{i(1+i)^n}$ $(p/a)^2_n$	n
60	3.281	0.3048	0.00877	0.02877	114.052	34.761	60
65	3.623	0.2761	0.00763	0.02763	131.126	36.197	65
70	4.000	0.2500	0.00667	0.02667	149.978	37.499	70
75	4.416	0.2265	0.00586	0.02586	170.792	38.677	75
80	4.875	0.2051	0.00516	0.02516	193.772	39.745	80
85	5.383	0.1858	0.00456	0.02456	219.144	40.711	85
90	5.943	0.1683	0.00405	0.02405	247.157	41.587	90
95	6.562	0.1524	0.00360	0.02360	278.085	42.380	95
100	7.245	0.1380	0.00320	0.02320	312.232	43.098	100

2½%

n	To find F, given P: $(1+i)^n$ $(f/p)^{21/2}_n$	To find P, given F: $\frac{1}{(1+i)^n}$ $(p/f)^{21/2}_n$	To find A, given F: $\frac{i}{(1+i)^n-1}$ $(a/f)^{21/2}_n$	To find A, given P: $\frac{i(1+i)^n}{(1+i)^n-1}$ $(a/p)^{21/2}_n$	To find F, given A: $\frac{(1+i)^n-1}{i}$ $(f/a)^{21/2}_n$	To find P, given A: $\frac{(1+i)^n-1}{i(1+i)^n}$ $(p/a)^{21/2}_n$	n
1	1.025	0.9756	1.00000	1.02500	1.000	0.976	1
2	1.051	0.9518	0.49383	0.51883	2.025	1.927	2
3	1.077	0.9386	0.32514	0.35014	3.076	2.856	3
4	1.104	0.9060	0.24082	0.26512	4.153	3.762	4
5	1.131	0.8839	0.19025	0.21525	5.256	4.646	5
6	1.160	0.8623	0.15655	0.18155	6.388	5.508	6
7	1.189	0.8413	0.13250	0.15750	7.547	6.349	7
8	1.218	0.8207	0.11447	0.13947	8.736	7.170	8
9	1.249	0.8007	0.10046	0.12546	9.955	7.971	9
10	1.280	0.7812	0.08926	0.11426	11.203	8.752	10
11	1.312	0.7621	0.08011	0.10511	12.483	9.514	11
12	1.345	0.7436	0.07249	0.09749	13.796	10.258	12
13	1.379	0.7254	0.06605	0.09105	15.140	10.983	13
14	1.413	0.7077	0.06054	0.08554	16.519	11.691	14
15	1.448	0.6905	0.05577	0.08077	17.932	12.381	15
16	1.485	0.6736	0.05160	0.07660	19.380	13.055	16
17	1.522	0.6572	0.04793	0.07293	20.865	13.712	17

2½%

n	To find F, given P: $(1+i)^n$	To find P, given F: $\frac{1}{(1+i)^n}$	To find A, given F: $\frac{i}{(1+i)^n-1}$	To find A, given P: $\frac{i(1+i)^n}{(1+i)^n-1}$	To find F, given A: $\frac{(1+i)^n-1}{i}$	To find P, given A: $\frac{(1+i)^n-1}{i(1+i)^n}$	n
	$(f/p)_n^{21/2}$	$(p/f)_n^{21/2}$	$(a/f)_n^{21/2}$	$(a/p)_n^{21/2}$	$(f/a)_n^{21/2}$	$(p/a)_n^{21/2}$	
18	1.560	0.6412	0.04407	0.06967	22.386	14.353	18
19	1.599	0.6255	0.04176	0.06676	23.946	14.979	19
20	1.639	0.6103	0.03915	0.06415	25.545	15.589	20
21	1.680	0.5954	0.03679	0.06179	27.183	16.185	21
22	1.722	0.5809	0.03465	0.05965	28.863	16.765	22
23	1.765	0.5667	0.03270	0.05770	30.584	17.332	23
24	1.809	0.5529	0.03091	0.05591	32.349	17.885	24
25	1.854	0.5394	0.02928	0.05428	34.158	18.424	25
26	1.900	0.5262	0.02777	0.05277	36.012	18.951	26
27	1.948	0.5134	0.02638	0.05138	37.912	19.494	27
28	1.996	0.5009	0.02509	0.05009	39.860	19.965	28
29	2.046	0.4887	0.02389	0.04689	41.856	20.454	29
30	2.098	0.4767	0.02278	0.04778	43.903	20.930	30
31	2.150	0.4651	0.02174	0.04674	46.000	21.395	31
32	2.204	0.4538	0.02077	0.04577	48.150	21.849	32
33	2.259	0.4427	0.01986	0.04486	50.354	22.292	33
34	2.315	0.4319	0.01901	0.04401	52.613	22.724	34
35	2.373	0.4214	0.01821	0.04321	54.928	23.145	35
40	2.685	0.3724	0.01464	0.03984	67.403	25.103	40
45	3.038	0.3292	0.01227	0.03727	81.516	26.833	45
50	3.437	0.2909	0.01026	0.03526	97.484	28.362	50
55	3.889	0.2572	0.00865	0.03365	115.551	29.714	55
60	4.400	0.2273	0.00735	0.03235	135.992	30.909	60
65	4.978	0.2009	0.00628	0.03128	159.118	31.965	65
70	5.632	0.1776	0.00540	0.03040	185.284	32.898	70
75	6.372	0.1569	0.00465	0.02965	214.888	33.723	75
80	7.210	0.1387	0.00403	0.02903	248.383	34.452	80
85	8.157	0.1226	0.00349	0.02849	286.279	35.096	85
90	9.229	0.1084	0.00304	0.02804	329.154	35.666	90
95	10.442	0.0958	0.00265	0.02765	377.664	36.169	95
100	11.814	0.0846	0.00231	0.02731	432.549	36.614	100

3%

n	To find F, given P: $(1+i)^n$ $(f/p)^3_n$	To find P, given F: $\frac{1}{(1+i)^n}$ $(p/f)^3_n$	To find A, given F: $\frac{i}{(1+i)^n-1}$ $(a/f)^3_n$	To find A, given P: $\frac{i(1+i)^n}{(1+i)^n-1}$ $(a/p)^3_n$	To find F, given A: $\frac{(1+i)^n-1}{i}$ $(f/a)^3_n$	To find P, given A: $\frac{(1+i)^n-1}{i(1+i)^n}$ $(p/a)^3_n$	n
1	1.030	0.9709	1.00000	1.03000	1.000	0.971	1
2	1.061	0.9426	0.49261	0.52261	2.030	1.913	2
3	1.093	0.9151	0.32353	0.35353	3.091	2.829	3
4	1.126	0.8885	0.23903	0.26903	4.184	3.717	4
5	1.159	0.8626	0.18835	0.21835	5.309	4.580	5
6	1.194	0.8375	0.15460	0.18460	6.468	5.417	6
7	1.230	0.8131	0.13051	0.16051	7.662	6.230	7
8	1.267	0.7894	0.11246	0.14246	8.892	7.020	8
9	1.305	0.7664	0.09843	0.12843	10.159	7.786	9
10	1.344	0.7441	0.08723	0.11723	11.464	8.530	10
11	1.384	0.7224	0.07808	0.10808	12.808	9.253	11
12	1.426	0.7014	0.07046	0.10046	14.192	9.954	12
13	1.469	0.6810	0.06403	0.09403	15.618	10.635	13
14	1.513	0.6611	0.05853	0.08853	17.086	11.296	14
15	1.558	0.6419	0.05377	0.08377	18.599	11.938	15
16	1.605	0.6232	0.04961	0.07961	20.157	12.561	16
17	1.653	0.6050	0.04595	0.07595	21.762	13.166	17
18	1.702	0.5874	0.04271	0.07271	23.414	13.754	18
19	1.754	0.5703	0.03981	0.06981	25.117	14.324	19
20	1.806	0.5537	0.03722	0.06722	26.870	14.877	20
21	1.860	0.5375	0.03487	0.06487	28.676	15.415	21
22	1.916	0.5219	0.03275	0.06275	30.537	15.937	22
23	1.974	0.5067	0.03081	0.06081	32.453	16.444	23
24	2.033	0.4919	0.02905	0.05905	34.426	16.936	24
25	2.094	0.4776	0.02743	0.05743	36.459	17.413	25
26	2.157	0.4637	0.02594	0.05594	38.553	17.877	26
27	2.221	0.4502	0.02456	0.05456	40.710	18.327	27
28	2.288	0.4371	0.02329	0.05329	42.931	18.764	28
29	2.357	0.4243	0.02211	0.05211	45.219	19.188	29
30	2.427	0.4120	0.02102	0.05102	47.575	19.600	30
31	2.500	0.4000	0.02000	0.05000	50.003	20.000	31
32	2.575	0.3883	0.01905	0.04905	52.503	20.389	32
33	2.652	0.3770	0.01816	0.04816	55.078	20.766	33
34	2.732	0.3660	0.01732	0.04732	57.730	21.132	34
35	2.814	0.3554	0.01654	0.04654	60.462	21.487	35
40	3.262	0.3066	0.01326	0.04328	75.401	23.115	40
45	3.782	0.2644	0.01079	0.04079	92.720	24.519	45
50	4.384	0.2281	0.00887	0.03887	112.797	25.730	50

3%

n	To find *F*, given *P*: $(1+i)^n$	To find *P*, given *F*: $\frac{1}{(1+i)^n}$	To find *A*, given *F*: $\frac{i}{(1+i)^n-1}$	To find *A*, given *P*: $\frac{i(1+i)^n}{(1+i)^n-1}$	To find *F*, given *A*: $\frac{(1+i)^n-1}{i}$	To find *P*, given *A*: $\frac{(1+i)^n-1}{i(1+i)^n}$	*n*
	$(f/p)_n^3$	$(p/f)_n^3$	$(a/f)_n^3$	$(a/p)_n^3$	$(f/a)_n^3$	$(p/a)_n^3$	
55	5.082	0.1968	0.00735	0.03735	136.072	26.774	55
60	5.892	0.1697	0.00613	0.03613	163.053	27.676	60
65	6.830	0.1464	0.00515	0.03515	194.333	28.453	65
70	7.918	0.1263	0.00434	0.03434	230.594	29.123	70
75	9.179	0.1089	0.00367	0.03367	272.631	29.702	75
80	10.641	0.0940	0.00311	0.03311	321.363	30.201	80
85	12.336	0.0811	0.00265	0.03265	377.857	30.631	85
90	14.300	0.0699	0.00226	0.03226	443.349	31.002	90
95	16.578	0.0603	0.00193	0.03193	519.272	31.323	95
100	19.219	0.0520	0.00165	0.03165	607.288	31.599	100

4%

n	To find *F*, given *P*: $(1+i)^n$	To find *P*, given *F*: $\frac{1}{(1+i)^n}$	To find *A*, given *F*: $\frac{i}{(1+i)^n-1}$	To find *A*, given *P*: $\frac{i(1+i)^n}{(1+i)^n-1}$	To find *F*, given *A*: $\frac{(1+i)^n-1}{i}$	To find *P*, given *A*: $\frac{(1+i)^n-1}{i(1+i)^n}$	*n*
	$(f/p)_n^4$	$(p/f)_n^4$	$(a/f)_n^4$	$(a/p)_n^4$	$(f/a)_n^4$	$(p/a)_n^4$	
1	1.040	0.9615	1.00000	1.04000	1.000	0.962	1
2	1.082	0.1246	0.49020	0.53020	2.040	1.886	2
3	1.125	0.8190	0.32035	0.36035	3.122	2.775	3
4	1.170	0.8548	0.23549	0.27549	4.246	3.630	4
5	1.217	0.8219	018463	0.22463	5.416	4.452	5
6	1.265	0.7903	0.15076	0.19076	6.633	5.242	6
7	1.316	0.7599	0.12661	0.16661	7.898	6.002	7
8	1.369	0.7307	0.10853	0.14853	9.214	6.733	8
9	1.423	0.7026	0.09449	0.13449	10.583	7.435	9
10	1.480	0.6756	0.08329	0.12329	12.006	8.111	10
11	1.539	0.6496	0.07416	0.11415	13.486	8.760	11
12	1.601	0.6246	0.06655	0.10655	15.026	9.385	12
13	1.665	0.6006	0.06014	0.10014	16.627	9.986	13
14	1.732	0.5775	0.05467	0.09467	18.292	10.563	14
15	1.801	0.5553	0.04994	0.08994	20.024	11.118	15
16	1.873	0.5339	0.04582	0.08582	21.825	11.652	16
17	1.948	0.5134	0.04220	0.08220	23.698	12.166	17

(Continued)

4%

n	To find F, given P: $(1+i)^n$ $(f/p)^4_n$	To find P, given F: $\frac{1}{(1+i)^n}$ $(p/f)^4_n$	To find A, given F: $\frac{i}{(1+i)^n-1}$ $(a/f)^4_n$	To find A, given P: $\frac{i(1+i)^n}{(1+i)^n-1}$ $(a/p)^4_n$	To find F, given A: $\frac{(1+i)^n-1}{i}$ $(f/a)^4_n$	To find P, given A: $\frac{(1+i)^n-1}{i(1+i)^n}$ $(p/a)^4_n$	n
18	2.026	0.4936	0.03899	0.07899	25.645	12.659	18
19	2107	0.4746	0.03614	0.07614	27.671	13.134	19
20	2.191	0.4564	0.03358	0.07358	29.778	13.590	20
21	2.279	0.4388	0.03128	0.07128	31.969	14.029	21
22	2.370	0.4220	0.02920	0.06920	34.248	14.451	22
23	2.465	0.4057	0.02731	0.06731	36.618	14.857	23
24	2.563	0.3901	0.02559	0.06559	39.083	15.247	24
25	2.666	0.3751	0.02401	0.06401	41.646	15.622	25
26	2.772	0.3607	0.02257	0.06257	44.312	15.983	26
27	2.883	0.3468	0.02124	0.06124	47.084	16.330	27
28	2.999	0.3335	0.02001	0.08001	49.968	16.683	28
29	3.119	0.3207	0.01888	0.05888	52.966	16.984	29
30	3.243	0.3083	0.01783	0.05783	58.085	17.292	30
31	3.373	0.2965	0.01686	0.05686	59.328	17.588	31
32	3.508	0.2851	0.01595	0.05595	62.701	17.874	32
33	3.648	0.2741	0.01510	0.05510	68.210	18.148	33
34	3.794	0.2636	0.01431	0.05431	69.858	18.411	34
35	3.946	0.2534	0.01358	0.05358	73.652	18.665	35
40	4.801	0.2083	0.01052	0.05052	95.026	19.793	40
45	5.841	0.1712	0.00826	0.04826	121.029	20.720	45
50	7.107	0.1407	0.00655	0.04855	152.667	21.482	50
55	8.646	0.1157	0.00523	0.04523	191.159	22.109	55
60	10.520	0.0951	0.00420	0.04420	237.991	22.623	60
65	12.799	0.0781	0.00339	0.04339	294.968	23.047	65
70	15.572	0.0642	0.00275	0.04275	364.290	23.395	70
75	18.945	0.0528	0.00223	0.04223	448.631	23.680	75
80	23.050	0.0434	0.00181	0.04181	551.245	23.915	80
85	28.044	0.0357	0.00146	0.04140	676.090	24.109	85
90	34.119	0.0293	0.00121	0.04121	827.983	24.267	90
95	41.511	0.0241	0.00099	0.04099	1012.785	24.398	95
100	50.505	0.0198	0.00081	0.04081	1237.624	24.505	100

5%

n	To find F, given P: $(1+i)^n$	To find P, given F: $\frac{1}{(1+i)^n}$	To find A, given F: $\frac{i}{(1+i)^n-1}$	To find A, given P: $\frac{i(1+i)^n}{(1+i)^n-1}$	To find F, given A: $\frac{(1+i)^n-1}{i}$	To find P, given A: $\frac{(1+i)^n-1}{i(1+i)^n}$	n
	$(f/p)^5_n$	$(p/f)^5_n$	$(a/f)^5_n$	$(a/p)^5_n$	$(f/a)^5_n$	$(p/a)^5_n$	
1	1.050	0.9524	1.00000	1.05000	1.000	0.952	1
2	1.103	0.9070	0.48780	0.53780	2.050	1.859	2
3	1.158	0.8638	0.31721	0.36721	3.153	2.723	3
4	1.216	0.8227	0.23201	0.28201	4.310	3.546	4
5	1.276	0.7835	0.18097	0.23097	5.526	4.329	5
6	1.340	0.7462	0.14702	0.19702	6.802	5.076	6
7	1.407	0.7107	0.12282	0.17282	8.142	5.786	7
8	1.477	0.6768	0.10472	0.15472	9.549	6.463	8
9	1.551	0.6446	0.09069	0.14069	11.027	7.108	9
10	1.629	0.6139	0.07950	0.12950	12.578	7.722	10
11	1.710	0.5847	0.07039	0.12039	14.207	8.306	11
12	1.796	0.5568	0.06283	0.11283	15.917	8.863	12
13	1.886	0.5303	0.05646	0.10646	17.713	9.394	13
14	1.980	0.5051	0.05102	0.10102	19.599	9.899	14
15	2.079	0.4810	0.04634	0.09634	21.579	10.380	15
16	2.183	0.4581	0.04227	0.09227	23.657	10.838	16
17	2.292	0.4363	0.03870	0.08870	25.840	11.274	17
18	2.407	0.4155	0.03555	0.08555	28.132	11.690	18
19	2.527	0.3957	0.03275	0.08275	30.539	12.085	19
20	2.653	0.3769	0.03024	0.08024	33.066	12.462	20
21	2.786	0.3589	0.02800	0.07800	35.719	12.821	21
22	2.925	0.3418	0.02597	0.07597	38.505	13.163	22
23	3.072	0.3256	0.02414	0.07414	41.430	13.489	23
24	3.225	0.3101	0.02247	0.07247	44.502	13.799	24
25	3.386	0.2953	0.02095	0.07095	47.727	14.094	25
26	3.556	0.2812	0.01956	0.06956	51.113	14.375	26
27	3.733	0.2678	0.01829	0.06829	54.669	14.643	27
28	3.920	0.2551	0.01712	0.06712	58.403	14.898	28
29	4.116	0.2429	0.01605	0.06605	62.323	15.141	29
30	4.322	0.2314	0.01505	0.06505	66.439	15.372	30
31	4.538	0.2204	0.01413	0.06413	70.761	15.593	31
32	4.765	0.2099	0.01328	0.06328	75.299	15.803	32
33	5.003	0.1999	0.01249	0.06249	80.064	16.003	33
34	5.253	0.1904	0.01176	0.06176	85.067	16.193	34
35	5.516	0.1813	0.01107	0.06107	90.320	16.374	35
40	7.040	0.1420	0.00828	0.05828	120.800	17.159	40
45	8.985	0.1113	0.00626	0.05626	159.700	17.774	45

(Continued)

5%

n	To find F, given P: $(1+i)^n$ $(f/p)_n^5$	To find P, given F: $\frac{1}{(1+i)^n}$ $(p/f)_n^5$	To find A, given F: $\frac{i}{(1+i)^n-1}$ $(a/f)_n^5$	To find A, given P: $\frac{i(1+i)^n}{(1+i)^n-1}$ $(a/p)_n^5$	To find F, given A: $\frac{(1+i)^n-1}{i}$ $(f/a)_n^5$	To find P, given A: $\frac{(1+i)^n-1}{i(1+i)^n}$ $(p/a)_n^5$	n
50	11.467	0.0872	0.00478	0.05478	209.348	18.256	50
55	14.636	0.0683	0.00367	0.05367	272.713	18.633	55
60	18.679	0.0535	0.00283	0.05283	353.584	18.929	60
65	23.840	0.0419	0.00219	0.05219	456.798	19.161	65
70	30.426	0.0329	0.00170	0.05170	588.529	19.343	70
75	38.833	0.0258	0.00132	0.05132	756.654	19.485	75
80	49.561	0.0202	0.00103	0.05103	971.229	19.596	80
85	63.254	0.0158	0.00080	0.05080	1245.087	19.684	85
90	80.730	0.0124	0.00063	0.05063	1594.607	19.752	90
95	103.035	0.0097	0.00049	0.05049	2040.694	19.806	95
100	131.501	0.0076	0.00038	0.05038	2610.025	19.848	100

6%

n	To find F, given P: $(1+i)^n$ $(f/p)_n^6$	To find P, given F: $\frac{1}{(1+i)^n}$ $(p/f)_n^6$	To find A, given F: $\frac{i}{(1+i)^n-1}$ $(a/f)_n^6$	To find A, given P: $\frac{i(1+i)^n}{(1+i)^n-1}$ $(a/p)_n^6$	To find F, given A: $\frac{(1+i)^n-1}{i}$ $(f/a)_n^6$	To find P, given A: $\frac{(1+i)^n-1}{i(1+i)^n}$ $(p/a)_n^6$	n
1	1.080	0.9434	1.00000	1.06000	1.000	0.943	1
2	1.124	0.8900	0.48544	0.54544	2.060	1.833	2
3	1.191	0.8396	0.31411	0.37411	3.184	2.673	3
4	1.262	0.7921	0.22859	0.28859	4.375	3.465	4
5	1.338	0.7473	0.17740	0.23740	5.637	4.212	5
6	1.419	0.7050	0.14336	0.20336	6.975	4.917	6
7	1.504	0.6651	0.11914	0.17914	8.394	5.582	7
8	1.594	0.6274	0.10104	0.16104	9.897	6.210	8
9	1.689	0.5919	0.08702	0.14702	11.491	6.802	9
10	1.791	0.5584	0.07587	0.13587	13.181	7.360	10
11	1.898	0.5268	0.08679	0.12679	14.972	7.887	11
12	2.012	0.4970	0.05928	0.11928	16.870	8.384	12
13	2.133	0.4688	0.05296	0.11296	18.882	8.853	13
14	2.261	0.4423	0.04758	0.10756	21.015	9.295	14
15	2.397	0.4173	0.04296	0.10296	23.276	9.712	15
16	2.540	0.3936	0.03895	0.09895	25.673	10.106	16

6%

n	To find F, given P: $(1+i)^n$ $(f/p)^6_n$	To find P, given F: $\frac{1}{(1+i)^n}$ $(p/f)^6_n$	To find A, given F: $\frac{i}{(1+i)^n-1}$ $(a/f)^6_n$	To find A, given P: $\frac{i(1+i)^n}{(1+i)^n-1}$ $(a/p)^6_n$	To find F, given A: $\frac{(1+i)^n-1}{i}$ $(f/a)^6_n$	To find P, given A: $\frac{(1+i)^n-1}{i(1+i)^n}$ $(p/a)^6_n$	n
17	2.693	0.3714	0.03544	0.09544	28.213	10.477	17
18	2.854	0.3503	0.03236	0.09236	30.906	10.828	18
19	3.026	0.3305	0.02962	0.08962	33.760	11.158	19
20	3.207	0.3118	0.02718	0.08718	36.786	11.470	20
21	3.400	0.2942	0.02500	0.08500	39.993	11.764	21
22	3.604	0.2775	0.02305	0.08305	43.392	12.042	22
23	3.820	0.2618	0.02128	0.08126	48.996	12.303	23
24	4.049	0.2470	0.01968	0.07968	50.816	12.550	24
25	4.292	0.2330	0.01823	0.07823	54.865	12.783	25
26	4.549	0.2198	0.01690	0.07690	59.156	13.003	26
27	4.822	0.2074	0.01570	0.07570	63.706	13.211	27
28	5.112	0.1956	0.01459	0.07459	68.528	13.406	28
29	5.418	0.1846	0.01358	0.07358	73.640	13.591	29
30	5.743	0.1741	0.01265	0.07265	79.058	13.765	30
31	6.088	0.1643	0.01179	0.07179	84.802	13.929	31
32	6.453	0.1550	0.01100	0.07100	90.890	14.084	32
33	6.841	0.1462	0.01027	0.07027	97.343	14.230	33
34	7.251	0.1379	0.00960	0.06960	104.184	14.368	34
35	7.686	0.1301	0.00897	0.06897	111.435	14.498	35
40	10.286	0.0972	0.00646	0.06646	154.762	15.046	40
45	13.765	0.0727	0.00470	0.06470	212.744	15.456	45
50	18.420	0.0543	0.00344	0.06344	290.336	15.762	50
55	24.650	0.0406	0.00254	0.06254	394.172	15.991	55
60	32.988	0.0303	0.00188	0.06188	533.128	16.161	60
65	45.145	0.0227	0.00139	0.06139	719.083	16.289	65
70	59.076	0.0169	0.00103	0.06103	967.932	16.385	70
75	79.057	0.0126	0.00077	0.06077	1300.949	16.456	75
80	105.796	0.0095	0.00057	0.06057	1746.600	16.509	80
85	141.579	0.0071	0.00043	0.06043	2342.982	16.549	85
90	189.465	0.0053	0.00032	0.06032	3141.075	16.579	90
95	253.546	0.0039	0.00024	0.06024	4209.104	16.601	95
100	339.302	0.0029	0.00018	0.06018	5638.368	16.618	100

7%

n	To find F, given P: $(1+i)^n$ $(f/p)^7_n$	To find P, given F: $\frac{1}{(1+i)^n}$ $(p/f)^7_n$	To find A, given F: $\frac{i}{(1+i)^n-1}$ $(a/f)^7_n$	To find A, given P: $\frac{i(1+i)^n}{(1+i)^n-1}$ $(a/p)^7_n$	To find F, given A: $\frac{(1+i)^n-1}{i}$ $(f/a)^7_n$	To find P, given A: $\frac{(1+i)^n-1}{i(1+i)^n}$ $(p/a)^7_n$	n
1	1.070	0.9346	1.00000	1.07000	1.000	0.935	1
2	1.145	0.8734	0.48309	0.55309	2.070	1.808	2
3	1.225	0.8163	0.31105	0.38105	3.215	2.624	3
4	1.311	0.7629	0.22523	0.29523	4.440	3.387	4
5	1.403	0.7130	0.17389	0.24389	5.751	4.100	5
6	1.501	0.6663	0.13980	0.20980	7.153	4.767	6
7	1.606	0.6227	0.11555	0.18555	8.654	5.389	7
8	1.718	0.5820	0.09747	0.16747	10.260	5.971	8
9	1.838	0.5439	0.08349	0.15349	11.978	6.515	9
10	1.967	0.5083	0.07238	0.14238	13.816	7024	10
11	2.105	0.4751	0.06336	0.13336	15.784	7.499	11
12	2.252	0.4440	0.05590	0.12590	17.888	7943	12
13	2.410	0.4150	0.04965	0.11965	20.141	8.358	13
14	2.579	0.3878	0.04434	0.11434	22.550	8.745	14
15	2.759	0.3624	0.03979	0.10979	25.129	9.108	15
16	2.952	0.3387	0.03586	0.10586	27.888	9.447	16
17	3.159	0.3166	0.03243	0.10243	30.840	9.763	17
18	3.380	0.2959	0.02941	0.09941	33.999	10.059	18
19	3.617	0.2765	0.02675	0.09675	37.379	10.363	19
20	3.870	0.2584	0.02439	0.09439	40.995	10.594	20
21	4.141	0.2415	0.02229	0.09229	44.865	10.836	21
22	4.430	0.2257	0.02041	0.09041	49.006	11.061	22
23	4.741	0.2109	0.01871	0.08871	53.436	11.272	23
24	5.072	0.1971	0.01719	0.08719	58.177	11.469	24
25	5.427	0.1842	0.01581	0.08581	63.249	11.654	25
26	5.807	0.1722	0.01456	0.08456	68.676	11.826	26
27	6.214	0.1609	0.01343	0.08343	74.484	11.987	27
28	6.649	0.1504	0.01239	0.08239	80.698	12.137	28
29	7.114	0.1406	0.01145	0.08145	87.347	12.278	29
30	7.612	0.1314	0.01059	0.08059	94.461	12.409	30
31	8.145	0.1228	0.00980	0.07980	102.073	12.532	31
32	8.715	0.1147	0.00907	0.07907	110.218	12.647	32
33	9.325	0.1072	0.00841	0.07841	118.923	12.754	33
34	9.978	0.1002	0.00780	0.07780	128.259	12.854	34
35	10.677	0.0937	0.00723	0.07723	138.237	12.948	35
40	14.974	0.0668	0.00501	0.07501	199.635	13.332	40
45	21.002	0.0476	0.00350	0.07350	285.749	13.606	45
50	29.457	0.0339	0.00246	0.07246	406.529	13.801	50

7%

n	To find F, given P: $(1+i)^n$	To find P, given F: $\frac{1}{(1+i)^n}$	To find A, given F: $\frac{i}{(1+i)^n-1}$	To find A, given P: $\frac{i(1+i)^n}{(1+i)^n-1}$	To find F, given A: $\frac{(1+i)^n-1}{i}$	To find P, given A: $\frac{(1+i)^n-1}{i(1+i)^n}$	n
	$(f/p)_n^7$	$(p/f)_n^7$	$(a/f)_n^7$	$(a/p)_n^7$	$(f/a)_n^7$	$(p/a)_n^7$	
55	41.315	0.0242	0.00174	0.07174	575.929	13.940	55
60	57.946	0.0173	0.00123	0.07123	813.520	14.039	60
65	81.273	0.0123	0.00087	0.07087	1146.755	14.110	65
70	113.989	0.0088	0.00062	0.07062	1614.134	14.160	70
75	159.876	0.0063	0.00044	0.07044	2269.657	14.196	75
80	224.234	0.0045	0.00031	0.07031	3189.063	14.222	80
85	314.500	0.0032	0.00022	0.07022	4478.576	14.240	85
90	441.103	0.0023	0.00016	0.07016	6287.185	14.253	90
95	618.670	0.0016	0.00011	0.07011	8823.854	14.263	95
100	867.716	0.0012	0.00008	0.07008	12381.662	14.269	100

8%

n	To find F, given P: $(1+i)^n$	To find P, given F: $\frac{1}{(1+i)^n}$	To find A, given F: $\frac{i}{(1+i)^n-1}$	To find A, given P: $\frac{i(1+i)^n}{(1+i)^n-1}$	To find F, given A: $\frac{(1+i)^n-1}{i}$	To find P, given A: $\frac{(1+i)^n-1}{i(1+i)^n}$	n
	$(f/p)_n^8$	$(p/f)_n^8$	$(a/f)_n^8$	$(a/p)_n^8$	$(f/a)_n^8$	$(p/a)_n^8$	
1	1.080	0.9259	1.00000	1.08000	1.000	0.926	1
2	1.166	0.8573	0.48077	0.58077	2.080	1.783	2
3	1.260	0.7938	0.30803	0.38803	3.246	2.577	3
4	1.360	0.7350	0.22192	0.30192	4.506	3.312	4
5	1.469	0.6806	0.17046	0.25048	5.867	3.933	5
6	1.587	0.6302	0.13832	0.21832	7.336	4.623	6
7	1.714	0.5835	0.11207	0.19207	8.823	5.206	7
8	1.851	0.5403	0.09401	0.17401	10.837	5.747	8
9	1.999	0.5002	0.08008	0.16008	12.488	6.247	9
10	2.159	0.4632	0.06903	0.14903	14.487	6.710	10
11	2.332	0.4289	0.06008	0.14008	16.645	7.139	11
12	2.518	0.3971	0.05270	0.13270	18.877	7.536	12
13	2.720	0.3677	0.04652	0.12652	21.495	7.904	13
14	2.937	0.3405	0.04130	0.12130	24.215	8.244	14
15	3.172	0.3152	0.03883	0.11683	27.152	8.559	15
16	3.426	0.2919	0.03298	0.11298	30.324	8.851	16
17	3.700	0.2703	0.02983	0.10963	33.750	9.122	17
18	3.996	0.2502	0.02670	0.10670	37.450	9.372	18

(Continued)

8%

n	To find F, given P: $(1+i)^n$	To find P, given F: $\frac{1}{(1+i)^n}$	To find A, given F: $\frac{i}{(1+i)^n-1}$	To find A, given P: $\frac{i(1+i)^n}{(1+i)^n-1}$	To find F, given A: $\frac{(1+i)^n-1}{i}$	To find P, given A: $\frac{(1+i)^n-1}{i(1+i)^n}$	n
	$(f/p)_n^8$	$(p/f)_n^8$	$(a/f)_n^8$	$(a/p)_n^8$	$(f/a)_n^8$	$(p/a)_n^8$	
19	4.316	0.2317	0.02413	0.10413	41.446	9.604	19
20	4.661	0.2145	0.02185	0.10185	45.762	9.818	20
21	5.034	0.1987	0.01983	0.09983	50.423	10.017	21
22	5.437	0.1839	0.01803	0.09803	55.457	10.201	22
23	5.781	0.1703	0.01642	0.09842	60.893	10.371	23
24	6.341	0.1577	0.01498	0.09498	66.765	10.529	24
25	6.848	0.1460	0.01368	0.09368	73.106	10.675	25
26	7.396	0.1352	0.01251	0.09251	79.954	10.810	26
27	7.988	0.1252	0.01145	0.09145	87.351	10.935	27
28	8.627	0.1159	0.01048	0.09049	95.339	11.051	28
29	8.317	0.1073	0.00962	0.08962	103.966	11.158	29
30	10.063	0.0994	0.00883	0.08883	113.283	11.258	30
31	10.868	0.0920	0.00811	0.08811	123.348	11.350	31
32	11.737	0.0852	0.00745	0.08745	134.214	11.435	32
33	12.676	0.0789	0.00685	0.08685	145.951	11.514	33
34	13.690	0.0730	0.00630	0.08630	158.627	11.587	34
35	14.785	0.0676	0.00580	0.08580	172.317	11.655	35
40	21.725	0.0460	0.00386	0.08386	259.057	11.925	40
45	31.920	0.0313	0.00258	0.08259	386.506	12.108	45
50	46.902	0.0213	0.00174	0.08174	573.770	12.233	50
55	68.914	0.0145	0.00118	0.08118	848.923	12.319	55
60	101.257	0.0099	0.00080	0.08080	1253.213	12.377	60
65	148.780	0.0067	0.00054	0.08054	1847.248	12.416	65
70	218.606	0.0046	0.00037	0.08037	2720.080	12.443	70
75	321.205	0.0031	0.00025	0.08025	4002.557	12.461	75
80	471.955	0.0021	0.00017	0.08017	5886.935	12.474	80
85	693.456	0.0014	0.00012	0.08012	8655.706	12.482	85
90	1018.915	0.0010	0.00008	0.08008	12723.939	12.488	90
95	1497.121	0.0007	0.00005	0.08005	18701.507	12.492	95
100	2199.761	0.0005	0.00004	0.08004	27484.516	12.494	100

9%

n	To find F, given P: $(1+i)^n$ $(f/p)^9_n$	To find P, given F: $\frac{1}{(1+i)^n}$ $(p/f)^9_n$	To find A, given F: $\frac{i}{(1+i)^n-1}$ $(a/f)^9_n$	To find A, given P: $\frac{i(1+i)^n}{(1+i)^n-1}$ $(a/p)^9_n$	To find F, given A: $\frac{(1+i)^n-1}{i}$ $(f/a)^9_n$	To find P, given A: $\frac{(1+i)^n-1}{i(1+i)^n}$ $(p/a)^9_n$	n
1	1.090	0.9174	1.00000	1.09000	1.000	0.917	1
2	1.188	0.8417	0.47847	0.56847	2.090	1.759	2
3	1.295	0.7722	0.30505	0.39505	3.278	2.531	3
4	1.412	0.7084	0.21867	0.30867	4.573	3.240	4
5	1.539	0.6499	0.16709	0.25709	5.985	3.890	5
6	1.677	0.5963	0.13292	0.22292	7.523	4.486	6
7	1.828	0.5470	0.10869	0.19869	9.200	5.033	7
8	1.993	0.5019	0.09067	0.18067	11.028	5.535	8
9	2.172	0.4604	0.07680	0.16680	13.021	5.995	9
10	2.367	0.4224	0.06582	0.15582	15.193	6.418	10
11	2.580	0.3875	0.05695	0.14695	17.560	6.805	11
12	2.813	0.3555	0.04965	0.13965	20.141	7.161	12
13	3.066	0.3262	0.04357	0.13357	22.953	7.487	13
14	3.342	0.2992	0.03843	0.12843	26.019	7.786	14
15	3.642	0.2745	0.03406	0.12406	29.361	8.061	15
16	3.970	0.2519	0.03030	0.12030	33.003	8.313	16
17	4.328	0.2311	0.02705	0.11705	36.974	8.544	17
18	4.717	0.2120	0.02421	0.11421	41.301	8.756	18
19	5.142	0.1945	0.02173	0.11173	46.018	8.950	19
20	5.604	0.1784	0.01955	0.10955	51.160	9.129	20
21	6.109	0.1637	0.01762	0.10762	56.765	9.292	21
22	6.659	0.1502	0.01590	0.10590	62.873	9.442	22
23	7.258	0.1378	0.01438	0.10438	69.532	9.580	23
24	7.911	0.1264	0.01302	0.10302	76.790	9.707	24
25	8.623	0.1160	0.01180	0.10181	84.701	9.823	25
26	9.399	0.1064	0.01072	0.10072	93.324	9.929	26
27	10.245	0.0976	0.00973	0.09973	102.723	10.027	27
28	11.167	0.0895	0.00885	0.09885	112.968	10.116	28
29	12.172	0.0822	0.00806	0.09806	124.135	10.198	29
30	13.268	0.0754	0.00734	0.09734	136.308	10.274	30
31	14.462	0.0691	0.00669	0.09669	149.575	10.343	31
32	15.763	0.0634	0.00610	0.09610	164.037	10.406	32
33	17.182	0.0582	0.00556	0.09556	179.800	10.464	33
34	18.728	0.0534	0.00508	0.09508	196.982	10.518	34
35	20.414	0.0490	0.00464	0.09464	215.711	10.567	35
40	31.409	0.0318	0.00296	0.09296	337.882	10.757	40
45	48.327	0.0207	0.00190	0.09190	525.859	10.881	45

(Continued)

9%

n	To find F, given P: $(1+i)^n$	To find P, given F: $\frac{1}{(1+i)^n}$	To find A, given F: $\frac{i}{(1+i)^n-1}$	To find A, given P: $\frac{i(1+i)^n}{(1+i)^n-1}$	To find F, given A: $\frac{(1+i)^n-1}{i}$	To find P, given A: $\frac{(1+i)^n-1}{i(1+i)^n}$	n
n	$(f/p)_n^9$	$(p/f)_n^9$	$(a/f)_n^9$	$(a/p)_n^9$	$(f/a)_n^9$	$(p/a)_n^9$	n
50	74.358	0.0134	0.00123	0.09123	815.084	10.962	50
55	114.408	0.0087	0.00079	0.09079	1260.092	11.014	55
60	176.031	0.0057	0.00051	0.09051	1944.792	11.048	60
65	270.864	0.0037	0.00033	0.09033	2998.288	11.070	65
70	416.730	0.0024	0.00022	0.09022	4619.223	11.084	70
75	641.191	0.0016	0.00014	0.09014	7113.232	11.094	75
80	986.552	0.0010	0.00009	0.09009	10950.556	11.100	80
85	1517.948	0.0007	0.00006	0.09006	16854.444	11.104	85
90	2335.501	0.0004	0.00004	0.09004	25939.000	11.106	90
95	3593.513	0.0003	0.00003	0.09003	39917.378	11.108	95
100	5529.089	0.0002	0.00002	0.09002	61422.544	11.109	100

10%

n	To find F, given P: $(1+i)^n$	To find P, given F: $\frac{1}{(1+i)^n}$	To find A, given F: $\frac{i}{(1+i)^n-1}$	To find A, given P: $\frac{i(1+i)^n}{(1+i)^n-1}$	To find F, given A: $\frac{(1+i)^n-1}{i}$	To find P, given A: $\frac{(1+i)^n-1}{i(1+i)^n}$	n
n	$(f/p)_n^{10}$	$(p/f)_n^{10}$	$(a/f)_n^{10}$	$(a/p)_n^{10}$	$(f/a)_n^{10}$	$(p/a)_n^{10}$	n
1	1.100	0.9091	1.00000	1.10000	1.000	0.909	1
2	1.210	0.8264	0.47619	0.57619	2.100	1.736	2
3	1.331	0.7513	0.30211	0.40211	3.310	2.487	3
4	1.464	0.6830	0.21547	0.31547	4.641	3.170	4
5	1.611	0.6209	0.16380	0.26380	6.105	3.791	5
6	1.772	0.5645	0.12961	0.22961	7.716	4.355	6
7	1.949	0.5132	0.10541	0.20541	9.487	4.868	7
8	2.144	0.4665	0.08744	0.18744	11.436	5.335	8
9	2.358	0.4241	0.07364	0.17364	13.579	5.759	9
10	2.594	0.3855	0.06275	0.16275	15.937	6.144	10
11	2.853	0.3505	0.05396	0.15396	18.531	6.495	11
12	3.138	0.3186	0.04676	0.14676	21.384	6.814	12
13	3.452	0.2897	0.04078	0.14078	24.523	7.103	13
14	3.797	0.2633	0.03575	0.13575	27.975	7.367	14
15	4.177	0.2394	0.03147	0.13147	31.772	7.606	15
16	4.595	0.2176	0.02782	0.12782	35.950	7.824	16
17	5.054	0.1978	0.02466	0.12466	40.545	8.022	17

10%

n	To find F, given P: $(1+i)^n$ $(f/p)_n^{10}$	To find P, given F: $\frac{1}{(1+i)^n}$ $(p/f)_n^{10}$	To find A, given F: $\frac{i}{(1+i)^n-1}$ $(a/f)_n^{10}$	To find A, given P: $\frac{i(1+i)^n}{(1+i)^n-1}$ $(a/p)_n^{10}$	To find F, given A: $\frac{(1+i)^n-1}{i}$ $(f/a)_n^{10}$	To find P, given A: $\frac{(1+i)^n-1}{i(1+i)^n}$ $(p/a)_n^{10}$	n
18	5.560	0.1799	0.02193	0.12193	45.599	8.201	18
19	6.116	0.1635	0.01955	0.11955	51.159	8.363	19
20	6.727	0.1486	0.01746	0.11746	57.275	8.514	20
21	7.400	0.1351	0.01562	0.11562	64.002	8.649	21
22	8.140	0.1228	0.01401	0.11401	71.403	8.772	22
23	8.954	0.1117	0.01257	0.11257	79.543	8.883	23
24	9.850	0.1015	0.01130	0.11130	88.497	8.985	24
25	10.835	0.0923	0.01017	0.11017	98.347	9.077	25
26	11.918	0.0839	0.00916	0.10916	109.182	9.161	26
27	13.110	0.0763	0.00826	0.10826	121.100	9.237	27
28	14.421	0.0693	0.00745	0.10745	134.210	9.307	28
29	15.863	0.0630	0.00673	0.10673	148.631	9.370	29
30	17.449	0.0573	0.00608	0.10608	164.494	9.427	30
31	19.194	0.0521	0.00550	0.10550	181.943	9.479	31
32	21.114	0.0474	0.00497	0.10497	201.138	9.526	32
33	23.225	0.0431	0.00450	0.10450	222.252	9.569	33
34	25.548	0.0391	0.00407	0.10407	245.477	9.609	34
35	28.102	0.0356	0.00369	0.10369	271.024	9.644	35
40	45.259	0.0221	0.00226	0.10226	442.593	9.779	40
45	72.890	0.0137	0.00139	0.10139	718.905	9.863	45
50	117.391	0.0085	0.00086	0.10086	1163.909	9.915	50
55	189.059	0.0053	0.00053	0.10053	1880.591	9.947	55
60	304.482	0.0033	0.00033	0.10033	3034.816	9.967	60
65	490.371	0.0020	0.00020	0.10020	4893.707	9.980	65
70	789.747	0.0013	0.00013	0.10013	7887.470	9.987	70
75	1271.895	0.0008	0.00008	0.10008	12708.954	9.992	75
80	2048.400	0.0005	0.00005	0.10005	20474.002	9.995	80
85	3298.969	0.0003	0.00003	0.10003	32979.690	9.997	85
90	5313.023	0.0002	0.00002	0.10002	53120.226	9.998	90
95	8556.676	0.0001	0.00001	0.10001	85556.760	9.999	95
100	13780.612	0.0061	0.00001	0.10001	137796.123	9.999	100

12%

n	To find F, given P: $(1+i)^n$ $(f/p)_n^{12}$	To find P, given F: $\frac{1}{(1+i)^n}$ $(p/f)_n^{12}$	To find A, given F: $\frac{i}{(1+i)^n-1}$ $(a/f)_n^{12}$	To find A, given P: $\frac{i(1+i)^n}{(1+i)^n-1}$ $(a/p)_n^{12}$	To find F, given A: $\frac{(1+i)^n-1}{i}$ $(f/a)_n^{12}$	To find P, given A: $\frac{(1+i)^n-1}{i(1+i)^n}$ $(p/a)_n^{12}$	n
1	1.120	0.8929	1.00000	1.12000	1.000	0.893	1
2	1.254	0.7972	0.47170	0.59170	2.120	1.690	2
3	1.405	0.7118	0.29635	0.41635	3.374	2.402	3
4	1.574	0.6355	0.20923	0.32923	4.779	3.037	4
5	1.762	0.5674	0.15741	0.27741	6.353	3.605	5
6	1.974	0.5066	0.12323	0.24323	8.115	4.111	6
7	2.211	0.4523	0.09912	0.21912	10.089	4.564	7
8	2.476	0.4039	0.08130	0.20130	12.300	4.968	8
9	2.773	0.3606	0.06768	0.18768	14.776	5.328	9
10	3.106	0.3220	0.05698	0.17698	17.549	5.650	10
11	3.479	0.2875	0.04842	0.16842	20.655	5.938	11
12	3.896	0.2567	0.04144	0.16144	24.133	6.194	12
13	4.363	0.2292	0.03568	0.15568	28.029	6.424	13
14	4.887	0.2046	0.03087	0.15087	32.393	6.628	14
15	5.474	0.1827	0.02682	0.14682	37.280	6.811	15
16	6.130	0.1631	0.02339	0.14339	42.753	6.974	16
17	6.866	0.1456	0.02046	0.14046	48.884	7.120	17
18	7.690	0.1300	0.01794	0.13794	55.750	7.250	18
19	8.613	0.1161	0.01576	0.13576	63.440	7.366	19
20	9.646	0.1037	0.01388	0.13388	72.052	7.469	20
21	10.804	0.0926	0.01224	0.13224	81.699	7.562	21
22	12.100	0.0826	0.01081	0.13081	92.503	7.645	22
23	13.552	0.0738	0.00956	0.12956	104.603	7.718	23
24	15.179	0.0659	0.00846	0.12846	118.155	7.784	24
25	17.000	0.0588	0.00750	0.12750	133.334	7.843	25
26	19.040	0.0525	0.00665	0.12665	150.334	7.896	26
27	21.325	0.0469	0.00590	0.12590	169.374	7.943	27
28	23.884	0.0419	0.00524	0.12524	190.699	7.984	28
29	26.750	0.0374	0.00466	0.12466	214.582	8.022	29
30	29.960	0.0334	0.00414	0.12414	241.333	8.055	30
31	33.555	0.0298	0.00369	0.12369	271.292	8.085	31
32	37.582	0.0266	0.00328	0.12328	304.847	8.112	32
33	42.091	0.0238	0.00292	0.12292	342.429	8.135	33
34	47.142	0.0212	0.00260	0.12260	384.520	8.157	34
35	52.800	0.0189	0.00232	0.12232	431.663	8.176	35
40	93.051	0.0107	0.00130	0.12130	767.091	8.244	40
45	163.988	0.0061	0.00074	0.12074	1358.230	8.283	45
50	289.002	0.0035	0.00042	0.12042	2400.018	8.305	50

15%

n	To find F, given P: $(1+i)^n$	To find P, given F: $\frac{1}{(1+i)^n}$	To find A, given F: $\frac{i}{(1+i)^n-1}$	To find A, given P: $\frac{i(1+i)^n}{(1+i)^n-1}$	To find F, given A: $\frac{(1+i)^n-1}{i}$	To find P, given A: $\frac{(1+i)^n-1}{i(1+i)^n}$	n
	$(f/p)_n^{15}$	$(p/f)_n^{15}$	$(a/f)_n^{15}$	$(a/p)_n^{15}$	$(f/a)_n^{15}$	$(p/a)_n^{15}$	
1	1.150	0.8696	1.00000	1.15000	1.000	0.870	1
2	1.322	0.7561	0.46512	0.61512	2.150	1.626	2
3	1.521	0.6575	0.28798	0.43798	3.472	2.283	3
4	1.749	0.5718	0.20027	0.35027	4.993	2.855	4
5	2.011	0.4972	0.14832	0.29832	6.742	3.352	5
6	2.313	0.4323	0.11424	0.26424	8.754	3.784	6
7	2.660	0.3759	0.09036	0.24036	11.067	4.160	7
8	3.059	0.3269	0.07285	0.22285	13.727	4.487	8
9	3.518	0.2843	0.05957	0.20957	16.786	4.772	9
10	4.046	0.2472	0.04925	0.19925	20.304	5.019	10
11	4.652	0.2149	0.04107	0.19107	24.349	5.234	11
12	5.350	0.1869	0.03448	0.18448	29.002	5.421	12
13	6.153	0.1625	0.02911	0.17911	34.352	5.583	13
14	7.076	0.1413	0.02469	0.17469	40.505	5.724	14
15	8.137	0.1229	0.02102	0.17102	47.580	5.847	15
16	9.358	0.1069	0.01795	0.16795	55.717	5.954	16
17	10.761	0.0929	0.01537	0.16537	65.075	6.047	17
18	12.375	0.0808	0.01319	0.16319	75.836	6.128	18
19	14.232	0.0703	0.01134	0.16134	88.212	6.198	19
20	16.367	0.0611	0.00976	0.15976	102.444	6.259	20
21	18.821	0.0531	0.00842	0.15842	118.810	6.312	21
22	21.645	0.0462	0.00727	0.15727	137.631	6.359	22
23	24.891	0.0402	0.00628	0.15628	159.276	6.399	23
24	28.625	0.0349	0.00543	0.15543	184.168	6.434	24
25	32.919	0.0304	0.00470	0.15470	212.793	6.464	25
26	37.857	0.0264	0.00407	0.15407	245.711	6.491	26
27	43.535	0.0230	0.00353	0.15353	283.569	6.514	27
28	50.066	0.0200	0.00306	0.15306	327.104	6.534	28
29	57.575	0.0174	0.00265	0.15265	377.170	6.551	29
30	66.212	0.0151	0.00230	0.15230	434.745	6.566	30
31	76.143	0.0131	0.00200	0.15200	500.956	6.579	31
32	87.565	0.0114	0.00173	0.15173	577.099	6.591	32
33	100.700	0.0099	0.00150	0.15150	664.664	6.600	33
34	115.805	0.0086	0.00131	0.15131	765.364	6.609	34
35	133.176	0.0075	0.00113	0.15113	881.170	6.617	35
40	267.863	0.0037	0.00056	0.15056	1779.090	6.642	40
45	538.769	0.0019	0.00028	0.15028	3585.128	6.654	45
50	1083.657	0.0009	0.00014	0.15014	7217.716	6.661	50

20%

n	To find F, given P: $(1+i)^n$ $(f/p)_n^{20}$	To find P, given F: $\frac{1}{(1+i)^n}$ $(p/f)_n^{20}$	To find A, given F: $\frac{i}{(1+i)^n-1}$ $(a/f)_n^{20}$	To find A, given P: $\frac{i(1+i)^n}{(1+i)^n-1}$ $(a/p)_n^{20}$	To find F, given A: $\frac{(1+i)^n-1}{i}$ $(f/a)_n^{20}$	To find P, given A: $\frac{(1+i)^n-1}{i(1+i)^n}$ $(p/a)_n^{20}$	n
1	1.200	0.8333	1.00000	1.20000	1.000	0.833	1
2	1.440	0.6944	0.45455	0.65455	2.200	1.528	2
3	1.728	0.5787	0.27473	0.47473	3.640	2.106	3
4	2.074	0.4823	0.18629	0.38629	5.368	2.598	4
5	2.488	0.4019	0.13438	0.33438	7.442	2.991	5
6	2.986	0.3349	0.10071	0.30071	9.930	3.326	6
7	3.583	0.2791	0.07742	0.27742	12.916	3.605	7
8	4.300	0.2326	0.06061	0.26061	16.499	3.837	8
9	5.100	0.1938	0.04808	0.24808	20.799	4.031	9
10	6.192	0.1615	0.03852	0.23852	25.959	4.192	10
11	7.430	0.1346	0.03110	0.23110	32.150	4.327	11
12	8.916	0.1122	0.02526	0.22526	39.581	4.439	12
13	10.699	0.0935	0.02062	0.22062	48.497	4.533	13
14	12.839	0.0779	0.01689	0.21689	59.196	4.611	14
15	15.407	0.0649	0.01388	0.21388	72.035	4.675	15
16	18.488	0.0541	0.01144	0.21144	87.442	4.730	16
17	22.186	0.0451	0.00944	0.20944	105.931	4.775	17
18	26.623	0.0376	0.00781	0.20781	128.117	4.812	18
19	31.948	0.0313	0.00646	0.20646	154.740	4.843	19
20	38.338	0.0261	0.00536	0.20536	186.688	4.870	20
21	46.005	0.0217	0.00444	0.20444	225.025	4.891	21
22	55.206	0.0181	0.00369	0.20369	271.031	4.909	22
23	66.247	0.0151	0.00307	0.20307	326.237	4.925	23
24	79.497	0.0126	0.00255	0.20255	392.484	4.937	24
25	95.396	0.0105	0.00212	0.20212	471.981	4.948	25
26	114.475	0.0087	0.00176	0.20176	567.377	4.956	26
27	137.371	0.0073	0.00147	0.20147	681.853	4.964	27
28	164.845	0.0061	0.00122	0.20122	819.223	4.970	28
29	197.813	0.0051	0.00102	0.20102	984.068	4.975	29
30	237.376	0.0042	0.00085	0.20085	1181.881	4.979	30
31	284.851	0.0035	0.00070	0.20070	1419.257	4.982	31
32	341.822	0.0029	0.00059	0.20059	1704.108	4.985	32
33	410.186	0.0024	0.00049	0.20049	2045.930	4.988	33
34	492.223	0.0020	0.00041	0.20041	2456.116	4.990	34
35	590.668	0.0017	0.00034	0.20034	2948.339	4.992	35
40	1469.772	0.0007	0.00014	0.20014	7343.858	4.997	40
45	3657.258	0.0003	0.00005	0.20005	18281.331	4.999	45
50	9100.427	0.0001	0.00002	0.20002	45497.191	4.999	50

Bibliography

Economic Analysis of Oil and Gas Engineering Operations

CHAPTER 1

en.wikipedia.org › wiki › Oil

www.iea.org › reports › oil-information-overview

Oil, 2020 IEA

U S Energy Information Administration, March 2016

Global Energy Statistical Year Book, 2020

Coronavirus impact on crude oil price: *economictimes.indiatimes.com › markets › coronavirus.*

U.S. Information Technologies Corporation: www.usinfotech.com

CHAPTER 2

wildwell.com › Articles, Shut-in and well kill procedures - wild well control

en.wikipedia.org › wiki › Well_kill

Oilfield Technology magazine, March 2019.

Guerriero V., et al., A Permeability Model for Naturally Fractured Carbonate Reservoirs, Mar. Pet. Geol. 40: 115–134, 2012. doi:10.1016/j.marpetgeo.2012.11.002.

"Organic Hydrocarbons: Compounds made from carbon and hydrogen". Archived from the original on July 19, 2011. "petroleum", in the American Heritage Dictionary.

Speight, James G., *The Chemistry and Technology of Petroleum,* Marcel Dekker, 215–216, 1999. ISBN 978-0-8247-0217-5.

CHAPTER 3

Abdel-Aal, H. K., and M. AlSahlawi, Editors, *Petroleum Economics and Engineering*, 3rd Edition, Taylor & Francis Group, CRC Press, 2014.

Speight, James G., *The Chemistry and Technology of Petroleum*, Marcel Dekker, 215–216, 1999.

Odebunmi O., and Adeniyi A., Infrared and Ultraviolet Spectrophotometric Analysis of Chromatographic Fractions of Crude Oils and Petroleum Products, Bull. Chem. Soc. Ethiop, 21(1), 135–140, 2007.

www.researchgate.net › publication › 337905545_Classifi... ceng.tu.edu.iq › images › lectures › chem-lec › lec.1.pdf

Abdel-Aal, H. K., *Surface Petroleum Operations*, Saudi Publishing House, Jeddah, 1998.

Abdel-Aal, H. K., et al., *Petroleum and Gas Field Processing*, Marcel Dekker Inc., New York, NY, 2003.

COQG: Crude Oil Quality Group Informational Industry Paper Crude Oil, "Contaminants and Adverse Chemical Components and Their Effects on Refinery Operations", Presented on May 27, 2004 at the General Session Houston, Texas.

Doolan, P. C., and Pujado, P. R., Make Aromatics from LPG, *Hydrocarbon Processing*, Vol. 70, No. 9, www.ems.psu.edu › ~radovic › Chapter8.

Hua, Rulxlang, Wang, J., Kong, H., Liu, J., Lu, X., and Xu, G., *Analysis of Sulfur-Containing Compounds in Crude Oils by Comprehensive Two-Dimensional Gas Chromatography with Sulfur Chemiluminescence Detection*, WILEY-VCH Verlag GmbH & Co.KGaA, Weinheim, 2004.

CHAPTERS 4, 5, 6, 7, 8

Peters Max, Timmerhaus Klaus, and West Ronald, *Plant Design and Economics for Chemical Engineers*, 5th Edition, McGraw-Hill Education, 2003.

Abdel-Aal, H. K., and AlSahlawi, M., Editors, *Petroleum Economics and Engineering*, 3rd Edition, Taylor & Francis Group, CRC Press, 2014.

Crosson, S.V., and Needles, B.E., *Managerial Accounting*, 8th Edition, Houghton Mifflin Company, Boston, 2008.

Gamble, J., Strickland, A., and Thompson, A., *Crafting & Executing Strategy*, 15th Edition, McGraw-Hill, New York, NY, 2007.

Inkpen Andrew, and Moffett Michael H., *The Global Oil & Gas Industry: Management, Strategy and Finance*, Penn Well Corporation, 2011.

Masseron Jean, *Petroleum Economics*, Publisher, Editions OPHRYS, Paris, 1990.

Lerche Iran, and MacKay James A., *Economic Risk in Hydrocarbon Exploration*, 1st Edition, Imprint: Academic Press, 1990, ISBN: 9780124441651.

Pearson Education Limited, "Fundamentals of Financial Management", 2004, 12/e.

Noland, Thomas R., The Sum-of years' Digits Depreciation Method: Use by SEC Filers, Journal of Finance and Accountancy. 1997.

Berg, M., De Waegenaere, A., and Wielhouwer, J., Optimal Tax Depreciation with Uncertain Future Cash-Flows. European Journal of Operational Research. 132: 197–209, 2001.

Merino, D. N., Developing Economic and Non-Economic Models, Incentives to Select among Technical Alternatives, Engineering Economist, 34(4), 1989.

Valle-Riestra, J. R., *Project Evaluation in the Chemical Process Industries*, 1st Edition, McGraw-Hill Book Company, New York, NY, 1983.

Blank, L.T., Tarquini, A.J., and Iverson, S. - 2005 - lavoisier.fr

Lehtonen, R., and Pahkinen, E., *Practical Methods for Design and Analysis of Complex Surveys*, J Wiley, 2004.

DeFusco, Richard, A., CFA, McLeavey, Dennis, W., CFA, Pinto, Jerald, E., CFA, and Runkle, David E., CFA, *Quantitative Investment Analysis*, John Wiley & Sons, Jan, 2011.

Hazelrigg, George A., Validation of Engineering Design Alternative Selection Methods, *Engineering Optimization*, Taylor & Fran., Volume 35, Issue 2, pages 103–120, April 2003.

CHAPTERS 9, 10

Abdel-Aal, H. K., *Chemical Engineering Primer with Computer Applications*, Taylor & Francis Group, CRC Press, 2017.

Abdel-Aal, H. K., *Magnesium from Resources to Production*, Taylor & Francis Group, CRC Press, 2019.

Choate, M. S., *Professional Wikis*. Wiley Publishing, Inc., Indianapolis, 2008.

Hohman, J., and Saiedian, H., Wiki customization to resolve management issues in distributed software projects, Crosstalk. Aug, 2008. pp. 1–8. Available at: http://www.stsc.hill.af.mil/crosstalk/2008/08/0808HohmanSaiedian.html.

Wenger, E.C., and Snyder, W.M., Communities of practice: the organizational frontier, Harvard Business Review. Jan-Feb, 2000. pp. 139–5. Available at: http://www.stevens-tech.edu/cce/NEW/PDFs/commprac.pdf.

Noel, S., and Robert, J. M., How the web is used to support collaborative writing. [February 23, 2009]. Available at: http://charlie.res.crc.ca/~sylvie/Articles/BIT_02_05_2003.pdf.

CHAPTER 11

AspenTech Software, AspenTech Enables Faster Decision-Making for Oil & Gas Assets in New Release of aspen ONE® Software, Washington, D.C., www.aspentech.com/ThreeColumnLayout.aspxpageid…id, May 25, 2011.

Ben-Haim, Y., *Information-Gap Decision Theory: Decisions under Severe Uncertainty*, Academic Press, 2001.

Deng, H., Runger, G., and Tuv, E., "Bias of Importance Measures for Multi-valued Attributes and Solutions", Proceedings of the 21st International Conference on Artificial Neural Networks, 2011.

Macmillan, Fiona, "Risk, Uncertainty and Investment Decision-Making in the Upstream Oil and Gas 5- Industry" A thesis presented for the degree of Ph.D. at the University of Aberdeen, October 2000.

Lawrence, John A., and Lawrence, Barry, A., *Applied Management Science: a Computer-Integrated Approach for Decision making/with CD*, John Wiley & Sons, Inc., 2000.

Taghavifard et.al., "Decision Making Under Uncertain and Risky Situations, Copyright 2009 by the Society of Actuaries, 2009.

Hillier, F. S., and Lieberman, G. J., *Introduction to Operations Research*, 4th Edition, Holden-Day, Oakland, CA, 1989.

Hertz, D. D., *Practical Risk Analysis*, John Wiley & Sons, New York, NY, 1983.

Goodwin, P., and Wright, G., *Decision Analysis for Management Judgment,* 3rd Edition, Wiley, Chichester, 2004. ISBN 0-470-86108-8.

Clemen, Robert, and Reilly, T., *Making Hard Decisions,* 2nd Edition, Southwestern College Pub., Belmont, CA, 2004, ISBN 978-0-495-01508-6.

Leach, Patrick, *Why Can't You Just Give Me the Number? An Executive's Guide to Using Probabilistic Thinking to Manage Risk and to Make Better Decisions,* 2006. Probabilistic. ISBN 0-9647938-5-7.

CHAPTER 12

Oil and Gas Production Handbook, Havard Devolt, SRH Publisher, Feb, 2012 (ISBN 9781105538643 American Petroleum Institute, "Industry Sectors", http://www.api.org/aboutoilgas/sectors/ Retrieved 12 May 2008.

Ranked in Order of 2007 Worldwide Oil Equivalent Reserves as Reported in "OGJ 200/100", *Oil & Gas Journal*, September 15, 2008, http://www.petrostrategies.org/Links/Worlds_Largest_Oil_and_Gas_Companies_Sites.htm

Abdel-Aal, H. K., and AlSahlawi, M., Editors, *Petroleum Economics and Engineering*, 3rd Edition, Taylor & Francis Group, CRC Press, 2014.

API 2008 Basic Petroleum Data Book, First Edition.

Oil & Gas Production Handbook, Ed 2x3 AB.

CHAPTERS 13, 14

US Energy Information, www.eia.gov/dnav/pet/pet_crd_wellcost.

REM, Int. Eng. J. 72(4) Ouro Preto Oct./Dec. 2019 Epub Sep 16, 2019.

www.investopedia.com › Commodities › Oil.

www.wiley.com › en-us › Fundamentals+of+Sustainabl.

Hossain, and Al-Majed, 2015.

US Energy Information, www.eia.gov/dnav/pet/pet_crd_wellcost.

http://www.rigzone.com/data/dayrates.

US Energy Information, www.eia.gov/dnav/pet/pet_crd_wellcost.

20th Century Petroleum Statistics, DeGolyer and MacNaughton, Publisher: Dallas, TX. 1999.

Abdel-Aal, H. K., Aggour, M., and Fahim, M., *Petroleum and Gas Field Processing*, Marcel Dekker Inc, New York, NY, 2003.

Devolt, Havard, *Oil and Gas Production Book*, edition 2.3 Oslo, April 2010 ABB.

Abdel-Aal, H. K., *Surface Petroleum Operations*, Saudi Publishing & Distributing House, Jeddah, Saudi Arabia 1998.

Al-Ghamdi, Abdullah, and Kokal, Sunil, "Investigation of Causes of Tight Emulsions in Gas Oil Separation Plants", SPE Proceedings, Middle East Oil Show, 9-12 June 2003, Bahrain.

CHAPTER 15

Abdel-Aal, H. K., and AlSahlawi, M., Editors, *Petroleum Economics and Engineering*, 3rd Edition, Taylor & Francis Group, CRC Press, 2014.

Abdel-Aal, H. K., *Chemical Engineering Primer with Computer Applications*, Taylor & Francis Group, CRC Press, 2017.

Abdel-Aal, H. K., Aggour, M., and Fahim, M., *Petroleum and Gas Field Processing*, Marcel Dekker Inc, New York, NY, 2003.

Abdel-Aal, H. K., *Surface Petroleum Operations*, Saudi Publishing & Distributing House, Jeddah, Saudi Arabia 1998.

CHAPTER 16

en.wikipedia.org › wiki › Separator_(oil_production).

www.sciencedirect.com › science › article › pii.

Abdel-Aal, H. K., and AlSahlawi, M., Editors, *Petroleum Economics and Engineering*, 3rd Edition, Taylor & Francis Group, CRC Press, 2014.

Abdel-Aal, H. K., *Chemical Engineering Primer with Computer Applications*, Taylor & Francis Group, CRC Press, 2017.

Gas–Oil Separators part, AONG website *www.arab-oil-naturalgas.com › gas-oil-separators-part* www.saudienergy.net › PDF › Intro Oil.

Al-Ghamdi, Abdullah, and Kokal, Sunil, "Investigation of Causes of Tight Emulsions in Gas Oil Separation Plants", SPE Proceedings, Middle East Oil Show, 9-12 June 2003, Bahrain.

Kokal, Sunil, and Wingrove, Martin, "Emulsion Separation Index: From Laboratory to Field Case Studies", SPE Proceedings, Annual Technical Conference and Exhibition, 1-4 October 2000, Dallas, TX.

CHAPTER 17

SPE 124823 Paper - New Mixer Optimizes Crude Desalting Plant (Paper presented at the 2009 SPE Annual Technical Conference and Exhibition in New Orleans, Louisiana, U.S.A, 2009.)

Abdel-Aal, H. K., Aggour, M., and Fahim, M. A., *Petroleum and Gas Field Processing*, Marcel Dekker Inc., New York, NY, 2003.

Abdel-Aal, H. K., *Surface Petroleum Operations*, Saudi Publishing House, Jeddah, Saudi Arabia, 1998.

Lunsford, Kevin M, and Bullin, Jerry A, "Optimization of Amine Sweetening Units", Proceedings of National Meeting, 1996 AIChE Spring, New York, NY.

Aitani, A., Oil Refining and Products, In: *Encyclopedia of Energy*, C.J. Cleveland, Ed., Elsevier, v. 4, 715–729, 2004.

CHAPTER 18

Standard Handbook of Petroleum and Natural Gas Engineering (3rd Edition).

Abdel-Aal, H. K., Aggour, M., and Fahim, M. A., *Petroleum and Gas Field Processing*, Marcel Dekker Inc., New York, NY, 2003.

Abdel-Aal, H. K., *Surface Petroleum Operations*, Saudi Publishing House, Jeddah, Saudi Arabia, 1998.

Lunsford, Kevin M, and Bullin, Jerry A., "Optimization of Amine Sweetening Units", Proceedings of National Meeting.,1996 AIChE Spring, New York, NY.

Abdel-Aal, H. K., and Shalabi, M. A., Noncatalytic Partial Oxidation of Sour Natural Gas versus Catalytic Steam Reforming of Sweet Natural Gas. Ind. Eng. Chem. Res. 35: 1787, 1996.

Processing Natural Gas, *NaturalGas.org naturalgas.org › naturalgas › processing-ng.*

CHAPTERS 19, 20, 21

Abdel-Aal, H. K., and AlSahlawi, M., Editors, *Petroleum Economics and Engineering*, 3rd Edition, Taylor & Francis Group, CRC Press, 2014.

Abdel-Aal, H. K., *Chemical Engineering Primer with Computer Applications*, Taylor & Francis Group, CRC Press, 2017.

Petroleum Refining: Technology and Economics, Fourth Edition, James H. Gary *Colorado School of Mines, Golden, Colorado,* Glenn E. Handwerk, *Consulting Chemical Engineer, Golden, Colorado,* Copyright © 2001 by Marcel Dekker, Inc.

Moldauer, B., Editor, *All About Petroleum*, American Petroleum Institute: Washington, D.C., 1998.

Weinkauf, K., *The Many Uses of Petroleum from Discovery to Present*, Desk and Derrick Club of Tulsa: Tulsa, OK, 2003.

"World Oil Balance Data" International Petroleum Monthly, 2004.

Coffey, F., Layden, J., and Allerton, C., Editors, *America on Wheels: The First 100 Years*, General Publishing Group: Santa Monica, CA, 1996.

"UOP's History" About UOP, UOP LLC: Des Plaines, IL, 2003.

Meyers Robert, Editor, *Handbook of Petroleum Refining Processes*, 2nd Edition, McGraw-Hill, 1996.

True, W. R., and Koottungal, L., Global Capacity Growth Reverses; Asian, Mideast Refineries Progress, Oil Gas Journal. December 5, 2011.

Aitani, A., Oil Refining and Products, In: *Encyclopedia of Energy*, C.J. Cleveland, Ed., Elsevier, v. 4, 715–729, 2004.

OPEC, Oil Demand by Product, In: *World Oil Outlook*, 2011, OPEC, Vienna.

Kaiser, M. J., and Gary, J. H., Study Updates Refinery Investment Cost Curves, Oil Gas Journal, April 23, 2007.

Abdel-Aal, H. K., Crude Oil Processing, In: *Petroleum Economics and Engineering*, 2nd Edition, H. K. Abdel-Aal et al. Eds., CRC, 1992.

CHAPTER 22

Wikipedia, the free encyclopedia.

Europe pipeline Maps- Crude oil (petroleum pipe lines), theodora.com, pipelines- Europe and gas-and production.

Oil and Gas Pipelines Industry, Global, Trunk/Transmission Pipeline Length by Region are indicated in Fig 22,10, Sept 2019.

Midstream Analytics, GlobalData Oil and Gas © GlobalData

Abdel-Aal, H. K., and AlSahlawi, M., Editors, *Petroleum Economics and Engineering*, 3rd Edition, Taylor & Francis Group, CRC Press, 2014.

Abdel-Aal, H. K., *Chemical Engineering Primer with Computer Applications*, Taylor & Francis Group, CRC Press, 2017.

Abdel-Aal, H. K. et al., *Petroleum and Gas Field Processing*, Marcel Dekker Inc., New York, NY, 2003.

Economies of Scale Economies of Scale | Microeconomics - Reading courses.lumenlearning.com › chapter › economies-of-scale.

Oil and gas Construction Costs, global energy monitor, www.gimwiki, Oil and gas pipe line construction.

Index

Note: Locators in *italics* represent figures and **bold** indicate tables in the text.

D

K

L

M

N

O

P

R

S

T

U

V

W

Printed in the United States
By Bookmasters